This well written book has strong technical content that meets the current and future needs of technology professionals who do hands-on work to design, implement and evaluate RAG usage with the organization's LLM pipeline. It also has understandable explanations and insightful summaries that additionally meet the needs of the broader range of managers and executives for acquiring an accurate conceptual understanding of what is involved in deploying RAG in enterprise settings, and of the related capabilities and limitations of the various approaches.

The rapidly increasing use of AI Agents will not eliminate the use of RAG. On the contrary, Agentic RAG will increasingly be the mechanism for orchestrating RAG usage and will therefore increase RAG usage whenever an LLM needs to make use of proprietary internal organizational knowledge or publicly available external information. The combination of the technical and conceptual content of this book will provide useful support to a wide range of people across the enterprise involved in making both technical and managerial decisions about the use of RAG or Agentic RAG in the context of their overall enterprise LLM efforts. It is also a great educational resource.

Steven Miller

Professor Emeritus, Singapore Management University, Singapore
Former SMU Vice Provost for Research and Founding Dean, SMU
School of Computing and Information Systems

Jun Xu's book fills a real gap in the RAG literature. It moves past the usual toy examples and dives straight into the hard, practical problems that teams face when taking RAG systems into production: freshness, cost, evaluation, monitoring, and the dozens of smaller operational headaches that kill most deployments. The finance-oriented examples are especially useful, but the patterns and runbooks apply anywhere you need reliable, grounded LLM applications. A solid, no-nonsense reference for engineers who actually have to ship and maintain these systems.

David R. Hardoon

Global Head of AI Enablement, Standard Chartered Bank

In the ever-accelerating landscape of Artificial Intelligence, a small number of innovations stand out as true paradigm shifts. Retrieval-Augmented Generation (RAG) is unequivocally one of them. It represents the necessary evolution from Large Language Models (LLMs) that merely generate human-like text to systems that ground their output in verifiable, real-time knowledge.

It is precisely this critical juncture that Jun Xu has chosen to illuminate with this definitive work.

The rapid adoption of LLMs has exposed a central vulnerability: the models' tendency toward "hallucination" and their inherent lack of current, domain-specific information. This book, as its preface so clearly lays out, is not just a theoretical exploration but a strategic blueprint for overcoming these limitations.

Jun Xu moves beyond simple definitions, presenting a comprehensive, production-ready guide to integrating the power of dynamic knowledge retrieval with generative AI.

I wholeheartedly commend the author for several reasons that make this text essential reading:

* Innovative Clarity: Jun Xu has taken a complex, cutting-edge technology and rendered it accessible and actionable for a wide audience, from AI researchers to MLOps professionals and domain specialists in fields like healthcare and finance.
* The Pragmatic Real-World Approach: The preface's commitment to addressing "over thirty real-world challenges, such as data freshness, retrieval accuracy, and scalability" is a testament to the book's immense practical value. This is not simply a high-level overview; it is a battle-tested manual for deploying RAG effectively in production environments.
* Balanced Vision: Crucially, the author maintains a balanced perspective, acknowledging RAG's current limitations and its role as a "transitional" technology leading toward the next wave of agentic AI. This forward-looking context ensures that readers are not just mastering today's tools, but preparing for tomorrow's breakthroughs.

I highly recommend this book to anyone involved in building, deploying, or relying on modern AI solutions. If you are a practitioner looking to move past proof-of-concept and implement scalable, reliable, and context-aware LLM applications using both unstructured and structured data, this book is your guide. It offers the rare combination of deep theoretical understanding and detailed, practical implementation strategies, complete with the case studies and code examples necessary for success.

This book is a vital stepping stone toward creating the next generation of reliable, grounded, and truly collaborative AI systems.

I wish Jun the best in this endeavor.

Faraaz Ali
Global head of Frontline Platform, Standard Chartered Bank

This book stands out as one of the most practical and production-oriented treatments of Retrieval-Augmented Generation (RAG) available today. Rather than focusing narrowly on prompt engineering or toy examples, it covers the full scope of building, deploying, and operating RAG systems in enterprise environments.

The book walks the reader through the complete RAG lifecycle, from data parsing and chunking to retrieval strategies, response synthesis, evaluation, monitoring, and continuous improvement, all grounded in real-world constraints. The discussion of MLOps/LLMOps is not an afterthought; it is integral to the architecture, reflecting the need to treat RAG systems as long-lived, observable, and governable production systems rather than experimental demos.

A significant strength is its focus on real-world implementation challenges. The author addresses practical issues that teams routinely face in production, including degradation in retrieval quality, hallucination mitigation, orchestration bottlenecks, and serving-time trade-offs. The inclusion of concrete architectures, patterns (including graph-based, tabular, and agentic RAG), and runnable examples makes the material directly applicable to engineers and platform teams.

Overall, this is an excellent reference for anyone serious about building business-ready RAG systems. Engineers will value the concrete guidance, architects will appreciate the system-level thinking, and product and AI leaders will benefit from the pragmatic perspective on adoption and scalability.

Highly recommended for teams transitioning from experimentation to enterprise-grade AI delivery.

Carlos Alexandre Queiroz
Managing Director of AI, OCBC

Retrieval Augmented Generation in Production

Architecture, Patterns, and Runbooks

Retrieval Augmented Generation in Production

Architecture, Patterns, and Runbooks

Jun Xu

Standard Chartered, Singapore

NEW JERSEY · LONDON · SINGAPORE · BEIJING · SHANGHAI · HONG KONG · TAIPEI · CHENNAI · TOKYO

Published by

World Scientific Publishing Co. Pte. Ltd.

5 Toh Tuck Link, Singapore 596224

USA office: 27 Warren Street, Suite 401-402, Hackensack, NJ 07601

UK office: 57 Shelton Street, Covent Garden, London WC2H 9HE

Library of Congress Cataloging-in-Publication Data
Names: Xu, Jun (Research and Development Engineer) author
Title: Retrieval augmented generation in production : architecture, patterns, and runbooks /
    Jun Xu, Standard Chartered, Singapore.
Description: New Jersey : World Scientific, [2026] | Includes bibliographical references and index.
Identifiers: LCCN 2025049805 | ISBN 9789819827596 hardcover |
    ISBN 9789819829323 paperback | ISBN 9789819827602 ebook for institutions |
    ISBN 9789819827619 ebook for individuals
Subjects: LCSH: Generative artificial intelligence | Natural language processing (Computer science) |
    Information retrieval--Automation
Classification: LCC Q335 .X85 2026
LC record available at https://lccn.loc.gov/2025049805

British Library Cataloguing-in-Publication Data
A catalogue record for this book is available from the British Library.

For any available supplementary material, please visit
https://www.worldscientific.com/worldscibooks/10.1142/14714#t=suppl

Desk Editors: Soundararajan Raghuraman/Veronica Lee

Typeset by Stallion Press
Email: enquiries@stallionpress.com

Preface

The swift progression of artificial intelligence (AI) and machine learning (ML) has initiated a paradigm shift in how machines process and generate information. At the forefront of these developments are Large Language Models (LLMs), which have transformed how machines understand and interact with human language. Among the innovations stemming from LLMs, Retrieval-Augmented Generation (RAG) technology stands out as a pivotal advancement with broad practical applications. This book delves into the intricacies of LLM-based RAG technology and explores its transformative potential across industries.

RAG technology merges the generative capabilities of language models with external information retrieval systems. Traditional LLMs often struggled with accessing real-time or domain-specific knowledge, limiting their accuracy and relevance in complex tasks. RAG addresses these limitations by dynamically retrieving contextual data during the generation process, enabling outputs that are both informed and up-to-date.

The advent of RAG technology marks a significant leap in AI applications. By merging the capabilities of LLMs with dynamic knowledge retrieval, RAG opens the door to a new era of knowledge management, i.e., one where machines are not limited to generating human-like responses but can ground their outputs in actionable, contextually relevant data. This synergy promises benefits across domains such as healthcare, finance, and education, where accurate and context-sensitive responses are paramount. In healthcare, RAG systems support clinicians by synthesizing insights from medical literature, aiding in diagnostics and treatment planning. In finance, these models analyze real-time market data and historical trends to inform forecasts. Education leverages RAG for adaptive learning tools that tailor content to individual student needs. Customer service benefits from RAG-powered chatbots that resolve inquiries with real-time data, while legal professionals use them to efficiently navigate case law. Researchers, too, harness RAG to accelerate innovation by accessing the latest scientific findings. However, as we delve into these advantages, it is crucial to maintain a balanced perspective on its current limitations.

While RAG provides significant improvements over traditional LLMs, challenges persist. Retrieval mechanisms, though effective at reducing hallucinations, cannot entirely eliminate them. LLMs may still generate plausible but incorrect outputs, posing risks in high-stakes applications like legal analysis or medical diagnosis. Furthermore, RAG is a strategic solution to the limitations of today's LLMs, which lack self-directed learning, abstract reasoning, and adaptive behavior. Its role, while vital now, is inherently transitional. As AI systems evolve, frameworks such as agentic AI, capable of goal-driven learning and real-world interaction, could surpass traditional RAG by integrating retrieval into a broader spectrum of capabilities. These systems may address current shortcomings, achieving greater autonomy and dynamic problem-solving capabilities.

This book provides a comprehensive guide to LLM-based RAG technology for both unstructured and structured data, from foundational concepts to advanced implementations in production in terms of MLOps (ML Operations). It examines the theoretical principles of language models and retrieval systems, demonstrating how their synergy creates context-aware AI solutions. Unlike other publications, this work directly addresses over thirty real-world challenges, such as data freshness, retrieval accuracy, and scalability, and systematically resolves them through practical strategies. Detailed case studies and code examples equip readers to deploy RAG effectively in their fields.

Looking ahead, RAG stands as a key stepping stone, bridging the gap between present-day limitations and future breakthroughs. Its role in enhancing knowledge-driven AI systems provides valuable lessons in designing more reliable and grounded models. As research progresses, we can expect RAG to inspire the next generation of innovations, where retrieval, reasoning, and adaptability converge. By addressing current constraints while fostering continuous improvement, the path forward is filled with opportunities for creating AI systems that not only assist but meaningfully collaborate with humans across complex environments.

I am deeply grateful to Professor Emeritus Steven Miller for his rigorous, end-to-end review of this manuscript. His comments, often more extensive than the original passages, significantly elevated the clarity and depth of the work. Our constructive debates, including one on the reasoning capabilities of large language models, sharpened the arguments throughout. I also appreciate his gracious recommendation for this book.

I would also like to extend my thanks to friends and (ex-)colleagues who offered endorsements, including Carlos Queiroz, David R. Hardoon, and Faraaz Ali.

I am equally indebted to my colleagues Elgar Teo, Jason Cao, Yanhui Chen, and Yuyu Xiong for their steady support. They read early drafts, provided some suggestions, and helped verify some ideas in practice. Portions of the methodology were implemented in our RAG products, enabling us to validate both efficiency and effectiveness in real-world settings. I am indebted as well to many other colleagues, former colleagues, and friends, such as Ahsan Ijaz, Mohammed Rahim, Yong Xia, Thorston Neumann, Emma Johnson, Soo Kiat Chan, Terrence Tan, Songhua Zhang, Le Zhang, Jia Huang, Xue Zhao and many more. While it is impossible to name everyone individually, I remain sincerely appreciative of all their contributions.

My family has been indispensable. Encouraged by this work, my wife, a non-technical senior executive, adopted RAG in her daily workflows and received enthusiastic feedback. My sons, too, have been inspired to explore AI tools. Their patience and encouragement made this project possible.

Finally, I owe special thanks to my editors, Veronica Lee and Soundararajan Raghuraman, whose meticulous editorial guidance surfaced numerous issues and greatly improved readability.

Any remaining errors are mine alone.

About the Author

 **Dr Jun Xu** is an Executive Director of Machine Learning (ML) engineering in a top financial institution, and was previously with HSBC, Western Digital, Temasek Labs and A-Star. His working experiences across multiple disciplines include Artificial Intelligence (AI), ML, LLM, FinTech, Data Storage, Robotics, Internet of Things (IoT), Big Data, Cloud, Digitalization Transformation, System Architecture, Optimization and Modelling. He has published 4 books, 18 patents, and around 60 peer-reviewed papers.

He held multiple editorial positions with international journals and conferences. He is a senior member of the Institute of Electrical and Electronics Engineers (IEEE) and a certificate financial risk manager (FRM). He was a guest professor at South China University of Technology and an industrial PhD supervisor of Nanyang Technological University. He received his Bachelor of Science (BS) from Southeast University, China, and his PhD from Nanyang University (NTU), Singapore.

Contents

Chapter 1

Introduction

1.1 Introduction

Large Language Models (LLMs) sit at the heart of today's generative-AI (GenAI) revolution, powering systems that read, reason, and write with near-human fluency. This opening chapter grounds readers in the essentials: how the transformer architecture and its attention mechanism sift through oceans of text to learn rich linguistic patterns; how transfer learning and task-specific fine-tuning turn a single pre-trained model into a multitude of domain experts; and how reinforcement learning from human feedback keeps model behavior aligned with real-world values. We then situate these ideas in the pragmatic reality of modern AI engineering: training a frontier model from scratch can cost millions, financial institutions (FIs) and most other organizations stand on the shoulders of foundation models released by leading tech companies and labs. The chapter therefore drills into the everyday toolkit that allows practitioners to adapt these models safely and economically: prompt engineering, parameter-efficient fine-tuning, retrieval-augmented generation (RAG), and agent frameworks that orchestrate complex workflows. By the end, readers will have a clear mental map of the techniques that make LLMs both powerful and practical, setting the stage for a deeper dive into the RAG solutions that follow.

1.2 GenAI and LLM

The advent of AI, particularly GenAI and LLMs, is causing a transformative shift across industries. LLMs, such as OpenAI's GPT models, Google Gemini, and Anthropic Claude, have demonstrated remarkable capabilities in understanding, generating, and reasoning about natural language. GenAI is rapidly moving from being a form of experimental technology to a critical business asset, with significant organizational benefits. McKinsey research indicates a 65% adoption rate of GenAI, underscoring its rapid integration into financial services.[1]

Modern LLMs represent a leap forward in AI, combining massive data ingestion with complex neural architectures to mimic human communication patterns. These digital polymaths digest everything from Victorian literature to Python repositories through multi-layered deep learning systems, capable of producing remarkably human-like responses that often blur the line between human and machine authorship. The evolution of LLMs was fueled by three key factors: unprecedented computing power, internet-scale training data, and architectural breakthroughs like transformer mechanisms. While the original Transformer framework (2017) [1] laid the foundation, contemporary iterations demonstrate exponential gains in contextual understanding—progress that has been sparking intense debates about ethical implementation, particularly in sensitive fields like banking where a single misinterpreted phrase could trigger financial chaos.

1.2.1 *LLM Opportunities and Challenges*

LLMs are rapidly transforming various industries, driving innovation and enhancing efficiency in areas such as customer service, marketing, reporting, education, and system integration. The financial sector, which operates under strict regulatory supervision, is a key focus, as successful applications in this sector can often be adapted to other less regulated commercial sectors more easily. Fortunately, this book is built on this domain area, where I have deep expertise in.

Table 1.1 maps the expanding frontier of language model applications in financial services, revealing how these tools are reshaping everything

---

[1]McKensey, "The state of AI: How organizations are rewiring to capture value" https://www.mckinsey.com/capabilities/quantumblack/our-insights/the-state-of-ai, 2025, Accessed at 26/07/2025.

Table 1.1.   Applications in different departments of a financial institution.

Customer Service	Financial Consulting	Marketing	Risk Control and Compliance	Operations
– **Intelligent Outbound Calls (call center)**	– **Investment Consulting Assistant**	– **Marketing Content Review**	– **Public Opinion Analysis**	– **Contract Information Extraction**
– **Service Quality Inspection**	– **Investment Consulting Content Quality Inspection**	– **Marketing Content Writing**	– **Public Opinion Search**	– **Contract Writing**
– **Intelligent RM**	– **Investment Advice Report Writing**	– Market insights/ intelligence	– **Event Tagging**	– **Contract Review**
– Agent Training	– Investment Consulting Script Recommendations	– SEO optimization	– **Event Extraction**	– **Form Recognition**
– Loyalty program		– Social media management	– **Regulation Extraction**	– **Comprehensive Search**
			– **Regulation Search**	– NL2 Big Screen
			– KYC/AML	– NL2 Report
				– NL2SQL

(*Continued*)

Table 1.1.   (*Continued*)

Investment Research	Investment Banking	Quantitative Trading	IT
– **Research Report Writing**	– **Investment Banking Draft Generation**	– **Public Opinion Factor**	– **Software development reconstruction**
– **Dehydrated Research Report**	– **Investment management & advisor**	– **Instruction Recognition**	– **Data assets reconstruction**
– **Research Report Retrieval**	– Bank Transaction Flow Single Recognition	– End-to-End Trading	– **Knowledge & doc management**
– **Research Report Tagging**		– Dialogue (Intent Recognition, Similarity Calculation, NL2SQL)	

– Conversional chat (Intelligent Question Answering, intention identification, similarity computing, NL2SQL)

– Documentation processing: information extraction, assistant writing, trend monitoring, index tracking and real-time monitoring for contracts, reports, announcement and regulations

– Multimodal processing: multi-formatting parsing, refined table parsing, conversion between text and image/audio/video

Items in light color are less matured than others.

from back-office operations to client-facing innovation. At the core of this transformation lies natural language processing (NLP) capabilities that go far beyond simple chatbots, i.e., cutting-edge techniques now decode market sentiment, extract key entities from earnings calls, and even predict regulatory concerns before they escalate.

Take text analysis in investment firms for example: hedge funds employ AI-driven sentiment dissection of news cycles to time trades, while compliance departments use context-aware summarization to digest thousands of pages of legal documents into actionable bullet points. These aren't isolated experiments, as evidenced by recent industry analyses [2–4], such applications collectively form a new operational backbone for modern finance.

These technologies are pivotal in transforming financial services, from automating customer service inquiries and enhancing the accuracy of investment research to streamlining compliance and regulatory reporting. By leveraging LLMs for these tasks, financial institutions can not only achieve greater operational efficiencies but also gain deeper insights into market trends, risk factors, and customer needs, thereby delivering more personalized and effective financial products and services.

In general, LLM and GenAI offer significant benefits to enterprises beyond existing AI frameworks by enhancing productivity and streamlining operations. LLMs can automate routine linguistic tasks, integrate data across platforms, and speed up decision-making. By handling tasks like drafting emails and generating reports, LLMs allow employees to concentrate on more strategic activities, improving efficiency.

LLMs enable the creation of new value for customers and stakeholders in the main business departments through personalized interactions and insights into consumer behavior. They can power advanced chatbots for high-quality customer service and help tailor business offerings. Additionally, LLMs improve efficiency in various sectors by automating complex processes, such as supply chain communication and financial forecasting. They also improve the speed and capacity of customer and public interaction by automating responses to public inquiries.

Moreover, LLMs offer transformational potential in functional departments, e.g., HR and legal, by automating the tasks like resume/contract screening and providing personalized education. They can reduce manual documentation processing in risk and compliance to bridge educational gaps. This helps ensure consistency and accuracy in communication and processes. Beyond these mainstream applications, LLMs also hold notable promise in underserved regions and population segments where access to skilled professionals and quality resources is limited. By enabling scalable tutoring, affordable translation, and cost-effective automation of administrative tasks, LLMs can help bridge gaps in education, compliance, and communication, ultimately expanding opportunities and reducing disparities in access to essential services.

In conclusion, LLMs empower organizations by improving operational efficiencies, enhancing customer interactions, and addressing challenges in various sectors and regions. Their ability to analyze data, automate complex tasks, and personalize interactions is essential for organizations looking to stay competitive.

While LLMs promise to revolutionize many industries, e.g., finance, health and education, their implementation resembles navigating a high-stakes obstacle course, e.g., every potential breakthrough comes with hidden tripwires. Let's unpack these hurdles through the lens of industry practitioners wrestling with AI adoption.

1. Technical Constraints: Complexity, Hallucinations, and Bias

LLMs require sophisticated architecture, substantial computational resources, and deep expertise in NLP and machine learning (ML). Organizations lacking such specialized proficiencies must invest heavily in talent development and technical infrastructure, presenting a significant adoption barrier.

A central technical challenge is the phenomenon of hallucination where LLMs generate plausible but inaccurate or fabricated information, such as fictitious corporate earnings or invented regulatory citations. In financial services, such errors have been empirically shown to occur frequently. To address this, practitioners emphasize rigorous data curation, the use of RAG, and debiasing techniques during training.[2]

---

[2] The main sources of LLM hallucination are data-driven issues. These include discrepancies and misalignments between training sources and reference materials, as well as inconsistencies and biases in the data—factors that directly undermine the fidelity of the outputs. Moreover, models trained on static datasets suffer from temporal limitations, rendering them prone to fabricating facts about events beyond their knowledge cutoff.

Another key contributor is the probabilistic nature of LLMs, combined with imperfect representation and decoding mechanisms. As statistical models predicting the next token, they may generate outputs that are plausible in form but wrong in substance. Additional vulnerabilities stem from encoder weaknesses, decoding strategies like sampling or beam

Moreover, algorithmic bias, e.g., stemming from skewed or unrepresentative training corpora, can result in discriminatory or systematically inaccurate outputs. Financial contexts are particularly sensitive, as biased model outputs may propagate unfair credit evaluations or risk assessments. Mitigation strategies often integrate rule-based engines or deterministic data models to counteract bias and promote output fairness.

2. Strategic Alignment: Integration with Planning, Infrastructure, and Customization

The rapid evolution of LLM technology often outpaces traditional enterprise planning cycles, which commonly span one to two years. This mismatch can result in outdated strategies or misaligned operational deployment, necessitating more agile investment frameworks.

The choice between on-premises, cloud-based, or AI-as-a-service deployments significantly influences cost, scalability, and compliance posture. Each model presents trade-offs with respect to resource allocation, maintenance overhead, and integration complexity. Furthermore, the scarcity of enterprise-grade LLM solutions that can be seamlessly embedded in legacy systems compounds integration challenges.

Effective utilization of LLMs demands deep customization to align with domain-specific jargon, existing IT architectures, and compliance obligations. Such customization efforts are often labor-intensive and resource-intensive; for example, adapting an LLM to account for complex financial instruments or regulatory changes may require retraining with updated corpora and integration logic.

3. Regulatory and Governance Hurdles

Financial institutions face a stringent regulatory landscape that includes frameworks such as the EU AI Act, GDPR, and Singapore's Model AI

---

search, and attention misdirections. Furthermore, issues such as tokenizer artifacts, ambiguous or complex prompts, and a general lack of real-world grounding can exacerbate hallucinations, prompting models to invent or misconstrue information.

Lastly, intrinsic architectural and statistical limitations play a role. Hallucinations are not merely fixable bugs but are embedded in the mathematical and probabilistic framework of LLMs themselves, which makes them fundamentally inevitable to some degree. This is compounded by phenomena like model collapse—where recursive training on synthetic outputs degrades reliability—and the "stochastic parrot" effect, where models mimic patterns without understanding, further risking the propagation of falsehoods.

Governance Framework. These regulations impose requirements on explainability, fairness, data provenance, and auditability. The United States and China also have their own legal and regulatory frameworks that are evolving. In China, there are multiple rules released recently, e.g., Generative AI's interim measures and labeling rules, Data Annotation Security Specification, etc. The US previously announced a U.S. AI Bill of Rights in 2022, though the next president and administration are creating a different oversight and regulatory approach.

Governance-related challenges must also be overcome, which include ensuring steerability (guiding model behavior), ensuring interpretability (understanding decision logic), observability (monitoring outputs), and testability (evaluating under diverse conditions). Additionally, data privacy concerns, such as avoiding cross-client leakage, and protection of intellectual property demand stringent frameworks and secure data pipelines.

AI-specific cybersecurity threats, such as adversarial perturbations, prompt injection, denial-of-service attacks, and theft of model assets, require enhanced safeguards beyond traditional information security protocols. Building these protections incurs both technical and regulatory costs.

4. Organizational and Human-Centered Considerations

LLMs have the potential to disrupt workplace culture by altering social dynamics and interpersonal trust. Overreliance on automated agents may foster detachment or anxiety among employees and clients, who may find AI-generated interlocutions unsettling, or feel surveilled by persistent monitoring systems within the workflow.

Persistent use of AI to replace or critique human decision-making can undermine executive confidence and damage the emergence of the "ghost analyst" phenomenon, wherein employees resist AI without understanding its functioning. Tools must therefore preserve human accountability and transparency, with carefully considered approaches to where and how to retain human oversight (including various forms of human-in-the-loop) and quality control.

The emergence of "shadow AI", where employees adopt unsanctioned LLM tools, introducing further governance risks regarding security, data consistency, and compliance. Establishing clear enterprise policies, embedding ethical guardrails, and cultivating shared cultural norms are essential to sustaining organizational cohesion and employee trust.

In general, full-scale integration of LLMs in FIs hinges upon addressing complexity, ensuring strategic alignment, complying with evolving regulations, and preserving the human element. Systematic attention to each domain is indispensable for transitioning LLMs from promising prototypes to dependable, enterprise-grade systems.

1.2.2 *GenAI/LLM Adoption*

The successful adoption of robust GenAI solutions in enterprise production environments necessitates a systematic and methodical approach encompassing technology selection, solution development, and deployment strategies. A foundational aspect of this framework involves enhancing the performance of LLMs through four core techniques: prompt engineering, fine-tuning, RAG and LLM-based agents. These are outlined as follows:

1. Prompt Engineering: Structured Interaction Design for LLMs
Prompt engineering refers to the strategic crafting of input queries to guide LLMs in generating more accurate and contextually relevant outputs. Conceptually, a prompt serves as a contextual cue for a pre-trained language model, facilitating the alignment between human intent and machine response. Technically, prompt engineering involves augmenting the input with tailored textual structures that better exploit the latent knowledge encoded in the model.

Numerous prompt-based strategies exist, ranging from simple heuristics such as "Think step-by-step" to sophisticated approaches like graph-of-thought prompting. These methods can be classified into zero-shot, few-shot, and template-based categories. For instance, in financial analytics, prompt-based tool learning enables models to dynamically invoke external APIs (e.g., Alpha Vantage) by generating well-formed function calls based on natural language inputs, incorporating parameters such as ticker symbol, date, and price type.

Real-world applications illustrate its impact: a global asset manager redesigned its equity research process using dynamic prompt chains. By replacing generic queries with structured analytical frameworks (e.g., sector-specific DuPont analyses),[3] the firm achieved a significant reduction in research time. Prompt engineering proves particularly effective in time-sensitive contexts such as fraud detection, where natural language interfaces rapidly adapt model behavior without modifying underlying code. This concept has been extended to context and harness engineering, covering the whole LLM application pipeline (see Chapter 12).

2. Fine-Tuning: Specialization Through Data-Centric Model Adaptation

Fine-tuning involves re-training a pre-trained LLM on a domain-specific dataset to improve task performance, particularly in specialized applications such as finance, healthcare, or legal analysis. This technique not only enhances accuracy but also imbues the model with contextual understanding of industry-specific language, data structures, and compliance requirements.

Empirical studies underscore the efficacy of fine-tuned models. For instance, domain-specific adaptations of LLMs for financial sentiment analysis and stock prediction demonstrate superior performance relative to general-purpose models. Notably, research has shown that even smaller fine-tuned models (e.g., RoBERTa variants) can outperform general LLMs like ChatGPT in supervised tasks. Moreover, instruction tuning on limited labeled data can yield state-of-the-art results, suggesting cost-effective pathways for domain customization.

Nevertheless, challenges persist. Fine-tuning large models for multiple tasks often entails maintaining separate model instances, imposing significant storage and compute burdens, especially in edge environments. Parameter-efficient fine-tuning techniques, such as adapter modules or low-rank updates, offer practical alternatives by modifying only a subset of parameters.

A notable case study involves a FinTech startup that was fine-tuning LLaMA, a family of LLMs, for regional credit assessments. By incorporating local dialects, informal invoice formats, and non-standard

---

[3] The DuPont analysis is a powerful framework that decomposes Return on Equity (ROE) into its key drivers: profitability, efficiency, and leverage. The general formula is:

$$ROE = \text{Net Profit Margin} \times \text{Asset Turnover} \times \text{Equity Multiplier}$$

When applied to different sectors, the relative weight and interpretation of each component differ. Below is a structured explanation by sector:

communication styles, the adapted model significantly improved interpretability for underbanked demographics. With open-source tools reducing fine-tuning costs dramatically in recent years, the key consideration now lies in assessing the return on investment (ROI) based on data uniqueness and use-case specificity.

3. RAG: Enhancing Contextual Fidelity

RAG systems integrate external knowledge retrieval with language generation, enabling LLMs to access and synthesize up-to-date or domain-specific information. A standard RAG pipeline involves three stages: (1) retrieving relevant documents via embedding-based similarity search, (2) refining results through reranking or augmentation, and (3) generating responses conditioned on the retrieved content.

While RAG frameworks mitigate hallucination and introduce temporal freshness, they also introduce complexity. For example, embedding models face limitations in capturing semantic nuances due to insufficient sample diversity, noisy training signals, and uniform treatment of easy and hard cases. As a result, enhanced variants such as advanced RAG or agentic modular RAGs have emerged.

A salient illustration would be LLMs that integrate real-time data feeds into their responses, using this current information to generate subsequent responses. These real-time feeds may come from a wide range of sources such as recent SEC filings or recent satellite imagery. Examples of such LLMs are Perplexity, and the Research Assistants from OpenAI, Anthropic, Manus (from a Chinese start up) and other LLM providers. Here, models like GPT-4o/5 or Claude 3.7/4 are leveraged for their reasoning capabilities, while RAG systems ensure factual grounding in live data streams.

However, the implementation of RAG is non-trivial. Data preprocessing, encompassing normalization, security, and format alignment, can account for the majority of development effort. Building a robust RAG infrastructure thus requires not only technical sophistication but also organizational commitment to data integration and governance.

4. LLM-Based Agents: Autonomous and Context-Aware AI Systems

LLM agents represent an evolution from passive question-answering models to autonomous systems capable of planning, memory management, action, and tool integration. These agents typically comprise components such as short-term and long-term memory modules, task planning units, and interfaces for invoking external tools (e.g., APIs, databases, or code interpreters).

Memory modules enable contextual continuity, allowing agents to retain knowledge from past interactions and adapt their behavior accordingly. Planning modules decompose complex user queries into sequential subtasks, which are then executed using appropriate tools. Advanced reasoning techniques, such as ReAct (Reasoning and Acting) and self-reflection, further empower agents to iteratively improve their decision-making processes.

In practical terms, LLM agents are transforming enterprise workflows. Goldman Sachs's GS AI Assistant and DBS's CSO Assistant, for instance, coordinates across systems to verify compliance, query risk models, and draft communications, all while adapting to institutional knowledge and prior user interactions.

Nevertheless, integrating such agents into legacy IT environments is often fraught with hidden complexity. Connecting to mainframe systems or outdated APIs can require extensive reverse engineering and long-term infrastructural investment. Yet, the payoff is substantial: LLM agents serve as intelligent intermediaries that unify disparate systems into a cohesive, context-aware enterprise intelligence layer.

How to design a proper implementable strategy and architecture is a key question, considering the above-mentioned techniques have their advantages and disadvantages; Each offers distinct methods to enhance LLMs' capabilities and applicability across various tasks. Table 1.2 shows a brief comparison.

Figure 1.1 illustrates the four techniques in x-y axis in terms of knowledge optimization and LLM optimization, which is further explained in Figure 1.2 with decision paths and the index of a general procedure. Essentially, it outlines a structured decision-making process for applying or augmenting an LLM within various scenarios. Initially, the model is applied with general prompt engineering (instruct tuning). A key step involves assessing the current performance of the LLM against the desired objectives. If there's a gap in knowledge completeness and freshness, RAG is recommended, which utilizes external up-to-date data to enhance the model's responses, making it ideal for scenarios requiring up-to-date and context-rich information. If the query is complex and involves multiple steps or pre-defined actions, we may engage agent-based techniques. If we recognize that existing tools such as search engines, financial database APIs, and financial metric calculators can enhance task performance, it is highly recommended to

Table 1.2.   Comparison of typical LLM extensions.

Technique	Description	Advantages	Suitable Scenarios
Prompt Engineering	Involves crafting specific input prompts that guide the LLM to generate desired outputs without modifying the underlying model.	– Requires no retraining – Quick to implement – Flexible and creative use of model capabilities	– Rapid prototyping – Tasks requiring quick turnaround and adaptability – Situations where model updates are not feasible
Fine-Tuning	Refers to retraining the LLM on a specific dataset (usually not very big) to adapt its responses to particular needs or domains.	– Customizes the model to specific requirements – Improves performance on targeted tasks	– Domain-specific applications – Tasks needing deep customization – When a dedicated dataset for training is available
RAG	Combines the generative capabilities of LLMs with real-time data retrieval to enrich responses with external, factual content.	– Provides up-to-date and factually accurate information – Enhances responses with external data	– Information-intensive tasks requiring accuracy – Dynamic content generation – Fact-checking and knowledge-based applications
Agent Techniques	Employs LLMs within an agent framework to enable complex decision-making, multi-step reasoning, and interaction with other systems or agents.	– Handles complex workflows and tasks – Integrates with multiple systems and APIs – Adaptable to a wide range of environments	– Multi-agent systems – Complex problem-solving environments – Interactive applications requiring ongoing learning and adaptation

employ the techniques with a focus on utilizing agents with tools through prompt engineering to shape their outputs.[4] Conversely, if the

---

[4]Static workflow with tools might perform better than the LLMs in some cases. For example, to calculate capital asset pricing model (CAPM) or conditional value at risk (cVaR), we can simply use agents to call the corresponding functions, which can provide exact results, instead of using LLM's reasoning capability.

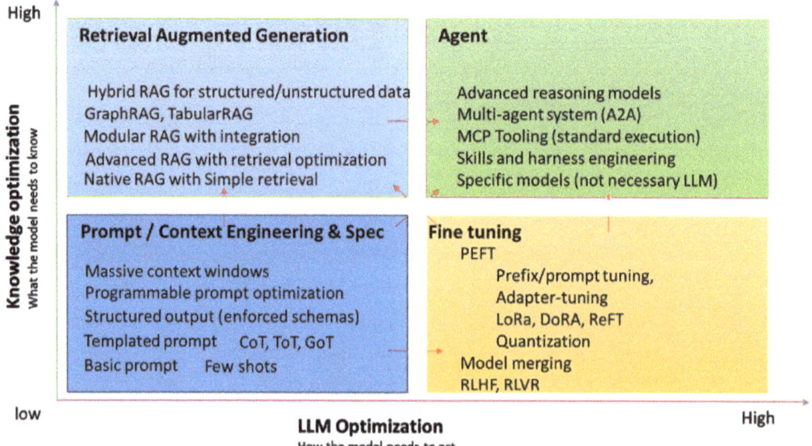

Figure 1.1. Tech comparison.

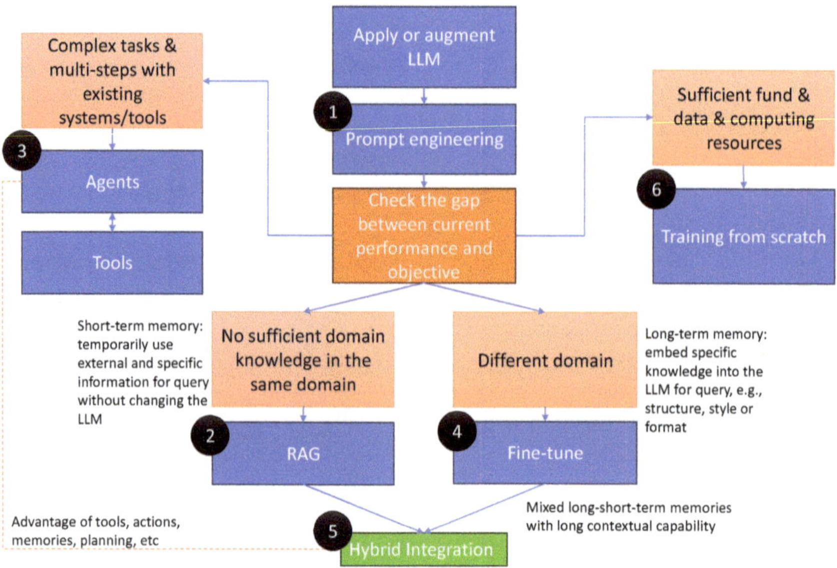

Figure 1.2. Decision path for suitable techniques.

application involves a different domain than the model's initial training, fine-tuning is suggested to adapt the model to the specific characteristics of the new domain, enhancing its accuracy and relevance in specialized contexts.

Although these techniques have distinct features and applicable scenarios, they may not be exclusive to each other. In fact, they can still merge and integrate to improve the overall performance. For example, modular RAG can use Agents to improve the search and reasoning capabilities. Meanwhile, a fine-tuned smaller but application-specific LLM might play a big role for mathematical reasoning inside an agent. A recent idea on agentic context engineering (ACE) treats prompts/memories as evolving playbooks rather than one-shot summaries, directly countering brevity bias and context collapse, which might replace the fine-tuning process of models [5].[5] This usage of those combinations will be reflected in the later chapters.

Note that organizations should only train their proprietary LLMs only if all four techniques cannot meet the targets and they have sufficient funds and resources for it. This systematic approach ensures that the LLM can be effectively tailored to meet diverse and specific needs.

1.3 A Close Look at RAG

Recent surveys of more than 400 banks show that 80% are already piloting or expanding generative-AI projects, with RAG flagged as a top-three enabling pattern for compliance, risk and research.[6] Market analysts project that the global RAG tooling market will grow at 49% CAGR, jumping from around USD 1–2 bn in 2024 to over USD 11 billion by 2030, confirming sustained investor confidence.[7] IDC goes further, predicting that by 2028, 80% of production-grade gen-AI features will

---

[5]Methodologically, it runs a modular loop (generation → reflection → curation) to incrementally update, organize, and preserve detailed strategies that scale to long-context models.

ACE optimizes both offline (system prompts) and online (agent memory) contexts using natural execution feedback instead of labels, yielding gains while reducing adaptation latency and rollout cost. On AppWorld, it matches the top production agent overall and surpasses it on the harder test-challenge split despite using a smaller open-source model.

[6]Temenos, "Survey: Majority of financial institutions deploying generative AI", ABA Banking Journal, https://bankingjournal.aba.com/2025/05/survey-majority-of-financial-institutions-deploying-generative-ai/, May 2025, Accessed on 28/07/2025.

[7]Retrieval Augmented Generation Market Summary, https://www.grandviewresearch.com/industry-analysis/retrieval-augmented-generation-rag-market-report, 2025, Accessed on 28/07/2025.

embed RAG under the hood, making it the de-facto retrieval layer for enterprise LLMs.[8]

Given the practical popularity and versatility of RAG, this book continues by focusing on the use of this technique. The RAG workflow is structured around three main steps: retrieving relevant external information based on the input, fusing or augmenting this information with the input and prompt or intermediate stages using selected fusion techniques, and generating outputs based on the enriched input and retrieval data. Thus, RAG, as its name suggests, comprises three main modules as follows:

Retriever Module: This module typically includes an encoder for transforming inputs into embeddings, an indexing mechanism for efficient approximate nearest neighbor search, and a datastore that maintains external knowledge in key-value pairs. The primary challenge here is balancing retrieval efficiency, how swiftly relevant information can be accessed, which involves optimizing encoding, indexing, and batch queries-and retrieval quality, how accurately the needed information is retrieved, which depends on sophisticated chunk representation learning and search algorithms.

Augmenter Module: It is sometimes also called retrieval fusion, retrieval enhancement/refinement, or retrieval integration/combiner. This component utilizes the information fetched by the retriever to enhance the generation process through various fusion techniques: query-based fusion, latent fusion, and logits-based fusion. Query-based fusion integrates retrieved data with input before processing in the generator. Latent fusion incorporates retrieval data directly into the generator's latent representations, subtly enhancing performance. Logits-based fusion manipulates the output logits from the generator by integrating retrieval logits, aiming to stabilize and enrich the output. This process usually involves prompt engineering techniques that ensure the additional information complements and enhances the original query, effectively transforming it into a more informative input for the generator.

---

[8]"IDC FutureScape: Worldwide Artificial Intelligence and Automation 2025 Predictions" https://my.idc.com/research/viewtoc.jsp?containerId=US51666724, Oct 2024, Accessed on 28/07/2025.

Generator Module: This module is tasked with crafting coherent and contextually appropriate responses based on both the augmented prompt and its pre-existing knowledge. It takes the enriched input provided by the augmenter and processes it through one or few LLMs, which synthesizes information from multiple sources to produce human-like text. This integration allows the generator to leverage both real-time data retrieved by the retriever and its extensive training on diverse language patterns, resulting in outputs that are informative and engaging. However, challenges remain, such as ensuring factual accuracy and managing potential biases in generated content, which necessitate ongoing refinement and evaluation of both retrieval and generation processes.

In real scenarios, there are various types of RAG structure for different applications, depending on the user requirements, data availability, LLM capacity, etc. As summarized by [5–6], the RAG foundations are divided into four types as shown in Figure 1.3, and its query complexity is classified into four levels listed in Table 1.3. The taxonomy categorizes the different ways in which a retriever can augment a generator within the RAG framework.

- **Query-based RAG:** Dominating current implementations, this approach functions like a meticulous research assistant cross-referencing notes mid-analysis. It integrates the user's query with insights from the information retrieved during the process, directly into the initial

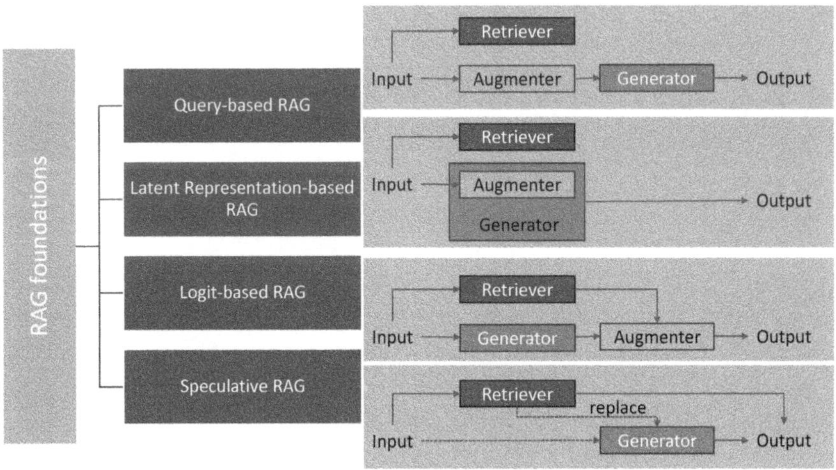

Figure 1.3. RAG foundations.

Table 1.3.   Four levels of queries.

Level	Features	Challenges
1. Explicit Facts	Query asks for facts directly present in data; task is location and extraction. LLM uses retrieved info directly.	Main challenge is correct data retrieval for accuracy; evaluating RAG system performance.
2. Implicit Facts	Query is factual, but answer is not in a single passage; requires combining facts with common-sense reasoning.	Adaptive retrieval volume (avoiding noise/insufficient info); coordinating reasoning and retrieval.
3. Interpretable Rationales	Query demands understanding and applying domain-specific rationales explicitly provided (plain text guidelines, structured instructions like workflows or pseudocode). Requires aligning with expert thinking.	Integrating domain rationales comprehensively and understandably.
4. Hidden Rationales	Rationales are not explicitly documented but inferred from patterns and outcomes in external data (in-domain data, preliminary knowledge); represents implicit domain expertise.	Logical retrieval difficulty (standard methods struggle); decoding and leveraging latent wisdom in disparate data.

stage of the language model's input, i.e., it creates a dialogic context where external knowledge actively reshapes the generation pathway. It is widely adopted and involves merging the retrieved content with the original user query to create a composite input sequence for the generator.

- **Latent Representation-based RAG:** In this structure, retrieved objects are incorporated into generative models as latent representations, e.g., mirroring how human experts subconsciously synthesize disparate concepts, this paradigm embeds retrieved content as compressed neural activations. This improves the model's comprehension abilities and the quality of the generated content by fusing information from multiple retrieved paragraphs in the decoder. Models like Fusion-in-Decoder (FiD)[9] and Memorizing Transformers[10] employ such techniques to facilitate nuanced integration of retrieved information.

---

[9] https://github.com/facebookresearch/FiD.

[10] https://github.com/lucidrains/memorizing-transformers-pytorch.

- **Logit-based RAG:** This method combines retrieval information through logits during the decoding process, typically by summing or integrating them to influence the probability distribution of generated tokens, e.g., assessing the confidence of the logits produced by the generation models. A typical instance is to operate like an editorial team voting on sentence completion, and this method computationally aggregates probabilities from both parametric knowledge and retrieved evidence. This approach requires careful calibration to avoid conflicting gradient signals during training.
- **Speculative RAG index:** It employs a strategy where the model alternates between retrieval and generation steps, allowing for dynamic incorporation of external information during the generation process. This can lead to accelerated generation in certain contexts, though it may introduce trade-offs in accuracy for novel cases. Another implementation might use small-large models, where the small model acts for drafting with extracting from data sources, while the large model acts verification to generate the final answer based on the drafting results.

The actual RAG implementation can be one of the four structures or some extended variants.

Finally, it's important to address the potential issues in each step of RAG process. Organizations must be prepared to address issues like data quality, privacy, model interpretability, and system reliability, which are all significant barriers to adoption. LLM hallucinations can be mitigated using high-quality data, real-time data, RAG, and unlearning techniques. Algorithmic bias can be addressed by combining LLMs with data models and rule engines, while model and data drift can be addressed with quality control measures and algorithm updates. Additionally, RAG can address LLM limitations such as context windows and lack of transparency, and RAG systems can be improved by implementing filtering and ranking mechanisms to address information overload. The complexity of RAG components also poses a challenge that needs to be managed.

1.3.1 *Implementation Challenges and Governance Imperatives*

Imagine equipping an AI with both a library card and a fact-checker: that's RAG in essence. This hybrid approach transforms language models from encyclopedic memorizers into dynamic researchers, but like any powerful tool, it demands skilled handling. Let's dissect its double-edged nature through the lens of real-world deployment.

1. Source Credibility and Verification: The principal advantage of RAG, i.e., its ability to ground answers in external evidence, collapses when the retrieval layer surfaces material of dubious provenance. Empirical analyses show that omissions or low-quality documents in the knowledge base are a dominant cause of erroneous outputs. Industry guidance therefore recommends multi-tier verification pipelines that grade documents on authority, recency, and consistency, analogous to peer review in scientific publishing. Without such controls, even a single obsolete regulation or retracted study can corrupt an otherwise correct synthesis.

2. Context Interpretation and Temporal Fit: A RAG system must not only locate pertinent sources but also interpret the user query with sufficient precision to retrieve temporally and semantically aligned evidence. Recent work on generative query rewriting and advanced semantic ranking demonstrates material gains in relevance and latency, yet it also highlights failure modes when queries are underspecified or ambiguously scoped. Organizations have begun introducing "context gates," algorithmic checkpoints that validate retrieved passages against domain-specific credibility metrics before generation proceeds, thereby reducing misalignment between query intent and retrieved content.

3. Hallucination Containment: Although RAG is frequently promoted as an antidote to LLM hallucination, some recent benchmarking across four public datasets shows that faulty retrieval can amplify fictitious assertions by providing spurious but well-formatted context [6]. Effective defenses include bias-detection layers that flag anomalous passages, real-time block-lists of untrusted domains, and calibrated confidence scores that expose conflicts among sources. These mechanisms collectively reduce the probability of confidently incorrect statements escaping into user-visible outputs.

4. Dynamic Knowledge Maintenance: Unlike static fine-tuned models, RAG architecture is expected to reflect the current state of the world. Achieving this goal demands newsroom-grade ingestion pipelines with strict version control for every document update. Case studies in biomedical fact-checking during the COVID-19 pandemic reveal that in the absence of rigorous timestamping and archival policies, systems continued to retrieve pre-pandemic economic or clinical assumptions well into 2020, producing dangerously outdated recommendations [7]. Continuous monitoring of data freshness therefore remains a non-negotiable operational requirement.

5. Transparency versus Trustworthiness: RAG lends itself to "explainable AI" because it can expose the passages on which an answer is based. However, apparent transparency is illusory if all citations originate from a single biased source. Recent evaluations of RAG pipelines emphasize the necessity of diversity audits that measure not only the number of supporting documents but also the independence and epistemic quality of those documents [6]. Traceability thus constitutes a necessary but insufficient condition for trustworthiness.

6. Multimodal Retrieval Complexity: Modern applications increasingly demand retrieval across text, tables, imagery, and sensor data. Multimodal RAG surveys catalogue unique obstacles, including modality selection, cross-modal alignment, and unit corruption during transformation pipelines. For instance, numerical values extracted from engineering schematics may lose their associated units when converted to plain text, thereby converting "5 mm tolerance" into an ambiguous magnitude that risks unsafe recommendations. Robust schema validation across modalities remains essential to preserve semantic fidelity.

7. End-to-End Pipeline Orchestration: A production-grade RAG system comprises discrete yet tightly coupled stages: intent rewriting, semantic retrieval, dynamic ranking, and constrained text generation. Studies comparing end-to-end pipelines show that errors introduced at any stage propagate multiplicatively through the system. Organizations that achieve near-human precision typically incorporate automated fact-checking against ground-truth databases and mandate human review for high-stakes outputs. Modular architectures that allow hot-swapping of individual components further enable incremental improvements without wholesale redeployment.

Collectively, these seven non-exhausted considerations reframe RAG implementation as a discipline of continuous knowledge governance rather than a one-off engineering task. Only by embedding verification, temporal validation, hallucination checks, transparency audits, multi-modal safeguards, and modular orchestration into the development life-cycle can AI practitioners deliver systems that evolve reliably alongside the domains they serve. To systematically discuss the RAG lifecycle, we will consider the MLOps (ML operations), LLMOps (LLM Operations) and RAGOps (RAG Operations) concepts in Chapter 2. We will further dissect war stories to map the path from theoretical potential to operational excellence in Chapter 3.

1.4 Book Organization

RAG isn't just an AI technique: it's a paradigm shift in machine knowledge management. Once mastered, it creates systems that learn and adapt like seasoned experts. However, when mismanaged, it builds automated misinformation factories. The difference lies in recognizing that RAG implementation isn't an IT project, but an ongoing knowledge governance discipline in a DevOps (Development and Operations) pipeline. Therefore, we describe this pipeline in a systematic way in this book, as illustrated in Figure 1.4.

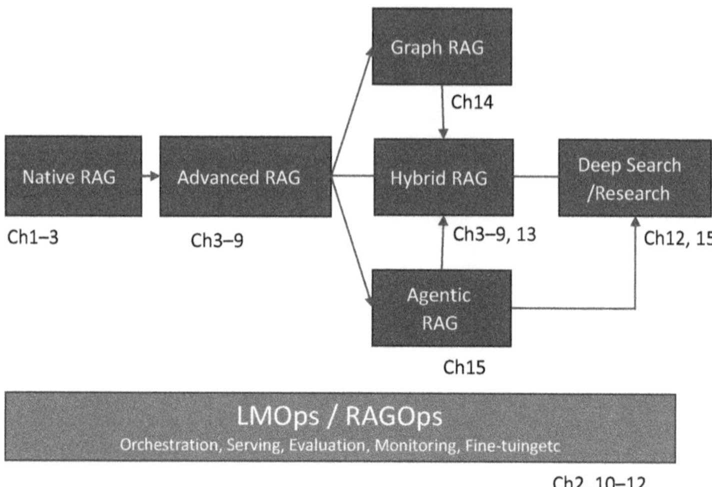

Figure 1.4.   Book structure.

- **Chapter 2** establishes a foundation by surveying the modern ML life-cycle and its operational counterpart, MLOps, to gain a full picture of AI/ML production. We then generalize these principles to LLMOps for governing LLM systems, and finally, to RAGOps, the specialized discipline of productionizing RAG architectures.
- **Chapter 3** maps the principal pain points that arise when RAG is pushed from prototype to production. It serves as an annotated table of contents for the twelve technical chapters that follow, each of which dives into advanced algorithms, implementation patterns, and runnable code, not a surficial introduction.
- **Chapter 4** inaugurates the RAG pipeline with data acquisition and preprocessing, covering ingestion, aggregation, cleansing, parsing, normalization and parallelism.
- **Chapter 5** addresses the text embedding technology, e.g., chunking strategies, embedding models, vector-store selection, and metadata governance, i.e., the building blocks for any high-recall retriever. Over 10 chunking techniques are compared with different applicable scenarios. Tens of embedding models and vector databases are discussed for their pros and cons.
- **Chapter 6** turns to query transformation and prompt engineering, showing how careful reformulation boosts both retrieval accuracy and generative coherence. The typical techniques of query rewriting, decomposition, iterative and routing techniques are illustrated with charts and codes.
- **Chapter 7** operationalizes retrieval itself, demonstrating how to surface the most pertinent vectors from the stores engineered in Chapter 5, guided by the queries refined in Chapter 6. The pre-, in- and post-retrieval techniques are differentiated with focus on the in-retrieval phase. The sparse, dense, advanced and hybrid approaches are discussed.
- **Chapter 8** advances retrieval quality augmentation through reranking, context-window management (e.g., context compression), and hybrid search tactics. For example, the reranking technique, as one of the most important steps in post-retrieval, can be used to significantly improve the content relevancy.
- **Chapter 9** presents controlled-generation methods that ground the language model firmly in retrieved evidence, thereby mitigating hallucinations and improving factual fidelity. The output format, response synthesis and model context are intensively discussed.

- **Chapter 10** introduces rigorous evaluation frameworks in both component-level and end-to-end with over 10 commonly used metrics, while **Chapter 11** translates those metrics into live serving and monitoring practices suitable for production traffic.
- **Chapter 12** explores the auxiliary components in the RAG pipeline such as caching layers, memory stores, and task-specific fine-tuning, illustrating how their orchestration can lift overall performance beyond any single technique.
- **Chapters 13 through 15** push the frontier: Chapter 13 adapts RAG for natural-language-to-SQL systems; Chapter 14 integrates knowledge-graph reasoning via GraphRAG techniques; and Chapter 15 synthesizes agent workflows, i.e., Agentic RAG, that coordinates specialized agents and tools to solve enterprise-grade tasks across structured and unstructured data.
- **Chapter 16** closes the volume with a forward-looking synthesis. RAG is framed not as a terminus but as the current waypoint in the evolution of AI-driven knowledge management. The chapter surveys active research threads from self-improving retrievers, multimodal grounding, edge-native deployment, to autonomous agent coordination, and outlines how these advances may converge into an impending "RAG 2.0" paradigm. In doing so, it provides readers with a strategic lens for anticipating the next wave of techniques and for future-proofing the systems they design today.

Taken together, these chapters offer a complete, hands-on blueprint for AI and ML professionals intent on deploying RAG pipelines that are robust, explainable, and ready for real-world scale. Each chapter also concludes with a concise overview that spotlights promising techniques in its domain, yet we do not examine them in detail because of space constraints. It also sketches forward-looking directions to guide further inquiry.

To support the practical implementation of the solutions discussed, the book is complemented by extensive sample code, available at https://github.com/junxu-ai/RAG_Codes. *The same repository also hosts an appendix, which serves as a reference for common tools, libraries, and additional materials. Readers are warmly invited to contribute their own insights and code via this GitHub repository or through the book's official website at* https://www.worldscientific.com/worldscibooks/ 10.1142/14714.

References

[1] A. Vaswani and *et al.*, *Attention Is All You Need*, Advances in Neural Information Processing Systems 30 (NIPS 2017), 2017.

[2] J. Xu and *et al.*, *The Future and FinTech: ABCDI and Beyond*, J. Xu, Ed., World Scientific Press, 2022.

[3] F. Xing, *Designing heterogeneous LLM agents for financial sentiment analysis,* arXiv:2401.05799, 2024.

[4] A. W. Lo and J. Ross, Generative AI from theory to practice: A case study of financial advice, *An MIT Exploration of Generative AI,* 2024.

[5] Y. Ding and *et al.*, *A Survey on RAG Meets LLMs: Towards Retrieval-Augmented Large Language Models*, arxiv:2405.06211v1, 2024.

[6] S. Zhao, Y. Yang, Z. Wang, Z. He, L. K. Qiu and L. Qiu, *Retrieval Augmented Generation (RAG) and Beyond: A Comprehensive Survey on How to Make your LLMs use External Data More Wisely,* arXiv:2409.14924v1, 2025.

[7] H. Li, J. Huang, M. Ji , Y. Yan and R. An, Use of retrieval-augmented large language model for Covid-19 fact-checking: Development and usability study, *Journal of Medical Internet Research,* 2025.

Chapter 2

MLOps, LLMOps and RAGOps
for Production

This chapter explores the foundational concepts and practical implementations of machine learning operations (MLOps), large language model operations (LLMOps), and retrieval-augmented generation operations (RAGOps) within the context of AI lifecycle management and production deployment. It begins by introducing the AI lifecycle, emphasizing the iterative stages of model development, deployment, and maintenance, alongside the challenges associated with scalability, reproducibility, and governance.

It then delves into the key principles of MLOps, highlighting its core components such as version control, continuous integration/continuous deployment/continuous training (CI/CD/CT), monitoring, and retraining pipelines. Building on this foundation, it introduces LLMOps as a specialized domain, focusing on the operationalization of LLMs. Key considerations include handling large-scale model training, fine-tuning, serving, and optimizing inference costs while ensuring fairness and robustness. Similarly, RAGOps is presented as an emerging framework for integrating retrieval systems with generative models to enhance performance, reduce hallucinations, and improve real-world applicability in domains requiring factual accuracy.

The chapter concludes with an overview of practical frameworks and tools used in these disciplines for LLMOps and RAGOps. By combining theoretical insights with practical implementation guidance, this chapter equips readers with the knowledge to operationalize AI systems at scale, enabling robust, efficient, and responsible AI-driven solutions.

2.1 AI/ML Lifecycle and MLOps

The lifecycle of AI and machine learning (ML) is a multifaceted journey that encompasses several critical phases [1, 2]. For clarity, we break down the ML lifecycle into five primary stages, as depicted in Figure 2.1:

- **Stage 0. Project Formulation and Feasibility Study:** In this initial phase, the AI or ML project is conceptualized and evaluated for feasibility. The project's objectives and scope are defined clearly, establishing a solid foundation for all future efforts. During this stage, the team identifies the essential ML infrastructure and resources needed to achieve the project's goals. Additionally, comprehensive documentation is prepared, and necessary approvals are secured from stakeholders and regulatory bodies, such as those governing data and infrastructure.
- **Stage 1. Data Preparation and Preprocessing:** Once the project receives approval, attention turns to gathering and cleaning data. This stage, primarily managed by developers, involves collecting relevant data from various sources and performing initial cleaning and validation to ensure high quality. Developers employ exploratory data analysis (EDA) techniques to gain insights into key statistical metrics, such as counts, maximum and minimum values, averages, and distributions, and to format the data appropriately for model development.
- **Stage 2. Model Development and System Design:** In this stage, developers focus on transforming raw data into features that the model can effectively use. They engage in feature engineering, select the most predictive variables, and choose suitable ML algorithms. The model is then iteratively trained and fine-tuned, with techniques such as cross-validation employed to evaluate performance. Concurrently, system design ensures that the model integrates seamlessly into the broader application architecture. Sandbox environments might be used for some further system integration tests when necessary.
- **Stage 3. Model Deployment and Validation:** Following development, the model is deployed into various environments, including staging and production. This phase is dedicated to ensuring that the model operates

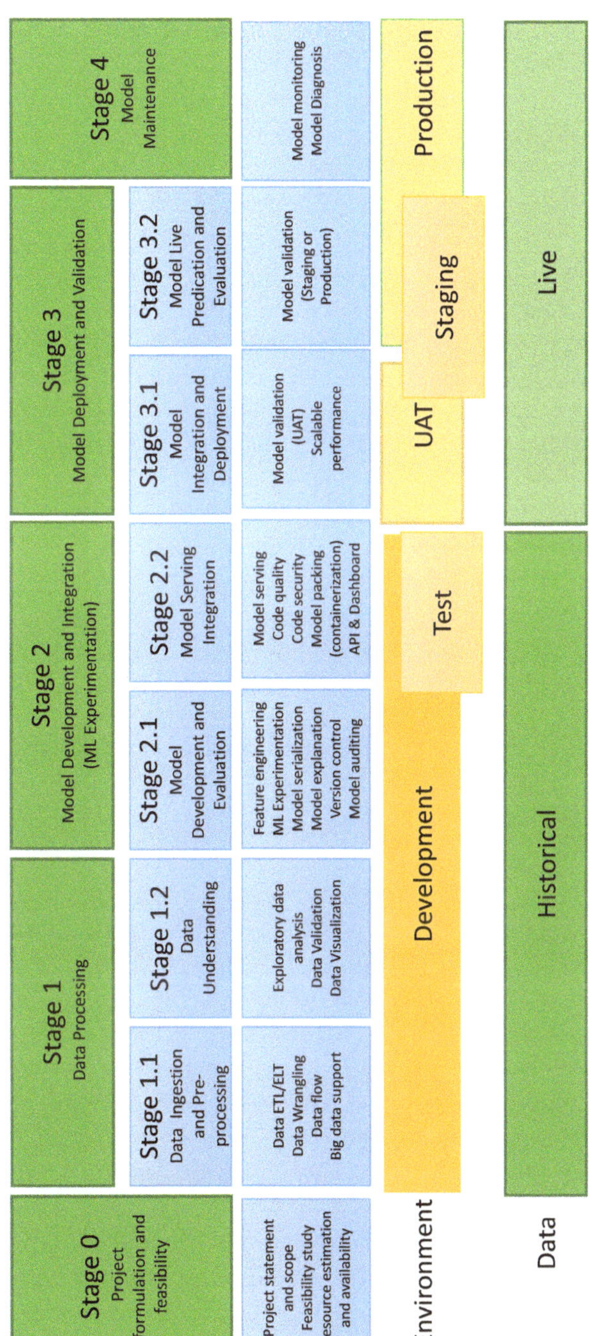

Figure 2.1. Five stages of the ML lifecycle.

reliably in real-world scenarios and meets all operational criteria. Rigorous testing (e.g., UAV, UVT) validates that the model's predictions are both accurate and dependable, all while adhering to pertinent regulations and standards.

- **Stage 4. Monitoring, Maintenance, and Diagnosis:** The final stage focuses on continuous monitoring, maintenance, and troubleshooting to preserve the model's performance over time. Ongoing tracking systems are established to evaluate the model's effectiveness, promptly address any emerging issues, and update the model as new data or conditions arise. Feedback is consistently collected to drive further improvements.

By thoughtfully navigating these five stages, organizations can effectively manage the complexities of the AI and ML lifecycle. This systematic approach not only paves the way for successful project outcomes but also ensures sustained value from AI initiatives. To streamline this process further, it is crucial to automate lifecycle pipelines using DevOps methodologies, which blend software development with IT operations.

MLOps [1–4] was therefore proposed and has become a rapidly growing field that combines the best practices of software development, data science, and DevOps to streamline the lifecycle. MLOps is a set of practices and tools that enable data scientists and machine learning engineers to develop, deploy, and maintain machine learning models in a production-ready environment. It involves the entire ML lifecycle, from data preparation, model development and integration to model deployment and monitoring, as illustrated in Figure 2.1. MLOps aims to bridge the gap between data science and software engineering, ensuring that ML models are reliable, reproducible, and scalable. MLOps is crucial in today's data-driven world, where ML models are being used to make critical decisions in various industries. MLOps enhances data quality and ethical management in LLM training, mitigating biases to produce reliable and compliant models. It simplifies AI development, fostering innovation, increasing developer engagement, and accelerating time-to-market. MLOps streamlines collaboration and standardization across teams, boosting productivity and efficiency. It provides tools for risk mitigation and ensures ethical AI use and compliance, especially in regulated sectors. Additionally, it improves scalability and model performance through systematic evaluation and monitoring.

Almost all major cloud service providers sell their MLOps platforms, e.g., Azure Machine Learning, AWS SageMaker, Google Vertex, and Aliyun PAI. Other popular platforms include Databricks (MLflow), DataRobot, Weights & Biases (W&B), Valohai, Comet ML, and Kubeflow. Choosing or building a suitable MLOps platform involves a thorough evaluation of several critical factors tailored to the organization's specific needs and existing infrastructure. First, assess the cloud and technology strategy to ensure compatibility with current tools and frameworks, as well as the expertise of the team, which dictates the platform's usability and learning curve. Consider the integration capabilities with existing data sources and DevOps tools to facilitate seamless workflows. It's also essential to define key use cases that the platform must support, ensuring it can effectively address the organization's primary business challenges. Additionally, evaluate commercial details such as pricing models and vendor support arrangements, alongside the platform's scalability to accommodate future growth. Finally, look for an active user community and a clear product roadmap to ensure ongoing support and development alignment with organizational goals.

2.2 From MLOps to LLMOps

With the recent emergence of LLMs, the operations on these have become another interesting topic. This is where LLMOps,[1] a specialized domain focusing on the lifecycle management of applications powered by LLMs, come into play. Figure 2.2 highlights the LLM functionalities and components associated with LLMOps; while not exhaustive, it captures the primary or predominant elements.

Note that although LLMOps is a subset of MLOps, LLMOps differs from the general MLOps mainly in handling LLMs, focusing on efficient data usage, specialized experimentation, and distinct evaluation challenges. Experimentation in LLMOps focuses on prompt engineering versus fine-tuning, requiring different setups, and evaluation often relies on A/B testing because of the nuanced outputs of LLMs. Costs are higher

---

[1]Large model operations (LMOps) is a more general case than LLMOps, when the former contains both large language models (LLM) and vision language models (VLM) or even some general pre-trained models.

Service management

Scenario management

Configuration	Task creation
Knowledge management	Application management
pipeline	Automated test

Online services

Security/persona Auth	Monitoring/Logging/cost reporting	audit trail
guardrails	Validation evaluation	
QC	API gateway	

Model/data management

Metadata/data annotation	Collaboration	Content moderation
Data mining & synthesis	Security and privacy: PII detection & masking	

Agent/prompt management

Prompt studio/recipe	Agent studio/template	Tool management
NLP recipe	Safety and compliance	

Engineering

Prompt engineering

Zero-short/few-short	Chain/tree/graph-of-thought
Pattern-based template: costar/automate	self-criticism/role-playing

RAG engineering

Query rewriting	Routing	Rerank
Auto-merging	Recursive	Hybrid fusion
compression	Dense x	

Fine-tuning engineering

LoRa/QLoRa	quantization	SFT/RLHF
distillation	ReFT	Adaptor
PEFT	ZeRo/DeepSpeed	

Agent engineering

ReAct	self-criticism	memory
tooling	planning	decision
workflow	collaboration	

Technology foundation

LLM (generation)

OpenAI	Gemini/gemma	Claudra
ChatGLM	Llama	Mistral/Mixtral
Phi-MoE	Yi/Qwen	WizardLM
deepseek	Kimi	Blossom

Embedding

BGE	Bert/FinBert FinGPT
OpenAI	M3E/ERNIE

Core algorithms

Context fusion	memory
multimodal	Long context

Vector database

pinecore	Chroma	pgvector
Faiss	Milvus	Qdrant

Framework

Langchain	Lammaindex
GPTflow	graphGPT
AutoAgent	HF Agents
Langroid	Haystack
MetaGPT	RAGflow

Feature store

Feathr	Databricks	Feast
Vertex AI FS	Hopsworks	AWS FS

Figure 2.2.   LLM functionalities.

during the inference in LLMOps, especially when managing long prompts, and LLMs demand robust computational resources and optimization techniques like model compression. Latency is more pronounced in LLMOps, impacting both development speed and user experience, and there's a greater emphasis on transfer learning, human feedback loops, and prompt engineering to mitigate risks like model hallucinations and prompt hacking. Additionally, LLMOps uses different performance metrics such as BLEU, ROUGE, trustworthiness and relevance, and specialized tools for pipeline construction, highlighting the distinct nature of LLMOps compared to MLOps.

Therefore, the lifecycle of an LLMOps is even more comprehensive than a general MLOps, starting from problem formulation, data management to model selection, iterative prompt management, testing and evaluation, deployment, and continuous monitoring, although we can still fit those steps into five stages. Note that each step presents its own set of challenges and requires specialized tools and methodologies. In the following, the unique features of LLMOps are briefly described:

- **Problem Statement and Downstream Definition:** We shall have a clear problem description and then properly define the corresponding downstream tasks. Unlike traditional MLOps, which often deal with structured outputs or classification tasks, LLMOps must define downstream tasks with a focus on language generation, understanding, and nuanced contextual interpretation.
- **Data Management:** In LLMOps, robust data management is crucial, encompassing both the extensive training data and the more specialized fine-tune datasets. For training data, the process involves handling vast, heterogeneous text corpora, entailing rigorous cleaning, preprocessing, annotation, and storage. This phase demands advanced techniques to organize and version-control large volumes of diverse language inputs, ensuring data quality and consistency at scale. For fine-tuning data, the emphasis shifts toward curating domain-specific datasets that enhance the model's performance on targeted tasks. This includes additional layers of annotation and refinement to capture nuanced contextual information and mitigate biases. Effective version control is imperative to track changes and improvements in these smaller, yet critical, datasets, ensuring that iterative updates align with evolving business requirements and model objectives. Together, these strategies underline the dual focus in LLMOps: managing expansive, general-purpose

training data alongside meticulously curated fine-tuning data, each requiring tailored processes to optimize the overall performance and reliability of language models.

- **Model Selection:** Choosing the right pre-trained model(s) is a critical step.[2] It involves considering the model's size, architecture, performance, and suitability for the task (alignment) at hand. Once a foundation model is chosen, it can be accessed via its application programming interface (API) for adaptation to downstream tasks. Interacting with LLM APIs may initially seem unfamiliar, as the relationship between input prompts and output completions is not always predictable; the API generates responses based on learned patterns, which may lead to unexpected results. Therefore, iterative prompt engineering and thorough testing are recommended to achieve the desired outcomes. In rare cases, training a model from scratch might be considered, particularly when unique data requirements or specific customization needs arise. However, this approach is resource-intensive and typically undertaken by organizations with substantial resources or for specialized research purposes. For most applications, leveraging and fine-tuning existing pre-trained models is a more practical and efficient strategy.
- **Vector Databases (Embedding Storage) and/or Feature Store:** After post-processing, the model may return more than just plain text responses. Advanced applications may require embeddings—high-dimensional vectors that represent semantic content, e.g., RAG, sentiment analysis and recommendation systems. These embeddings can be stored or provided as a service, enabling quick retrieval or comparison of semantic information, thus enriching the ways in which model functionalities are utilized beyond just text generation. A feature store may also be required when the structured data is used in the application too.
- **Prompt Engineering (Iterative Prompt Management):** Crafting the right prompts to guide the LLM's responses is an art. It requires iteration and fine-tuning to ensure the model's outputs are accurate and relevant.
- **Fine-tuning Pre-trained Models:** It can significantly enhance their performance on specific tasks. While this process requires additional training efforts, it often leads to more efficient inference, thereby reducing operational costs. LLM API costs are typically influenced by

---

[2]Depending on the application and the system architecture, multiple models might be used, e.g., embedding model, generation model, and reasoning/schedule/planning model.

the length of input and output sequences; therefore, minimizing the number of input tokens (by making them clearer and more concise) can effectively lower these expenses. By internalizing recurrent prompts during fine-tuning, the need for lengthy inputs is diminished, resulting in faster inference and cost savings.

- **RAG:** It is an AI framework that enhances the capabilities of LLMs by integrating external information retrieval mechanisms. In this architecture, when a user submits a query, the system retrieves relevant data from external sources (such as databases, documents, or the internet) and combines this information with the LLM's internal knowledge to generate a more accurate and contextually relevant response. This approach allows LLMs to access up-to-date information without requiring retraining, thereby improving response quality and reducing the likelihood of generating outdated or incorrect information.

- **AI Agent:** In the context of AI agent systems, the term "Agent-Tool" refers to the collaborative interaction between the agent and external tools or systems. Within this framework, the "Agent" is responsible for interpreting user queries, determining the appropriate external tools or data sources required, and orchestrating the activation of these tools to fulfill the task effectively. The "Tool" represents the external system or resource that provides specific functionalities or data, such as APIs, databases, or software applications. For instance, an AI agent designed for customer support might utilize a knowledge base tool to retrieve relevant information in response to user inquiries. This dynamic selection and activation of tools enable AI agents to perform a wide range of tasks by leveraging external resources, thereby enhancing their versatility and effectiveness.

- **Testing and Evaluation:** Assessing the model's performance is complex due to the qualitative nature of language tasks. It involves developing a diverse set of metrics and testing strategies. For responsible AI (RAI), we shall also satisfy ethics and legal compliance, i.e., ensure that the use of LLMs adheres to ethical standards and complies with legal norms, especially concerning privacy, bias, and fairness.[3] Key measures include selecting cloud storage locations compliant with local data residency laws, securing data transmission through isolated network environments, and employing auditing and monitoring tools like AWS CloudTrail and GCP's Cloud Audit Logs. Additionally,

---

[3]https://github.com/junxu-ai/LLM_fairness.

protecting against prompt injection attacks and model inversion attacks is critical, with techniques such as input validation and differential privacy enhancing security. Optionally, the playground component provides an environment where developers can iterate and test AI prompts, ensuring they perform optimally before being embedded into the application. Tools such as OpenAI, nat.dev, and Humanloop offer platforms for fine-tuning and testing LLM prompts.

- **Orchestration:** The integration step coordinates the various components and workflows within the LLM application, abstracting details, such as prompt chaining and interfacing with external APIs. These layers are crucial for maintaining memory across multiple LLM calls and ensuring smooth operation. Human-in-the-loop interruption might be embedded into the pipeline whenever necessary.

- **Deployment:** Deploying LLM requires addressing unique challenges, such as API integration for real-time language processing, managing latency issues inherent to high-compute operations, and scaling infrastructure to support variable workloads. In LLMOps, techniques like sampling multiple outputs and ensemble methods are critical to capture the inherent stochasticity of language generation. Additionally, model management in LLMOps involves a more rigorous version control system—not only for the models but also for their linguistic context—allowing for seamless updates with new data or performance improvements. This contrasts with general MLOps deployments, where such language-specific versioning and output management techniques are less pronounced.

- **Monitoring:** Continuous monitoring in LLMOps must extend beyond typical performance metrics to include evaluation of language quality, coherence, and user satisfaction. It is vital to track the linguistic behavior of the model over time and adapt to evolving language patterns and business needs. This may involve specialized monitoring for output diversity, error patterns in language generation, and computational resource utilization. In contrast, general MLOps monitoring is typically centered on performance, scalability, and system reliability without the added complexity of natural language evaluation.

The general steps in LLMOps closely resemble those in MLOps, but there are some distinct differences due to the unique nature of working with LLMs. For example, feature engineering, a key step in MLOps, shifts to prompt engineering in LLMOps, where the focus is on designing effective prompts to guide the model's responses. Similarly, the traditional training process in MLOps is often replaced by fine-tuning existing pre-trained models in LLMOps to adapt them for specific tasks. Additionally, model evaluation in LLMOps is more complex due to the open-ended nature of language generation tasks, which require nuanced metrics to assess performance beyond standard metrics like accuracy or loss. These differences highlight the specialized adaptations necessary for handling LLMs in production.

2.3 From LLMOps to RAGOps

When the RAG projects go into the production environments, we will expect a comprehensive architecture with some unique supporting components. Similar to the MLOps or LLMOps framework as discussed in the previous chapter, we can build a reference architecture for RAG Operations (RAGOps) as depicted in Figure 2.3 with five layers. This architecture incorporates additional modules and phases beyond the three primary modules previously mentioned. In the remainder of this section, we will describe these components in more detail.

Data Preparation: The phase is the foundation of the RAG system, as the quality of the retrieved and generated content depends heavily on the data used during this phase. In this step, the external knowledge base that the RAG system will draw from is curated. This involves collecting, cleaning, and structuring data to ensure it is accurate, relevant, and properly indexed. Depending on the use case, this data may come from a variety of sources, such as documents, databases, or real-time feeds. Techniques such as entity extraction, semantic indexing, and embedding generation are often employed to ensure that the data are in a format that can be efficiently retrieved by the system. Special attention must be given to the quality and freshness of the data to ensure that the RAG system produces reliable and timely responses.

This process is a meticulous orchestration of several steps: tokenization, encoding, quality filtering, and data deduplication. Tokenization meticulously segments text into discrete units known as tokens, enhancing the model's ability to process data efficiently. Positional encoding infuses these tokens with a sense of order, crucial for the model to comprehend the narrative flow within sequences. The sanctity of data is preserved through rigorous quality filtering and deduplication, removing the

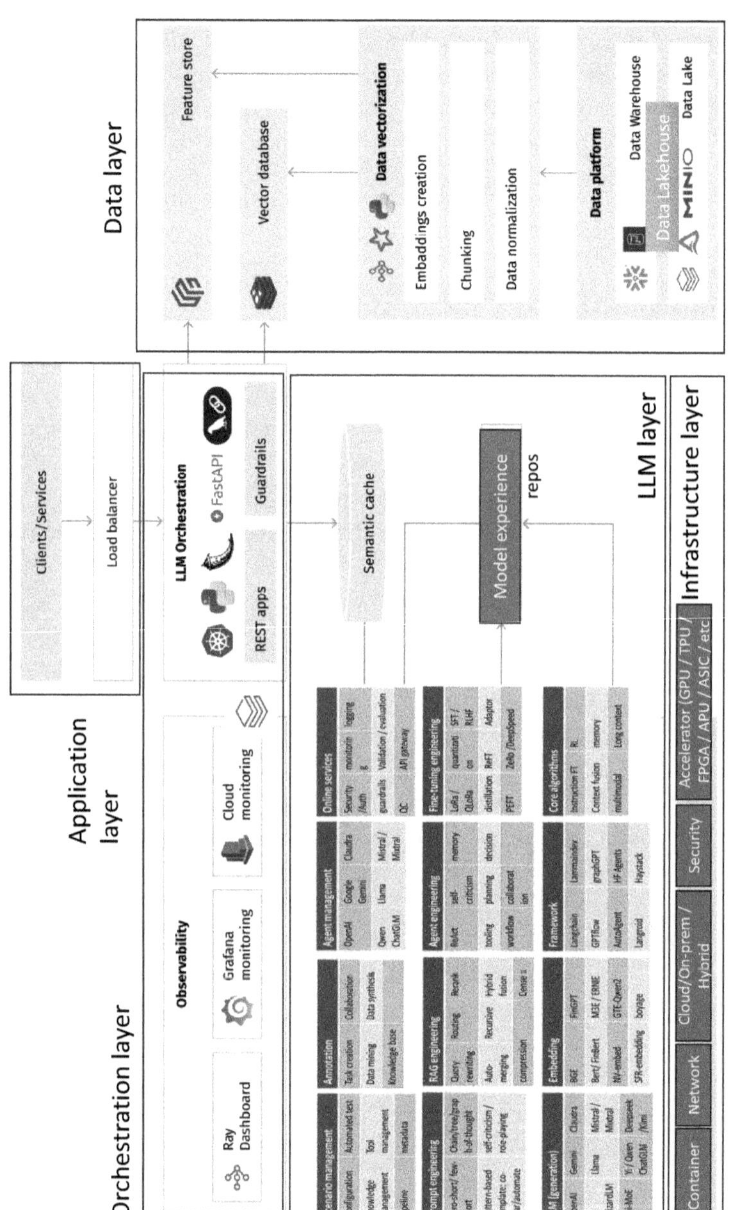

Figure 2.3.   A sample RAG reference architecture.

cacophony of noisy or redundant data. This cleansing ritual ensures that the data are pristine, structured, and optimally primed for training, significantly boosting the model's performance, reliability, and accuracy. Moreover, sophisticated processing techniques tackle the inherent complexities of training, enabling these computational behemoths to thrive even on constrained hardware resources through strategies like optimizer parallelism.

Quality Control: Vital for curating high-quality datasets, techniques such as classifier-based predictions and heuristic rules (involving language metrics, statistics, and keyword analyses) refine data, enhancing the model's efficacy in tasks requiring coding acumen and commonsense reasoning. In addition, we pay attention to the data diversity (e.g., format, contents) and freshness (e.g., data temporal shift and misalignment). In terms of instruction used for model fine-tuning, the importance of the quality and diversity is even higher than that used for the unsupervised phase.

Data Deduplication: This technique is crucial for excising redundancy at various data levels—sentences, documents, and datasets—thereby optimizing training efficiency and model accuracy through mechanisms like N-gram similarity assessments using MinHash.

Tokenization and Encoding: Tokenization dissects text into fundamental units, tokens, which could be characters, subwords, or symbols, depending on the tokenization technique employed. Encoding, particularly in the domain of transformers, involves supplementing token embeddings with positional encodings to imbue them with sequential intelligence, crucial for effective model training.

Alignment Filtering: Toxicity control refers to the filtering of the text content which is "rude, disrespectful, or unreasonable language that is likely to make someone leave a discussion." Meanwhile, we must consider the social bias, e.g., the marginalization of minority groups caused by data detoxifying, representational harms as well as excluded voices and identities in large web text corpora.

Query Refinement: This phase, often referred to as part of Prompt/ Context Engineering, is crucial for optimizing how the RAG system interacts with the LLM. This phase focuses on crafting precise, context-aware queries that can be effectively understood and processed by the model. Since the quality of the prompts directly affects the relevance and accuracy of the retrieval and generation process, prompt engineering involves experimenting with different query formulations, selecting the right keywords, and structuring queries in a way that maximizes the LLM's ability to understand and generate useful information. This iterative process ensures that the system retrieves the most relevant data from the external knowledge base and generates content that is accurate, coherent, and aligned with the user's intent.

RAG Evaluation: This phase assesses the performance of the RAG system by measuring its effectiveness across various metrics, such as accuracy, relevance, coherence, and response time during different phases. For example, in the retrieval phase, we shall evaluate accuracy, relevance and importance of the retrieved contents. In the generation phase, we might pay more attention to accuracy, coherence and fairness. This step often involves both automated testing and human-in-the-loop evaluations. For automated testing, classification metrics (e.g., precision, recall, and F1 scores) and NLP metrics (e.g., perplexity, BLEU, ROUGE and SPICE) and fairness metrics (e.g., WEAT, SEAT, DisCo and LPBS) [1] are commonly used to measure how well the system retrieves and generates information. Human evaluators may assess the quality of the output, checking whether the responses are factually correct, contextually appropriate, and useful for the task at hand. Continuous evaluation (with monitoring) is important for identifying performance bottlenecks, improving retrieval algorithms, and fine-tuning the LLM. This phase ensures that the RAG system remains reliable, robust, and capable of producing high-quality outputs. Then we have the phases in the main procedure, retrieval phase, augmentation phase and generation phase, as discussed in the previous chapter.

LLM Serving and Monitoring: This phase involves deploying the RAG system into a production environment where it can handle real-time queries from users. In phase is similar to the general LLMOps; however, it may involve multiple model deployments. During this step, the LLM is integrated with retrieval systems and deployed using scalable infrastructure. Continuous monitoring of system performance is essential to ensure that the LLM responds to user queries efficiently and with minimal latency. Monitoring tools are used to track key performance indicators (KPIs), such as the metrics aforementioned, response times, and error logs. This phase also includes managing resources such as server load, ensuring that the system can scale effectively to meet varying demand. Monitoring allows for real-time detection of issues, such as system slow-downs, inaccurate outputs, or model drift, enabling timely interventions to maintain optimal performance.

RAG Pipeline (Orchestration): The module focuses on the end-to-end management of the RAG system's workflow. Orchestration tools like Apache Airflow or Kubernetes are often used to manage the multiple components involved in the RAG pipeline, including data ingestion, query processing, retrieval, and generation. This phase ensures that each part of the system, from data retrieval to response generation, works seamlessly together. The pipeline is responsible for ensuring that queries flow through the system efficiently, retrieving the necessary information from the knowledge base, feeding it into the LLM, and generating the final response in a timely manner. Orchestration also handles tasks like error recovery, retries, and automated scaling, ensuring that the system remains robust and resilient under different workloads. By streamlining the entire workflow, the RAG pipeline ensures high efficiency, scalability, and fault tolerance.

Note that the primary goal of LLMOps is to streamline the process of bringing LLM-based applications to production efficiently and effectively, and focuses on optimizing the performance of the LLM itself, emphasizing inference speed and resource consumption. RAGOps, on the other hand, is specifically focused on managing systems that combine retrieval mechanisms with language generation models; thus, it optimizes both the retrieval process (e.g., search speed, accuracy), argumentation and the generation process, balancing them for overall system performance. The following chapter will describe the key points encountered in each step. Another important point is that the steps outlined here illustrate a typical RAGOps pipeline commonly used in current production scenarios. However, they might not apply in certain situations, such as a unified model that integrates RAG functionality directly into a single component.

2.4 Concluding Remarks

Successful adoption of RAG via LLMOps and RAGOps requires a deliberate upgrade of both IT infrastructure and application-layer design. We therefore introduce a reference logical architecture organized into five functional strata, as shown in Figure 2.4: the application layer, where end-user workflows reside; the orchestration layer, which coordinates agents, retrieval pipelines, and safety guardrails; the LLM layer, encapsulating foundation models and domain-specific adapters; the data layer, spanning vector stores, knowledge graphs, and governance services; and the infrastructure layer, comprising compute, storage, networking, and observability primitives. More details can be found in the supplementary material.

Compared with classical MLOps stacks, this infrastructure now assumes a more central role. Modern platforms must expose low-latency retrieval APIs, streaming in-context learning buffers, and scalable inference gateways, in effect turning what were once bespoke model-development tasks into platform capabilities. Consequently, activities traditionally performed by data scientists, such as feature engineering or bespoke model fine-tuning, are increasingly supplanted by prompt- and context-engineering workflows executed by ML engineers. These workflows can, in turn, be exposed through low-code or no-code interfaces, enabling business analysts and product owners with minimal programming expertise to compose sophisticated RAG and agentic pipelines. The architecture thus re-balances responsibilities across roles while raising the strategic importance of a robust, extensible infrastructure foundation.

We also discuss two types of implementations in the supplementary material: cloud-dependent and cloud-agnostic. In cloud-dependent designs, hyperscale providers (such as AWS, Azure, and Google Cloud) supply deeply integrated services for every stage of the LLMOps and RAGOps lifecycle. Native offerings such as SageMaker, Azure Machine Learning, and Vertex AI deliver turnkey capabilities for model training, versioning, continuous integration and delivery, observability, and elastic inference, thereby streamlining automation and performance tuning for LLM or RAG workloads. Cloud-agnostic architectures, by contrast, prioritize portability. They are constructed from interoperable building blocks, including open-standard orchestration engines, vendor-neutral observability stacks, pluggable vector databases, and registry-based model repositories, so that the same pipeline can run unchanged across

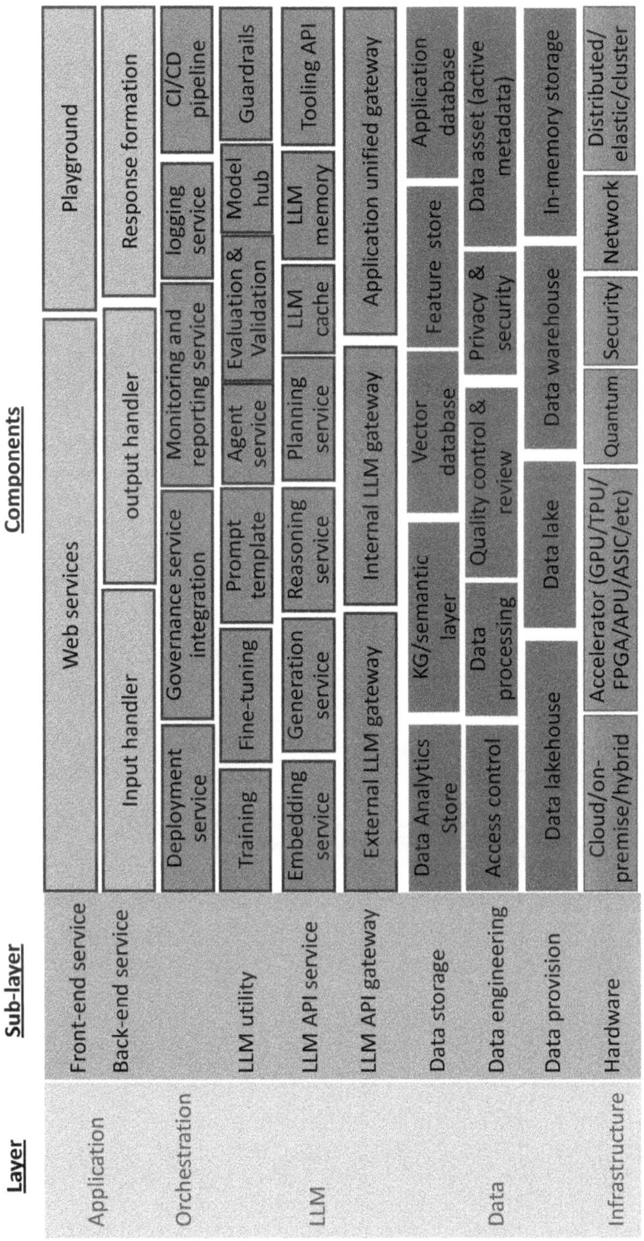

Layer	Sub-layer	Components
Application	Front-end service	Web services; Playground
Application	Back-end service	Input handler; output handler; Response formation
Orchestration		Deployment service; Governance service integration; Monitoring and reporting service; logging service; CI/CD pipeline
LLM	LLM utility	Training; Fine-tuning; Prompt template; Agent service; Evaluation & Validation; Model hub; Guardrails
LLM	LLM API service	Embedding service; Generation service; Reasoning service; Planning service; LLM cache; LLM memory; Tooling API
LLM	LLM API gateway	External LLM gateway; Internal LLM gateway; Application unified gateway
Data	Data storage	Data Analytics Store; KG/semantic layer; Vector database; Feature store; Application database
Data	Data engineering	Access control; Data processing; Quality control & review; Privacy & security; Data asset (active metadata)
Data	Data provision	Data lakehouse; Data lake; Data warehouse; In-memory storage
Infrastructure	Hardware	Cloud/on-premise/hybrid; Accelerator (GPU/TPU/FPGA/APU/ASIC/etc); Quantum; Security; Network; Distributed/elastic/cluster

Figure 2.4. The LLM stack.

multiple clouds or on-premises clusters. This approach safeguards against lock-in while preserving scalability and operational consistency. The supplementary material concludes with a comparative analysis of prominent RAG frameworks and support tools, such as Langchain and LlamaIndex, detailing how each aligns with the capabilities and trade-offs inherent in these two infrastructure strategies.

References

[1] J. Xu *et al.*, *The Future and FinTech: ABCDI and Beyond*, J. Xu, Ed., World Scientific Press, 2022.

[2] M. Treveil and Dataiku team, *Introducing MLOps: How to Scale Machine Learning in the Enterprise*, O'Reilly, 2020.

[3] S. Alla and S. K. Adari, *Beginning MLOps with MLFlow: Deploy Models in AWS SageMaker, Google Cloud, and Microsoft Azure*, Apress, 2021.

[4] A. McMahon, *Machine Learning Engineering with Python: Manage the Lifecycle of Machine Learning Models using MLOps with Practical Examples*, 2nd ed., Packt, 2023.

Chapter 3

RAG Challenges and Solutions

3.1 Introduction

In the high-stakes arena of AI innovation, RAG has emerged as a bridge between the static intellect of traditional language models and the pulsating heartbeat of real-world data. Imagine transforming an encyclopedia-bound scholar into a field researcher with a live satellite feed: this is RAG's core transformative capability. By dynamically marrying external data streams (think SEC filings, peer-reviewed journals, or IoT sensor networks) with the reasoning prowess of LLMs,[1] RAG doesn't just generate text; it curates knowledge [1, 2].

Yet beneath this promise lies an implementation minefield: the system's credibility hinges entirely on its data sources, as seen when an AI chatbot's symptom-checker accidentally cited retracted studies or unreliable sources [3] or Google's AI Overviews pulled a Reddit joke and recommending glue on pizza,[2] while latency and consistency challenges force trade-offs between speed and accuracy.

---

[1]LLMs exhibit notable fluency and pattern recognition, which can mimic reasoning in familiar contexts. However, some critiques highlight that current LLMs often rely on statistical correlations rather than genuine deductive or causal reasoning theoretically, which might be non-robust via small vibrations, and prone to hallucination, affecting reliability, in some scenarios.

[2]Kylie Robison, Google promised a better search experience—now it's telling us to put glue on our pizza, https://www.theverge.com/2024/5/23/24162896/google-ai-overview-hallucinations-glue-in-pizza, May 2024, Accessed on 4/11/2025.

The technical tightrope walk involves balancing precision and context, e.g., Goldman Sachs' research tool mastered semantic paragraph clustering to extract financial insights without losing narrative flow, while Amazon's Q&A system juggles pre-indexed common queries with real-time niche searches. However, even flawless retrieval can't eliminate embedded biases; a credit-scoring RAG might impeccably cite current regulations while perpetuating historical prejudices hidden in court records. Version-controlled knowledge snapshots, like those in Norton Rose Fulbright's legal bot, and bias-aware filters are becoming essential safeguards, yet most implementations still overlook these layers. Ultimately, RAG's industrial adoption demands more than just technical prowess—it requires rigorous data governance and ethical scrutiny.

In this chapter, we will provide a comprehensive discussion on the challenges that we may encounter during the development and delivery of RAG products in production environments. It also serves as the outlines of the RAG solutions and methods of this book.

3.2 RAG Lifecycle

Developing and deploying a RAG system involves navigating a complex lifecycle that encompasses several critical phases, as discussed in Chapter 2. In Chapter 1, we can roughly divide its main procedure into three phases. However, in this chapter, we provide fine-granularity frameworks within RAGOps that comprise eight phases.[3] Each phase presents unique obstacles and difficulties that can have an impact on the system's overall performance and effectiveness. For example,

1. **Data Preparation Phase:** Challenges arise in collecting, cleaning, and structuring vast amounts of data in various formats. Ensuring data quality and relevance is crucial for accurate retrieval and generation.
2. **Query Refinement Phase:** Understanding user intent and crafting precise queries or prompts is essential. Difficulties here involve handling ambiguous or poorly formulated queries and optimizing them for better retrieval results. Similar techniques can be applied to augmentation and generation phases too.

---

[3] In terms of five stages in MLOps, Phase 1 is in Stage 1, Phases 2–5 are in Stage 2, Phases 6 & 8 are across Stage 2 & 3, and Phase 7 is in Stage 4.

3. **Retrieval Phase:** Efficiently retrieving relevant information from large and possibly unstructured knowledge bases poses significant challenges, such as managing latency and ensuring the relevance, importance and diversity of retrieved documents.
4. **Augmentation and Enhancement Phase:** Enhancing retrieved data to improve the language model's responses can be problematic. Issues include reordering the contents, integrating diverse data sources and maintaining consistency in the augmented data.
5. **Model Generation Phase:** Generating accurate, coherent, and contextually appropriate responses is at the core of RAG systems. Challenges involve mitigating biases, preventing hallucinations, and ensuring the model's outputs are trustworthy.
6. **RAG Evaluation:** Assessing the system's performance requires robust evaluation metrics and methodologies. Difficulties include designing meaningful evaluation frameworks and interpreting results to inform improvements.
7. **LLM Serving and Monitoring:** Deploying the language model at scale while maintaining low latency and high availability is complex. Monitoring the model's performance in real-time to detect and address issues promptly is also a significant challenge.
8. **RAG Pipeline and Integration:** Orchestrating the end-to-end workflow efficiently—from data ingestion to response generation—requires careful planning. Challenges here include pipeline integration, error handling, and ensuring scalability and robustness.

By mapping the challenges onto their respective phases in the workflow, Figure 3.1 helps visualize how each obstacle impacts the system at different stages. This visual representation enables a clearer understanding of the interdependencies between phases and highlights critical points where focused efforts can significantly enhance system performance. We can view the data preparation and query refinement phases as the pre-retrieval stage, while part of the data augmentation and enhancement phase is considered as post-retrieval stage. The evaluation can be conducted across the whole lifecycle, as we will not only evaluate the final result but also the intermediate outcome, e.g., the relevance of the retrieved chunks, the quality of the rewritten prompt from the original query, and etc.

In the subsequent sections, we will delve into approximately 30 specific challenges that emerge throughout these phases. We will offer

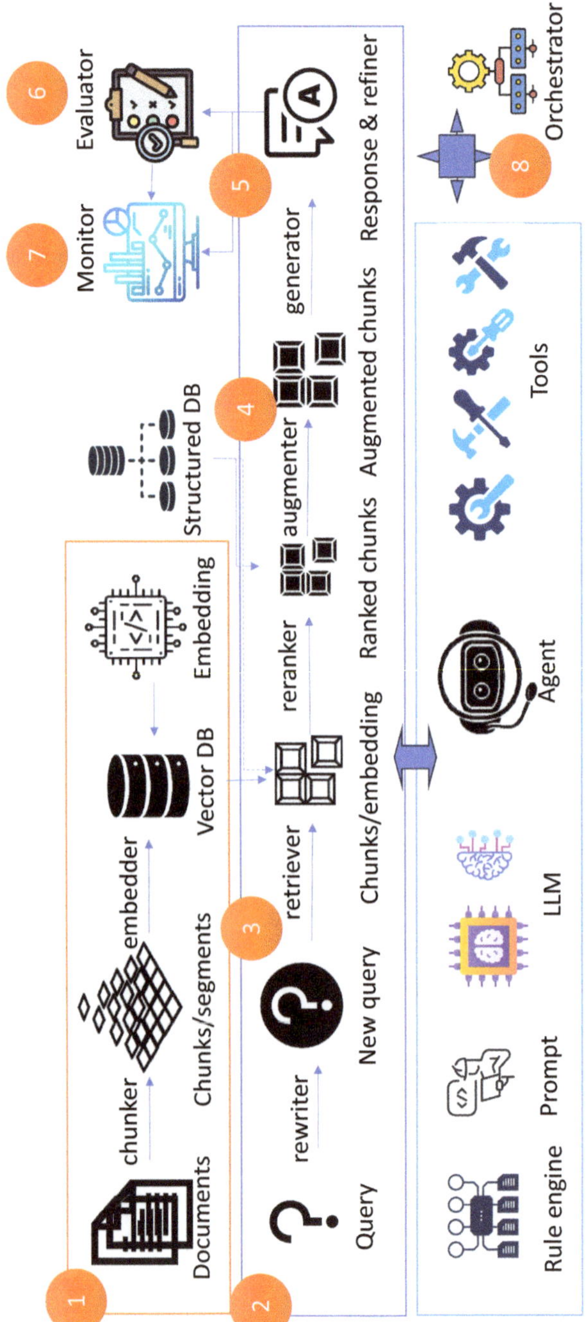

Figure 3.1.   Challenges inside a typical RAG workflow.

in-depth insights into each obstacle and discuss practical strategies to address them effectively. This structured approach ensures a logical flow of information and provides a solid foundation for readers to grasp the complexities involved in building and maintaining robust RAG systems.

3.3 Key Challenges and Concise Solutions

3.3.1 *Challenges in Data Preparation Phase*

3.3.1.1 *Challenge 1.1: Complexity in PDF parsing*

Pain point: PDFs are a cornerstone of modern data pipelines, yet they remain one of the most problematic formats to parse reliably. Unlike structured data formats such as JSON or XML, which are designed for machine readability, PDFs prioritize visual consistency, making automated extraction a persistent challenge. The difficulties stem from three core issues: unpredictable layouts, embedded multimedia elements, and unreliable text extraction. This situation may also apply to other formats, e.g., Word, PowerPoint, etc.

- **Non-Uniform Layouts:** PDFs rarely adhere to a single, predictable structure. A typical document might combine multi-column text, sidebars, footnotes, and embedded graphics—all within the same page. For instance, a research paper could switch between one and two-column formats, scatter superscripted footnotes throughout the text, or place critical data in marginal annotations. These variations force parsing systems to infer logical reading order from visual cues alone, a task that remains error-prone without sophisticated preprocessing.
- **Complex Embedded Elements:** Beyond raw text, PDFs often contain tables, forms, diagrams, and even scanned images, each requiring specialized handling. Extracting data from tables, for example, demands accurate detection of cell boundaries, which are often implied rather than explicitly marked. Meanwhile, charts and graphs require computer vision techniques to interpret, and multilingual documents introduce further complications with mixed character encodings and script directions.

- **Text Extraction Pitfalls:** Even seemingly straightforward text extraction is fraught with challenges. Font variability, such as mixed sizing, custom encodings, or legacy typefaces, can disrupt character recognition. What is worse is that some PDFs contain overlapping text layers, such as watermarks or annotations, which can obscure primary content. Additionally, the absence of semantic markup makes it difficult to distinguish between body text, headers, and captions, often leading to jumbled or out-of-sequence output.

Solution: We should leverage advanced AI and machine learning (ML) models, particularly OCR (optical character recognition), LLM and VLM (vision language model), which provide a robust approach to overcoming the complexities of PDF parsing, e.g.,

- **Preprocessing & Layout Analysis**
  - *Deep Learning-Enhanced OCR*: Advanced models (e.g., Tesseract 5+, Google Vision AI) handle diverse fonts and layouts.
  - *Layout Segmentation*: ML models classify regions (headers, tables, body text) to reconstruct document flow.
- **Content Extraction & Interpretation**
  - *Context-Aware LLMs/VLMs*: Preserve semantic coherence in non-linear layouts (e.g., footnotes reintegrated into main text).
  - Structured Data Extraction: NER models identify entities (dates, names), while specialized algorithms parse tables hierarchically.
- **Handling Multimedia & Multilingual Content**
  - *Visual Element Processing*: Computer vision extracts data from charts/diagrams; hybrid OCR-image models handle scanned text.
  - *Language-Agnostic Parsing*: Multilingual LLMs (e.g., mBERT, No Language Left Behind) manage mixed-language documents.
- **Validation & Continuous Improvement**
  - *Human-in-the-Loop*: Critical documents undergo manual review, with AI flagging ambiguities.
  - *Adaptive Learning*: Feedback loops refine models based on correction patterns.

These techniques leverage the strengths of LLMs and AI to transform the way we handle and extract data from PDFs, unlocking new possibilities for automation and data analysis. The detailed solutions with sample codes can be found under the Section "PDF Parsing" of Chapter 4.

3.3.1.2 *Challenge 1.2: Complexity in table parsing*

Pain point: Tables embedded in PDFs and Word documents often contain critical information needed to answer user queries, but extracting this data reliably remains a significant technical hurdle. Unlike structured datasets, document tables exhibit high variability in formatting, layout, and implicit semantics, making automated parsing exceptionally challenging.

- **Structural Variability:** Tables can have a wide variety of structures, ranging from simple grids to complex, nested arrangements with merged cells, multi-level headers, and varying cell spans. This diversity makes it difficult to develop a one-size-fits-all parsing solution. Tables may contain empty cells. When it is in a borderless form, column or row misalignment often occurs.
- **Contextual Dependencies:** Extracting meaningful information from tables often requires an understanding of the context in which the table is presented. For example, the significance of headers, the relationship between rows and columns, and the meaning of abbreviations or symbols used within the table, one table across multiple pages.
- **Data Heterogeneity:** A single table can mix text, numerical values, dates, currencies, and symbolic notations, each requiring distinct normalization. Formatting inconsistencies (e.g., "$1.5M" vs. "1,500,000 USD") compound interpretation difficulties, especially when locale-specific conventions come into play.
- **Noise and Ambiguity:** In reality, tables frequently contain typos, misaligned data, or missing values. Worse, some information is implied rather than stated. For example, a column labeled "Q1–Q4" might represent quarterly data without explicit headers for each quarter. Such nuances demand both pattern recognition and domain knowledge.

Solution: We should apply advanced preprocessing and postprocessing techniques, e.g., Implement sophisticated preprocessing algorithms to standardize table formats, handle merged cells, and normalize data types. The text content extracted from the table may be further verified by OCR, NLP and VLM for context correctness.

- **Preprocessing Standardization:** Advanced algorithms normalize tables by reconstructing merged cells, inferring borders in borderless designs, and flagging alignment ambiguities. OCR cross-validates text extraction, while data-type classifiers (e.g., distinguishing dates from IDs) reduce post-processing errors. Note that the final output to LLM must be in a proper format, e.g., markdown, XML and HTML.

- **Context-Aware Interpretation:** Fine-tuned language models analyze table captions, footnotes, and adjacent paragraphs to resolve abbreviations and infer relationships. For example, a model might link a column header like "ΔT" to surrounding text mentioning "temperature change" or recognize that a split table continues to the next page.
- **Hierarchical Structure Analysis:** Parsing occurs in layers: First, the system detects the table's global structure (e.g., header regions, data cells), then drills into row-column relationships, and finally extracts cell-level content. This phased approach improves accuracy for nested or irregular layouts.
- **Error Mitigation Techniques:** ML models impute missing values by analyzing patterns (e.g., interpolating missing dates in a time series) or flag inconsistencies for human review. Confidence scoring helps prioritize verification efforts.
- **Interactive Refinement:** Human-in-the-loop tools visualize parsed tables alongside the original document, allowing users to correct misalignments or validate ambiguous extractions. These corrections train the system to handle similar cases autonomously in the future.

By addressing these pain points with a combination of advanced preprocessing, context-aware models, hierarchical parsing, robust error handling, adaptive learning, and visualization tools, we can significantly improve its ability to parse and interpret complex tables accurately. The corresponding solutions can be found under the Section "Table Parsing" of Chapter 4.

3.3.1.3 *Challenge 1.3: Improper text segmentation and chunking*

Pain point: Improper text segmentation affects the accuracy and comprehensiveness of the retrieval content. The size of the text block division is directly related to the fit with the user's question during the query: too small a block may not cover all the relevant content of the question; on the contrary, too large may introduce redundant information.

Solution: Although most RAG libraries or tools extract content at the PDF page level by default, we can refine the process with the following methods:

- An iterative step to improve the segmentation or chunking size. We may treat the chunk size and segmentation methods as tuning parameters, then build an iteration over those parameters with evaluation datasets. See Section "Chunking" of Chapter 5.
- A hierarchical structure to contain two levels of information: a small chunk of raw text and a high-level summary of multiple chunks. See Section "Hierarchical" of Chapter 7.

3.3.1.4 *Challenge 1.4: Data ingestion scalability*

Pain point: The scalability challenges in a RAG pipeline's data ingestion phase manifest as prolonged processing times, system overload, data integrity risks, and reduced availability. These issues arise when the pipeline struggles to handle increasing data volumes, threatening real-time performance and reliability. Below lists some root causes:

1. **Sequential Processing Bottlenecks:** Linear ingestion workflows fail to leverage parallelism, leading to exponential delays as data grows.
2. **Resource Saturation:** Fixed hardware or poor resource allocation causes CPU/memory exhaustion during peak loads.
3. **Data Contention:** Concurrent read/write operations without proper locking mechanisms risk corruption or duplication.
4. **Monolithic Architecture:** Tightly coupled systems lack modular scaling, forcing entire pipelines to bear ingestion loads.

Solution: There are multiple ways to handle these ingestion issues.

- **Parallel Batch Processing with Map-Reduce:** Implementing parallel batch processing using the MapReduce paradigm involves dynamically partitioning data into logical units, such as by document type, customer ID, or timestamp, utilizing lightweight metadata tagging. For instance, partitioning loan agreements by region and year can enhance query performance. Worker orchestration is achieved by deploying parallel workers, like Kubernetes pods or AWS Lambda functions, each responsible for processing a specific data partition. Tools such as Apache Airflow or Prefect manage dependencies between these batches. In the Map phase, embeddings and metadata are extracted concurrently, while the Reduce phase merges results in tiered indexes, such as hot or warm storage. Leveraging Apache Spark facilitates large-scale batch processing, and integrating PyTorch Distributed enables GPU-accelerated embedding generation. This approach can significantly reduce data ingestion times. For example, Optum reported roughly a 10× improvement in query performance after adopting Starburst (Trino) and related optimizations, which illustrate how choices such as partitioning strategy and runtime pruning materially affect throughput.[4]

- **Streaming Integration for Real-Time Updates:** For real-time data updates, integrating streaming mechanisms within a hybrid architecture is essential. Priority-based routing can be implemented using platforms like Apache Kafka to differentiate between time-sensitive data, such as market news, and bulk updates like historical filings. Micro-batching, defined as processing data streams in short intervals (e.g., 5-minute windows) with tools like Apache Flink, ensures near-real-time index-ing without overloading the system. A practical application of this strategy is observed in fintech firms that stream live SEC filings into a "hot" index for immediate access, while separately batch-processing annual reports to maintain system efficiency.

- **Tiered Caching Mechanism:** To optimize data retrieval, we might use the cache mechanism. A more advanced version might be a tiered cache. This involves an L1 cache for storing precomputed embeddings of frequently accessed documents in an in-memory data store like Redis, ensuring rapid access. The L2 cache holds document summaries and key entities within search engines like Elasticsearch, facilitating swift filtering operations. An L3 cache retains final outputs for recur-ring queries, such as "show Q3 FX exposures," reducing the need for repetitive computations. Implementing an eviction policy that com-bines Least Recently Used (LRU) principles with Time-To-Live (TTL) settings helps balance data freshness with cache hit rates, maintaining an effective and efficient caching system.

---

[4] Starbust, Optum Accelerates Queries by 10× with Starburst Enterprise, https://cdn. featuredcustomers.com/CustomerCaseStudy.document/starburst_optum_685489.pdf, 2021. Accessed on 03/11/2025.

See Sections "Parallelizing ingestion pipeline" of Chapter 4, "Caching" of Chapter 15 and "Hierarchical Indexing" of Chapter 5 for the detailed solutions.

3.3.1.5 *Challenge 1.5: Vector space representation*

Pain point: In high-dimensional vector spaces, the distinction between closely related and unrelated items can become minimal. This phenomenon, often referred to as the "curse of dimensionality," may cause retrieval systems to misinterpret unrelated results as relevant, thereby diminishing the effectiveness of similarity searches.

Solution: To address this challenge, implementing dimensionality reduction techniques can be effective. Methods such as Principal Component Analysis (PCA), t-Distributed Stochastic Neighbor Embedding (t-SNE) or the recent randomly transformed bi-valued vectors for quantizing data vectors (RaBitQ) [4] reduce the number of dimensions, thereby enhancing the distinction between related and unrelated items. Additionally, employing advanced indexing structures like Hierarchical Navigable Small World (HNSW) graphs[5] can improve the efficiency and accuracy of similarity searches in high-dimensional spaces. See Section "Vector Space Representation" of Chapter 5.

3.3.2 *Challenges in Query Refinement Phase*

3.3.2.1 *Challenge 2.1: Complex query with multiple answering points, multiple-hop, etc*

Pain point: RAG systems often falter when faced with complex, multi-part queries that demand synthesis of disparate data types or temporal reasoning. Below are some typical examples:

- **Structured Analytics Combined with Semantic Search:** Queries such as identifying the risk factors of the highest-performing ridesharing company in Singapore integrate data from different sources to provide comprehensive insights.

---

[5]It is a graph-based algorithm that allows for fast approximate nearest-neighbor searches in high-dimensional spaces. By organizing data points in a hierarchical, navigable structure, HNSW significantly reduces search time compared to brute-force methods.

- **Comparative Analysis Queries:** Users often seek a deeper understanding by comparing tools or services, which necessitates analyzing both pros and cons effectively. An example would be evaluating the trade-offs between tool A and tool B for a specific task.
- **Summarization of Extensive Documents:** Providing concise summaries of lengthy documents, such as annual reports from companies like HSBC, Apple and Standard Chartered, requires distilling key information without losing essential context or details.
- **Multi-Part Queries with Temporal Dimensions:** Longer-term and short-term strategies often involve planning over extended periods (more than 5 years versus less than 5 years). Users expect a structured presentation of this information along with a clear conclusion.

Solution: The proposal suggests developing iterative information retrieval using sub-queries, incorporating graph-based methods to capture relationships between information pieces. It also explores reasoning techniques like multi-step reasoning and chain-of-thought prompting to guide language models through complex, interconnected queries for improved coherence.

- **Iterative Sub-Query Decomposition:** Complex queries can be effectively managed through iterative sub-query decomposition, which involves breaking down intricate questions into simpler, atomic sub-queries executed sequentially to synthesize comprehensive results. For example, to identify risk factors of Singapore's leading ridesharing company, one might:

  1. Retrieve market share data of Singapore ridesharing firms.
  2. Determine the top performer based on revenue growth.
  3. Fetch risk disclosures from the identified company's SEC filings.

  Tools using LLM can generate these sub-queries and manage their interdependence using dependency graphs, ensuring a structured and efficient retrieval process. See Section "Recursive retrieval" of Chapter 7.

- **Knowledge Graph-Augmented Retrieval:** Integrating knowledge graphs into retrieval systems enhances the modeling of relationships between entities such as companies, financial metrics, and regulatory events. In banking, for instance, nodes might represent companies and financial metrics, while edges denote relationships like a company's ESG report linking to specific regulations. This structure enables contextual traversal for comparative queries, facilitating nuanced analyses such as evaluating tools for foreign exchange risk assessment by mapping feature trade-offs. See Chapter 14 on "RAG with Knowledge Graph."

- **Chain-of-Thought Orchestration:** Techniques like multi-step reasoning or string-of-thought that prompt the LM to reason through complex sentences are methods to guide the LM through the inter-sentential field of relationships toward the desired coherence.

Implementing guided multi-step reasoning through chain-of-thought orchestration mirrors analyst workflows and enhances complex query resolution. This approach involves:

1. Extracting key entities.
2. Retrieving relevant documents.
3. Cross-referencing sources.
4. Resolving conflicts.
5. Generating structured outputs.

For temporal queries, strategies can be divided into short-term and long-term buckets, each with distinct evidence pools. Document summarization can be layered, progressing from executive summaries to financial details and risk assessments, utilizing section-aware retrieval to ensure comprehensive analysis. See Chapter 6 on "Query Transformation and Prompt Engineering."

- **Hybrid Structured/Semantic Workflow:** A hybrid structured/semantic workflow combines structured queries with semantic analysis to provide a holistic data examination pipeline. For example, one might run SQL queries to identify the top three Singapore ridesharing companies by revenue and then feed these results into a semantic search to analyze qualitative aspects like risk factors. Dynamic ranking within this workflow weights sources based on freshness and authority, prioritizing information from annual reports over news articles to ensure the reliability and relevance of insights. See "Hybrid or Fusion Retrieval" of Chapter 7 and the pipeline methods in Chapter 12.

3.3.2.2 *Challenge 2.2: Handling ambiguous or underspecified queries*

Pain point: Modern retrieval systems, despite their sophistication, often falter when confronted with ambiguous or context-starved queries. Consider the deceptively simple question *"How to use LLM?"*: it might seek coding tutorials for developers, business implementation strategies for executives, or ethical guidelines for policymakers. Without clarifying intent, even advanced systems risk generating fragmented or irrelevant responses.

Solution: Apply query resolution methodologies (such as asking additional questions or suggesting different interpretations for the user to choose from) and use function call or model context protocol (MCP) can leverage the issues.

- **Dynamic Query Refinement**
  - *Interactive Disambiguation*: Instead of passive guesswork, systems can *engage users* with targeted follow-ups. For example: *"Are you exploring technical implementation or organizational integration?"* See "ReAct policy" in Chapter 15.
  - *Multi-Perspective Suggestions*: Displaying 2–3 plausible interpretations (e.g., *"LLM fine-tuning vs. API deployment"*) lets users steer the dialogue, reducing misalignment risks. See "Subquery" in Chapter 6.
  - *Adaptive Context Leverage*: Systems that track user behavior (e.g., a developer's frequent API-related queries) can *anticipate intent* dynamically. For implementation frameworks, consult the section on "Query Rewriting" in Chapter 6.
- **Structured Data Agents.** Certain requests, such as *"Retrieve Q3 client count"* or *"Fetch SCB's latest annual report,"* demand precision over generative text. Here, **function-calling or MCP agents** bypass language models partially, directly interrogating structured databases or vector store metadata. This method slashes latency while improving factual accuracy. The topic of technical architectures is detailed in Chapter 15 under "Agentic RAG."

3.3.3 *Challenges in Data Retrieval Phase*

3.3.3.1 *Challenge 3.1: Insufficient or missing content*

Pain point: When essential information is absent from the knowledge base, a RAG system may produce plausible but incorrect answers instead of admitting it doesn't know. This misleads users and leads to frustration due to the dissemination of inaccurate information.

Solution: We may start with data cleaning, e.g., build a proper data cleaning mechanism to improve the data quality before building the RAG pipeline. Some typical steps to clean data include:

- **Remove Noise and Irrelevant Information:** Eliminate special characters, stop words (e.g., common words like "the," "an," "a," "I," and "we"), and HTML tags to reduce clutter and improve data clarity.
- **Identify and Correct Errors:** Use tools like spell checkers and language models to correct spelling mistakes, typos, and grammatical errors, ensuring the data is accurate.
- **Deduplicate Records:** Remove duplicate or similar entries that could bias the retrieval process and affect the system's responses.
- **Detect Harmful or Toxic Information:** Remove this information once identified.

For more advanced data cleaning functionalities, consider using tools like Unstructured.io, spacy.io, nltk, TextBlob, Cleanlab, Apache OpenNLP and their core libraries. See Section "Clean your data" of Chapter 4.

Meanwhile, we can enhance the prompts used with the RAG system to encourage more accurate answers, especially in situations where information might be lacking. Instruct the model to acknowledge its limitations and express uncertainty when appropriate by using prompts like "Tell me if you don't know when uncertain." While this doesn't guarantee absolute accuracy, better prompting is a valuable step toward improving the system's performance after data cleaning. See Chapter 6 on "Query Transformation and Prompt Engineering."

3.3.3.2 *Challenge 3.2: Relevance of retrieved information*

Pain point: Although vector retrieval, as the mainstream method, relies on similarity assessment, it also faces limitations in many aspects. When a user query involves specialized terms (e.g., "bi-directional pre-training mechanism of the BERT model"), vector retrieval relying only on cosine similarity is prone to miss documents with precise technical definitions, whereas keyword-only retrieval is unable to deal with semantically related, but non-overlapping, queries, such as "transformer architecture" and "self-attention mechanism." This contradiction is especially significant when the size of the knowledge base exceeds 10^6 documents, and the system recall (Recall@K) may drop by more than 30%.

- **Missed the Top Relevant Contents:** The initial retrieval missed the key information. Traditional vector retrieval often acts like a keyword-obsessed librarian – it might miss *"climate change impacts on coral reefs"* when searching for *"ocean warming effects on cnidarians,"* despite their semantic equivalence. This occurs because:
  - *Vocabulary Mismatch*: Queries and documents use different terminology (e.g., "AI" vs. "deep learning architecture").
  - *Semantic Understanding Deviation/Ambiguity*: Vector representation sometimes struggles to distinguish the subtle differences between near-synonymous concepts, which may lead to misunderstandings.
  - *Context Collapse*: Single vectors struggle to encode positional nuances (*"Apple stock"* vs. *"apple pie"*).
  - *Dimensional Tyranny*: High-dimensional spaces create false neighbors – documents sharing random feature overlaps rather than true relevance.
- **Magnitude vs. Direction:** The choice of different similarity methods may lead to large different results. Cosine similarity, a widely used metric, emphasizes the alignment (direction) of vectors rather than their magnitude. This can lead to results where vectors are directionally similar but semantically distant, potentially affecting relevance.
- **Imbalance in Dimension Consideration:** For example, cosine similarity focuses on the direction of the vector rather than its magnitude, which may lead to matches that are close in direction but far apart in meaning.

- **Insufficient Context Matching:** Existing vector search technology tends to judge overall similarity and may ignore higher similarity information in specific contexts or parts.
- **Difficulty in Retrieving Sparse Data:** Locating key fragments in a large-scale knowledge base is particularly challenging, especially when the required information is scattered across multiple documents.
- **Global vs. Local Similarities:** Traditional vector search mechanisms often prioritize global similarities, potentially overlooking local or contextual nuances. This limitation can result in the retrieval of items that are globally similar but lack relevance within specific contexts, thereby diminishing the effectiveness of search results.

Solution: Implementing advanced semantic search techniques, such as hybrid retrieval methods that mix sparse and dense representations, can significantly enhance search relevance. Additionally, fine-tuning retrieval models on domain-specific data further improves their effectiveness. Employing query expansion techniques captures different facets of user intent, leading to more accurate and comprehensive search results. Alternatively, Hyperparameter tuning in Chapter 12, such as chunk_size, overlap size, and similarity_top_k, or rerank techniques in Chapter 8, can also help.

- **Advanced Semantic Search Techniques:** Implementing advanced semantic search methods, such as dense vector retrieval, can significantly improve retrieval relevance. Dense vectors generated by deep learning models capture the semantic meaning of queries and documents more effectively than traditional sparse vectors. Hybrid retrieval methods that combine sparse and dense representations offer a balanced approach. By leveraging the strengths of both exact term matching (as in traditional keyword-based methods like BM25) and semantic similarity (through dense embeddings), hybrid methods can improve recall and precision. This combination ensures that documents containing exact query terms are retrieved while also capturing semantically related content that might use different wording. See Section "hybrid retrieval methods" and "Dense X Retrieval" of Chapter 7.
- **Query Expansion Techniques:** Employing query expansion techniques helps in capturing different aspects of the user's intent. By expanding the original query with synonyms, related terms, or

hierarchical concepts, the retrieval system can broaden its search to include documents that might not contain the exact query terms but are still relevant. Techniques such as pseudo-relevance feedback, where initial search results are used to identify additional relevant terms, can further enhance retrieval effectiveness. See Section "Query rewriting" of Chapter 6.

- **Hyperparameter Tuning:** Adjusting the chunk_size influences how documents are segmented, which affects the granularity of information retrieval. Modifying similarity_top_k determines the number of top similar documents retrieved, balancing between retrieving enough candidates for relevance and maintaining precision. See Section "Hyperparameter tuning" of Chapter 12.
- **Re-ranking techniques** involve applying a secondary model to reorder the initially retrieved documents based on more sophisticated criteria. For instance, using a **cross-encoder** model that jointly encodes the query and document can provide a more accurate relevance score compared to the initial retrieval based on individual embeddings. Re-ranking helps in promoting the most relevant documents to the top positions, increasing the likelihood that the system utilizes the best information available. See Section "Rerank techniques" of Chapter 8.
- **Utilizing Multi-Stage Retrieval Architectures:** Implementing multi-stage retrieval architectures can enhance retrieval performance, especially in large-scale knowledge bases. The initial stage retrieves a broad set of candidate documents using efficient but less precise methods. Subsequent stages apply more computationally intensive but accurate models to refine the results. This approach balances efficiency with effectiveness, ensuring that the system can handle large datasets without sacrificing relevance. See Section "Hierarchical Index Retrieval" and "Recursive retrieval" of Chapter 7.
- **Addressing Semantic Understanding and Context Matching:** To mitigate semantic understanding deviations, integrating contextual embeddings and models capable of capturing fine-grained meanings is essential. Techniques like contextualized topic modeling or incorporating knowledge graphs can enrich the semantic representation of documents and queries. Knowledge graphs, in particular, help in understanding entities and their relationships, which can improve the retrieval of contextually relevant information. For better context matching, methods such as attention mechanisms can be employed within the retrieval models to focus on the most relevant parts of

the documents. Additionally, segmenting documents into smaller, coherent passages and retrieving at the passage level rather than the whole document can enhance the matching of specific contexts. See Section "RAG with Knowledge Graph" of Chapter 14.

- **Dealing with Sparse Data Retrieval:** When facing the challenge of retrieving sparse data, techniques like entity linking and semantic indexing can help connect scattered pieces of information. By recognizing entities and concepts across documents, the retrieval system can aggregate related information, making it easier to locate key fragments. Moreover, employing approximate nearest neighbor (ANN) search algorithms, such as HNSW graphs, enables efficient retrieval in high-dimensional vector spaces, facilitating the discovery of relevant but infrequent information. See the content on "Sparse Data Retrieval" of Chapter 7. See also "Challenge 3.6: Latency and Scalability in Retrieval" in this chapter.
- **Fine-Tuning Domain-Specific Retrieval Models:** It allows the models to learn the particular language, terminology, and contextual cues relevant to a specific field. For example, in the medical domain, fine-tuning on medical literature enables the model to understand complex medical terminology and relationships between concepts. This specialization enhances the model's ability to retrieve documents that are more precisely aligned with the user's queries. See how to fine-tune retrieval models on domain-specific data to improve relevance in Chapter 12.

3.3.3.3 *Challenge 3.3: Inconsistencies in retrieved information*

Pain point: When a RAG system retrieves documents containing conflicting information, it may generate outputs that are inconsistent or contradictory. For example, if one source states that "RAG systems enhance accuracy by integrating external knowledge," while other claims that "RAG systems may introduce inaccuracies due to conflicting retrieved information," the generated response might confusingly present both statements without clarification. Additionally, if a user's query is highly specific but the database entries are on broader topics, the system might overgeneralize the response due to inconsistent information granularity.

Solution: Implementing fact verification modules, often referred to as retrieval evaluators, is essential for cross-referencing information across multiple sources to ensure accuracy. These modules typically involve a multi-step process: identifying claims, assessing their check-worthiness, generating queries for evidence retrieval, retrieving relevant evidence, and verifying the claims. This structured approach enhances the reliability of information by systematically validating claims against credible sources.

To address discrepancies in retrieved information, developing conflict resolution strategies is crucial. Techniques such as majority voting, where the most frequent information among sources is considered accurate, or assigning weights based on source credibility, can effectively resolve conflicts. For instance, prioritizing information from peer-reviewed journals over less authoritative sources ensures higher accuracy in the verification process.

Enhancing language models to recognize inconsistencies and explicitly highlight or explain them when detected further improves transparency and builds user trust. By integrating mechanisms that detect and address contradictions within the information, these models can provide users with a clearer understanding of the data's reliability. This proactive approach not only fosters confidence in the information presented but also encourages critical evaluation by users.

Incorporating these strategies into fact verification systems ensures a comprehensive approach to information validation, effectively managing conflicts and enhancing the overall trustworthiness of the data provided. See Section "Evaluation Frameworks" of Chapter 10.

3.3.3.4 *Challenge 3.4: Bias in retrieved information*

Pain point: The challenge of algorithmic bias in RAG systems mirrors a fundamental tension in modern AI: the gap between technical prowess and social responsibility. When retrieval systems ingest historical medical trial data skewed toward male participants, for instance, they might perpetuate diagnostic models that underperform for female patients – a real-world pattern documented in recent reviews of AI for cardiovascular care, which report demographic performance gaps and dataset bias; WHO's 2023 regulatory guidance similarly warns that biased training data threatens equity. This is not merely a data issue but a sociotechnical risk some scholars describe as "epistemic violence," where skewed retrieval and representation systematically sideline marginalized perspectives [5].

Solution: It's essential to implement bias detection and mitigation strategies at multiple stages of the RAG pipeline:

- **Bias Detection and Mitigation in Retrieval:**
  - *Develop Diverse and Representative Knowledge Bases*: Ensure that the knowledge base includes data from a wide range of sources representing different perspectives and demographics. Regularly audit the data to identify and fill gaps in representation.
  - *Fairness-Aware Ranking Algorithms*: Implement retrieval algorithms that consider fairness by weighting sources based on credibility and diversity. This helps provide a more balanced set of information during retrieval.
  - *Bias Detection Tools*: Utilize natural language processing techniques to analyze retrieved information for biased or discriminatory language, allowing for the filtering or flagging of such content.
- **Bias Mitigation in Generation:**
  - *Counterfactual Data Augmentation*: Enhance the training data with synthetic examples that counteract identified biases. This helps the model learn to produce more balanced and fair outputs.
  - *Ethical Guidelines and Prompting*: Incorporate prompts that instruct the language model to avoid biased language and encourage neutrality. Explicitly programming the model to recognize and correct for bias can improve output quality.
  - *Fine-Tuning with Bias Mitigation Objectives*: Adjust the model's training objectives to penalize biased outputs, encouraging the generation of fair and unbiased content.
- **Online Monitoring and Evaluation:**
  - *Continuous Evaluation*: Regularly assess the system's outputs using bias and fairness metrics to identify any ongoing issues.
  - *User Feedback Mechanisms*: Implement channels for users to report biased or unfair responses, allowing for iterative improvements.
  - *Cross-Validation*: Compare outputs across different subsets of data to detect inconsistencies or biases that may not be immediately apparent.

Note that this implementation usually requires cross-disciplinary collaboration. For example, it involves ethics and domain experts in ethics, sociology, and the relevant domain to identify potential biases and develop appropriate mitigation strategies. Organizational policies shall be established to define acceptable standards for bias and fairness in AI outputs, ensuring compliance with legal and ethical guidelines. By implementing these comprehensive strategies, organizations can significantly reduce the impact of biases in retrieved information within RAG systems. This not only enhances the fairness and reliability of the outputs but also fosters user trust and aligns such technology with ethical standards and societal values. See Chapter 10 on "Evaluation."

3.3.3.5 *Challenge 3.5: Context consistency across multiple turns/rounds dialogues*

Pain point: RAG systems often struggle to maintain and utilize the context from previous interactions in multi-round dialogues. They may lose track of the conversation history, leading to difficulties in choosing the required information for answering follow-up questions accurately. This results in responses that are irrelevant, disjointed, or lack the necessary continuity, which can frustrate users and diminish the effectiveness of the system.

Solution: Maintaining context is especially critical in complex conversations where each turn builds upon previous ones, such as in customer support, advisory services, or collaborative planning. Thus, implementing conversation history-aware retrieval practices is essential. Here are several strategies to enhance context maintenance in RAG systems:

- **Incorporate Conversation History in Retrieval:**
  - *Contextual Embeddings*: Include embeddings of previous dialogue turns when generating retrieval queries. This helps the system understand the current query in the context of the entire conversation.
  - *Session Awareness*: Design the retrieval mechanism to be session-aware, ensuring that it considers all relevant past interactions within the current conversation session.
- **Develop Dynamic Knowledge Graphs:**
  - *Real-Time Updates*: Build knowledge graphs that dynamically update as the conversation progresses, capturing new entities, relationships, and topics introduced by the user.

- o *Broader Contextual Understanding*: Utilize these evolving knowledge graphs to provide a wider and more accurate context for information retrieval, leading to more relevant and coherent responses.
- **Employ Retrieval-Based Memory Networks:**
  - o *Memory Augmentation*: Implement memory networks that store representations of past dialogue turns, allowing the system to reference previous interactions effectively.
  - o *Continuous Context Updating*: These networks can update and refine the context over time, ensuring that the system's understanding evolves with the conversation continuously.
  - o *Improved Relevance*: By accessing stored memories, the system can retrieve information that is directly pertinent to the user's current needs.
- **Advanced Dialogue Management Techniques:**
  - o *Anaphora Resolution*: Enhance the system's ability to resolve references to earlier mentioned entities or concepts (e.g., pronouns like "it" or "they").
  - o *Topic Tracking*: Keep track of ongoing topics or threads within the conversation to maintain coherence and relevance.
  - o *Contextual Prompts*: Use prompts that include summaries or key points from previous turns when generating responses.
- **Technical Implementation Considerations:**
  - o *Context Window Management*: Optimize the use of the model's context window to include as much relevant prior conversation as possible without exceeding token limits.
  - o *Hierarchical Encoding*: Use hierarchical models that encode dialogue at multiple levels (e.g., utterance-level and conversation-level) to capture both immediate and overarching context.
  - o *Latency Optimization*: Ensure that the incorporation of additional context does not significantly impact response times, maintaining a smooth user experience.

By implementing conversation history-aware retrieval practices, dynamic knowledge graphs, and retrieval-based memory networks, systems can significantly improve their contextual understanding. These strategies enable RAG systems to provide coherent, relevant, and accurate responses that enhance user engagement and satisfaction. Continuous advancements in dialogue management and contextual processing will further empower RAG systems to handle complex conversational scenarios with greater proficiency. See Sections "Prompt Compression" of Chapter 8, "RAG with Knowledge Graph" of Chapter 14 and "Chat Engine" of Chapter 12.

3.3.3.6 *Challenge 3.6: Latency and scalability in retrieval*

Pain point: In RAG systems, the retrieval component is crucial for fetching pertinent documents or data chunks that the language model uses to generate accurate and contextually relevant responses. However, as the volume of data increases, traditional retrieval methods may become inefficient and slow. The system may struggle to search through vast amounts of information quickly, leading to delays that are unacceptable in real-time applications such as customer support chatbots, virtual assistants, or any interactive services where prompt responses are expected.

Latency issues can degrade the overall performance of the system, leading to user frustration and reducing the effectiveness of the application. Scalability becomes a concern as well; the system may not be able to accommodate additional users or data without a proportional increase in resources, which is often impractical or cost prohibitive. This limitation hampers the ability of organizations to expand their services and meet growing demand.

Solution: Implement efficient indexing and retrieval techniques that can handle large-scale data with minimal computational overhead. One effective approach is the use of ANN search algorithms, such as HNSW graphs mentioned earlier. This makes it possible to retrieve relevant information quickly, even from large datasets. Besides HNSW, algorithms like Locality-Sensitive Hashing (LSH) and Product Quantization can also improve retrieval speeds by reducing the search space and focusing on the most relevant data points. We might consider customized retrieval algorithms for Optimization, e.g., tailor retrieval algorithms to the specific characteristics of the dataset and queries to enhance efficiency and apply adaptive thresholds to adjust the precision of retrieval dynamically based on the required speed and accuracy for each application scenario. Below are some other techniques.

- **Optimize Data Representations:**
  - *Dimensionality Reduction*: Apply techniques like PCA or t-SNE to reduce the dimensionality of data vectors without losing significant information. Lower-dimensional data requires less computational power to process. The solution is similar to the ones mentioned in Challenge 1.5 in this chapter.
  - *Vector Quantization*: Compress data representations to minimize memory usage and accelerate computations during retrieval. See more in Chapters 5 and 12.
- **Scalable Infrastructure:**
  - *Distributed Computing*: Leverage distributed systems and parallel processing to handle large-scale data across multiple machines or processing units. This approach divides the workload and accelerates retrieval times.
  - *Cloud-Based Solutions*: Utilize cloud infrastructure that can scale resources dynamically based on demand, ensuring that the system maintains performance during peak usage times.
- **Caching Mechanisms:**
  - *Result Caching*: Implement caching strategies to store frequently accessed data or recent query results. This reduces the need to perform full retrieval operations for repeated or similar queries, thereby decreasing latency.
  - *In-Memory Databases*: Use in-memory data storage for quick access to high-priority or commonly used data, improving response times.
- **Asynchronous Processing and Batching:**
  - *Asynchronous Retrieval*: Allow retrieval processes to occur independently of the main execution thread, preventing delays in response generation.
  - *Batch Processing*: Aggregate multiple retrieval requests and process them simultaneously to optimize resource utilization and throughput.
- **Load Balancing:**
  - *Distribute Workload*: Use load balancers to evenly distribute incoming requests across multiple servers or processing units, preventing any single component from becoming a bottleneck.
- **Data Sharding and Partitioning:**
  - *Segment Data*: Divide the knowledge base into smaller, more manageable partitions based on criteria like topic, date, or relevance.

This reduces the amount of data each retrieval operation needs to search through.

o *Geo-Distributed Databases*: Place data physically closer to users in different geographic locations to reduce network latency.

These strategies not only enhance the responsiveness of the RAG system but also improve its scalability, allowing it to maintain high levels of performance as it evolves. This results in a more robust, user-friendly application capable of delivering timely and accurate information, thereby increasing user satisfaction and supporting the organization's ability to expand its services. See Section "Prompt Compression" of Chapter 8 and "Caching" of Chapter 12 and some improved "Retrieval Techniques" of Chapter 7. Note that this issue may also happen in the generation phase. Some techniques, e.g., scalable infrastructure, caching mechanism and asynchronous processing and batching, are still applicable.

3.3.3.7 *Challenge 3.7: Handling temporal aspects*

Pain point: RAG systems often struggle to accurately answer questions that involve temporal dynamics, i.e., how information changes over time, or to provide time-sensitive data. The inability to manage temporal information can result in:

- **Outdated Responses:** Providing information that was once correct but is no longer valid.
- **Misinterpretation of Temporal Queries:** Failing to recognize when a question is about a specific time period or about changes over time.
- **Inconsistent Information:** Mixing data from different time periods without proper context, leading to confusing or misleading answers.

For instance, a user asking for the "current CEO of Company X" might receive an incorrect response if the system references outdated data. Similarly, questions about trends or historical changes, like "How have interest rates changed over the past decade?" require the system to understand and process temporal information, which standard RAG systems may not handle efficiently. Some questions also specify the time information, e.g., "What's the difference in revenue between 2022 and 2023 for Company X?".

Solution: It's essential to incorporate temporal awareness into the data processing, retrieval and generation components of RAG systems. In the data processing step, we shall include timestamp in the metadata of the chunks. Then the following strategies can enhance the system's ability to handle temporal aspects effectively:

- **Incorporate Document Timestamps into Retrieval:**
  - *Metadata Utilization*: Index documents with timestamps and temporal metadata, allowing the retrieval system to filter or prioritize information based on its relevance to the time frame specified or implied in the user's query.
  - *Time-Natural Language Processing Aware Retrieval Algorithms*: Modify retrieval methods to consider the temporal context, ensuring that the most recent or appropriately dated documents are retrieved. For example, implementing a decay function that boosts newer documents for queries requiring current information.
- **Develop Temporal Reasoning and Fact Updating Mechanisms:**
  - *Time Frame Assignment*: Utilize natural language processing techniques to assign temporal tags to facts within documents. This helps in understanding when certain information is valid.
  - *Temporal Information Extraction*: Extract and normalize temporal expressions (dates, times, periods) from both queries and documents to align the temporal dimensions accurately.
  - *Fact Updating Modules*: Implement systems that regularly update facts and figures in the knowledge base, ensuring that the most recent data is available for retrieval.
- **Create Dynamic Temporal Knowledge Graphs:**
  - *Temporal Knowledge Graphs (TKGs)*: Build knowledge graphs that represent entities and their relationships over time. TKGs capture the evolution of information, allowing the system to understand how facts change.
  - *Continuous Updating*: Regularly update the TKGs with new data, ensuring that temporal relationships remain accurate. This is particularly useful for tracking changes in leadership, pricing, regulations, etc.
  - *Temporal Queries Handling*: Adapt query processing to interpret and answer questions that involve temporal reasoning, such as trends, durations, and historical comparisons.

- **Implement Temporal Retrieval Models:**
  - *Time-Aware Language Models*: Fine-tune language models to handle temporal context by training them on datasets that include temporal annotations.
  - *Temporal Embeddings*: Incorporate temporal information into embeddings used for retrieval, enhancing the model's ability to match queries with time-relevant documents.
- **Utilize Time-Decay Functions in Ranking:**
  - *Recency-Based Ranking*: Apply time-decay functions to rank recent documents higher when freshness is crucial. This ensures that the most up-to-date information is prioritized in the retrieval process. This is applicable when the time is not specified.
  - *Balance with Relevance*: Ensure that the emphasis on recency does not override other relevance factors, maintaining a balance between timeliness and overall relevance.
- **Enhance User Queries with Temporal Context:**
  - *Query Expansion*: Automatically expand or reformulate user queries to include temporal terms when appropriate. For instance, adding "in 2023" to a query if the system infers that current information is desired.
  - *Clarification Dialogues*: If the temporal context is unclear, prompt the user for clarification. For example, "Are you asking about the current policies or those from a specific year?"
- **Temporal Consistency in Response Generation:**
  - *Maintain Temporal Coherence*: Ensure that generated responses are temporally consistent, avoiding mixing information from different time periods without proper context.
  - *Explicit Temporal References*: Clearly indicate the time frame of the information provided, such as "As of September 2023, the CEO of Company X is..."
- **Leverage Temporal Databases and Indexing Structures:**
  - *Temporal Indexes*: Implement indexing structures that are optimized for temporal data retrieval, such as time-based partitioning.
  - *Efficient Temporal Queries*: Optimize database queries to handle temporal conditions efficiently, reducing latency in retrieving time-specific information.
- **Employ Temporal Reasoning Techniques:**
  - *Temporal Logic Models*: Use models that understand and can reason about time, enabling the system to infer temporal relationships and answer complex temporal queries.

o *Event Sequencing*: Recognize and represent the sequence of events, which is essential for understanding narratives and causal relationships over time.

By effectively integrating temporal information into retrieval and generation processes, systems can offer more nuanced and precise responses that align with users' temporal needs. See Section "Indexing" and "Metadata" of Chapter 5, and "GraphRAG" in Chapter 14. Other tricks include the data version and update, which has been described in Section "Data Version and Update" of Chapter 4.

3.3.3.8 *Challenge 3.8: Extremely long content*

Pain point: RAG systems often struggle when tasked with processing extremely long content, such as summarizing an entire book, analyzing a full annual financial report, or comparing large sets of source code. These challenges arise due to inherent limitations in both the retrieval and generation components of the system.

- **Token Limitations with Language Model Constraints:** Most LLMs have a maximum token limit that restricts the amount of text they can process in a single prompt. For instance, the GPT-4o can process up to 128K tokens, while GPT-5 can further support 400,000 input tokens and 128,000 output tokens. However, several clients' comprehensive annual reports and ESG reports can easily exceed these limits, making it impossible to input all the content at once.
- **Information Overload, e.g., Cognitive Load:** Even if the model could process extremely long inputs, summarizing or comparing vast amounts of information presents a cognitive challenge for the AI. It may struggle to identify the most relevant points, leading to summaries that are either too general or miss critical details, e.g., context distraction, confusion and clashes.
  - o *Context Dilution*: With large inputs, important context can get diluted. The model might focus on irrelevant sections, leading to inaccurate or unhelpful outputs.
- **Computational Resources:**
  - o *Processing Time*: Handling large texts requires significant computational power and time, resulting in higher latency. Users may experience delays, which is detrimental to the user experience. Meanwhile,

limited by infrastructure resources, the full context length of LLMs may not be achieved due to limited VRAM required for both model weights, intermediate activations and KV cache during inference.

o *Cost Implications*: Increased computational requirements lead to higher operational costs, which may not be sustainable, especially for applications with many users or frequent requests.

- **Retrieval Challenges:**
  o *Segmented Data Retrieval*: Retrieving relevant information from massive documents is complex. Traditional retrieval methods may not efficiently pinpoint the necessary sections, leading to incomplete or fragmented outputs.

- **Applications Affected:**
  o *Summarization Tasks*: Users seeking concise summaries of lengthy documents receive incomplete or superficial overviews.
  o *Comparative Analysis*: Comparing large sets of source code or documents becomes impractical, as the system cannot process both inputs simultaneously within token limits.
  o *Legal and Medical Documents*: Professionals needing to analyze extensive legal briefs or medical records may find the RAG system inadequate for their needs.

Solution: A straightforward approach might be Divide and Conquer, i.e., when processing document segmentation, we can split the document into smaller chunks that fit within the model's token limit. Process each chunk independently, possibly in parallel, to improve efficiency. Meanwhile, we may utilize parallel computing resources to process multiple chunks simultaneously, reducing overall processing time. However, this approach may generally lose some information. Users may also interact with the RAG system, e.g., allow users to specify particular sections, chapters, or topics of interest within the document to narrow down the content that needs to be processed, or engage the user in a dialogue to iteratively refine the request, ensuring that the system focuses on the most relevant content. However, these interactions are generally manual and query-specific and hard to extend.

To overcome such issues, several strategies can be implemented to enable RAG systems to handle extremely long content effectively:

- **Prompt Compression Techniques:**
  - *Hierarchical Summarization*: We can apply **Stepwise Summarization or Recursive Summarization**. The former breaks the long content into smaller, manageable sections and generate summaries for each. Subsequently, combine these summaries to produce a comprehensive overview. The latter handles summaries in a hierarchical manner until the content fits within the token limit.
  - *Abstract Representation using Topic Modeling*: Use algorithms like Latent Dirichlet Allocation (LDA) to identify key topics and themes within the document, providing a high-level summary without processing the entire text through the language model.
  - *Key Point Extraction using Extractive Summarization*: Identify and extract the most significant sentences or paragraphs using statistical methods or ML models before inputting them into the LLM.
  - *Data Compression*: Apply compression techniques to reduce the size of the text data. However, this is often limited in effectiveness for natural language due to the need to maintain semantic meaning.
- **Re-Ranking (Ordering) Techniques:**
  - *Priority Sequencing, e.g., Relevance Ranking*: Rank sections of the document based on relevance to the user's query or the summarization task, processing the most important parts first.
  - *Dynamic Content Selection, e.g., Adaptive Content Selection*: Dynamically select which parts of the content to process based on initial analysis, ensuring that the most pertinent information is included within the token limit.
  - *Iterative Processing, e.g., Sliding Window Approach*: Move a window of allowable token size across the document, processing and summarizing each segment sequentially.
- **External Knowledge Integration:**
  - *Retrieval Systems Enhancement with Indexing and Metadata*: Enhance the retrieval component by indexing the document with detailed metadata, allowing the system to fetch only the most relevant sections in response to specific queries.
  - *Knowledge Graphs with Semantic Representation*: Convert the document into a knowledge graph that represents entities and their relationships, enabling the model to reason over the content without processing the raw text.

- **Model Optimization (structural modification, chunking and sliding techniques):**
  - *Advanced Model Architectures with Long-Range Transformers*: Utilize models designed to handle longer sequences, such as Gemini 2.5 Pro (up to 1M input tokens and 64K output tokens) or Llama 3.1 (up to 128K, while Llama3-Gradient extends from 8K to 1M+; Llama 4 herd supports up to 10M tokens theoretically.) or Mixtral (up to 32K), which incorporate hierarchical attention, advanced positional encoding techniques or Mixture-of-Experts (MoE) to manage extended contexts. LongRAG combining a long retriever and a long reader is a good example [6].
  - *Memory Networks with External Memory Modules*: Implement models with external memory components that can store and retrieve information over longer contexts.
  - *Efficient Tokenization*: Use more efficient tokenization methods to maximize the amount of content that fits within the token limit.
  - *StreamingLLM*: This technique addresses the "attention sink" phenomenon by merging window contexts with initial tokens, allowing LLMs trained on finite-length attention windows to effectively generalize infinite sequence lengths without needing extensive retraining [7].
  - *Compression of Representations using Knowledge Distillation*: Compress the knowledge from larger models or datasets into smaller, more efficient models that can handle longer inputs.
  - *SLED (SLiding Encoder and Decoder)*: This approach partitions long texts into overlapping chunks that are processed independently by a pretrained encoder. The outputs are then fused using a decoder, allowing for effective long-context understanding while maintaining low computational overhead [8].

We may also apply hybrid approaches, which leverage a combination of techniques to enhance the processing of long texts effectively. These methods often involve multistage processing, where the workflow begins with initial retrieval and extraction, followed by summarization and refinement stages. Additionally, algorithmic preprocessing plays a crucial role by utilizing algorithms to reduce data complexity before engaging the AI model. This step may include removing redundancies and irrelevant sections, ensuring that the model focuses on the most pertinent information. By integrating these strategies, hybrid approaches optimize performance and improve the overall efficiency of handling extensive content.

By implementing these strategies, RAG systems can more effectively handle extremely long content, providing accurate summaries, analyses, or comparisons even when dealing with large documents or datasets. These solutions help overcome the inherent limitations of token size and computational resources, enabling the system to deliver valuable insights without overwhelming the model or the user. See Sections on prompt compression techniques and rerank techniques of Chapter 8, GraphRAG of Chapter 14 and finetuning techniques and context engineering of Chapter 12.

3.3.4 *Challenges in Data Augmentation Phase (Reranking, Integration)*

3.3.4.1 *Challenge 4.1: Inappropriate ranking*

Pain point: Inappropriate ranking and prioritization of retrieved information fragments can lead to the LLM missing key points in its responses. When multiple relevant documents or paragraphs are retrieved, an ineffective ranking algorithm may fail to order them based on their true relevance and importance to the user's query. This misalignment causes the LLM to focus on less critical information, resulting in answers that are incomplete, inaccurate, or not aligned with user expectations. The ranking process must balance several factors:

- **Relevance:** How closely does the information relate to the user's query?
- **Importance:** How critical is the information in addressing the key aspects of the query?
- **Contextual Fit:** Does the information align with the user's intent and the context of the conversation?

Achieving this balance requires sophisticated ranking algorithms capable of understanding nuanced language and domain-specific knowledge. The challenge intensifies in specialized fields where domain expertise is necessary to discern the significance of different pieces of information. Inadequate ranking can lead to:

- **Missed Key Points:** Critical information may be overlooked if not properly ranked.
- **Information Overload:** Irrelevant or less important data may clutter the response.
- **User Frustration:** Users may lose trust in the system if it consistently provides subpar answers.

Solutions: We can generally improve the ranking quality by using advanced reranking algorithm or reducing the data noise.

- **Advanced Ranking Algorithms**
  - *Implement Learning-to-Rank Models*: The ML algorithms like LambdaMART, RankNet, or RankBoost are specifically designed for ranking tasks. For example, LambdaMART is a pairwise ranking method that evaluates the relevance differences between every pair of documents within a query group by computing a proxy gradient for each comparison. We can train a LambdaMART model on annotated data where documents are labeled based on relevance and importance to specific queries.
  - *Utilize Transformer-Based Models for Ranking*: We can employ models like BERT, RoBERTa, or ELECTRA fine-tuned for passage ranking to capture semantic nuances. For Example, we may Fine-tune BERT on a dataset of query-document pairs to improve semantic understanding in ranking. Meanwhile, we may consider similarity metrics other than cosine similarity.
  - *Cross-Encoders*: These models jointly encode the query and document pairs to generate precise relevance scores, allowing for a more nuanced understanding of how well each document matches the query.
  - *Lost-in-the-Middle Rankers*: This method strategically positions the most relevant documents at both the beginning and end of the ranked list while placing less relevant ones in the middle, optimizing for the model's context window.
  - *Diversity Rankers*: These techniques focus on maximizing the diversity of retrieved documents by selecting those that are least similar

to already chosen documents, thereby ensuring a broader range of perspectives.

- o *Reinforcement Learning Rerankers*: These models optimize rankings based on user interaction data and long-term rewards, adapting over time to improve user satisfaction and relevance.
- o *Hybrid Rerankers*: By combining multiple reranking strategies, such as LTR with various ML or deep learning models, hybrid approaches can leverage the strengths of different techniques for improved performance. It generally has two types. One is two-stage (or multi-stage), i.e., see an initial retrieval to gather a broad set of documents (such as keyword), then apply a more precise re-ranking. The other is Ensemble Methods, e.g., combine multiple ranking models to leverage different strengths. For example, use a weighted average of scores from both a semantic model and a keyword-based model.

- **Domain Knowledge Integration**
  - o *Knowledge Graphs*: Integrate domain-specific knowledge graphs to enhance understanding of relationships and entity importance. For example, use a medical knowledge graph to prioritize critical health information in healthcare applications.
  - o *Expert Annotations*: Incorporate feedback from domain experts to adjust ranking algorithms. For example, use expert-labeled datasets to train and validate the ranking model.

- **Feedback Mechanisms**
  - o *User Feedback Loops*: Collect user interactions and feedback to continuously improve ranking accuracy. For example, implement thumbs-up/down buttons for users to rate the helpfulness of responses.
  - o *Active Learning*: Utilize active learning to focus on the most informative data points for improving the ranking model. For example, prioritize retraining on queries where the system's confidence is low or user feedback indicates issues.

- **Personalization via User Profiling and Session Content:**
  - o We may adjust rankings for a user who frequently searches for advanced technical content by prioritizing more in-depth documents. If a user has been asking about a specific product, prioritize documents related to that product.

Note that we may encounter multi-objective optimization in this phase, when considering the balance Multiple Factors, such as relevance, importance, freshness, and diversity. Although relevance is diversity promotion ensures that the top-ranked documents provide a comprehensive view. We might implement a Maximal Marginal Relevance (MMR) approach to balance relevance and diversity. In addition, appropriate performance metrics, shall be carefully chosen, e.g., NDCG (Normalized Discounted Cumulative Gain), MAP (Mean Average Precision), or Precision@K (see the details in the supplementary material).

By enhancing the ranking component with advanced algorithms, semantic understanding, and domain-specific knowledge, RAG systems can more effectively prioritize retrieved information. Incorporating feedback mechanisms and multi-objective optimization ensures that the system remains aligned with user needs and domain requirements. These improvements lead to more accurate and relevant responses from the LLM, increasing user satisfaction and trust in the system. See Chapter 8 on augmentation and Chapter 12 for integration.

3.3.4.2 *Challenge 4.2: Content integration & deduplication*

Pain points: The seamless integration of retrieved context and the avoidance of redundancy in generated outputs. Firstly, if the context from retrieved passages is not smoothly integrated into the generation task, the output may appear disjointed or lack coherence. For example, if a retrieved passage provides in-depth information about "Python's history" and the generation task involves elaborating on "Python's applications," the output might overemphasize the history at the expense of applications. Secondly, when multiple retrieved passages contain similar information, the generation step might produce repetitive content. For instance, if several retrieved articles all mention "PyTorch's dynamic computation graph," the generated content might redundantly emphasize this point multiple times.

Solutions: For context integration, employing contextual retrieval methods can enhance the coherence of generated outputs. By adding specific contextual information to text chunks before they are embedded or indexed, the relationship with broader documents is preserved, ensuring that the retrieved information aligns seamlessly with the generation task.

In addition, the format of context shall be standardized, e.g., canonicalization, deduplication, ordering, truncation, summarization, safety & injection hardening, citation, etc. Regarding redundancy, utilizing techniques that promote diversity among retrieved passages can mitigate repetitive content. For instance, optimizing relevant information gain, e.g., a probabilistic measure of the total information pertinent to a query, ensures that each retrieved passage contributes unique and relevant insights, reducing redundancy in the generated output. By integrating these strategies, RAG systems can produce more coherent and concise outputs, effectively addressing the issues of disjointed content and redundancy. See reranking techniques in Chapter 8 and metadata in Chapter 5.

3.3.5 *Challenges in Large Model Selection & Generation Phase*

3.3.5.1 *Challenge 5.1: Domain-specific LLM*

Pain point: The Challenge of Specialized Knowledge in LLMs. While general-purpose language models like GPT-4 excel at broad tasks, they often falter in specialized domains such as medicine, law, or engineering. Imagine asking a brilliant general practitioner to perform neurosurgery: their broad medical knowledge, while impressive, lacks the precision required for such a niche task. Similarly, LLMs trained on diverse datasets struggle with domain-specific jargon, contextual nuances, and the tacit conventions governing professional discourse. This gap manifests in three critical ways:

1. **Terminology Pitfalls:** Technical terms like "tortious interference" (legal) or "neutropenia" (medical) may be misparsed or oversimplified, akin to a novice misreading a seasoned engineer's schematic.
2. **Conceptual Blind Spots:** Complex processes—say, pharmaceutical drug approval workflows or semiconductor fabrication protocols—often elude the model's grasp, leading to superficial or fragmented explanations.
3. **Inconsistent Outputs:** Like interns with varying training, different LLMs exhibit erratic performance; one might ace a patent law analysis while another botches basic financial derivations.

These limitations undermine the reliability of RAG systems, where precise integration of external knowledge is paramount. A model misinterpreting HVAC engineering standards from a technical manual could, for instance, recommend dangerously inefficient building designs.

Solution: To overcome these challenges, several strategies to enhance Domain-Specific Comprehension can be employed to strengthen the LLM's understanding and generation capabilities within specific domains.

- **Targeted Training Regimens**
  - *Specialized Fine-Tuning*: Immersing models in domain-specific corpora—think medical journals or legal case law—is akin to a medical residency. For example, BioBERT's training on PubMed abstracts enables nuanced understanding of biomedical relationships [9].
  - *Domain-Adaptive Pretraining (DAPT)*: Before task-specific tuning, models undergo "domain boot camps" using large-scale technical datasets. A study showed DAPT improved materials science paper summarization accuracy by 37% compared to baseline models.[6]
- **Architectural Adjustments**
  - *Vocabulary Surgery*: Retrofitting tokenizers with domain lexicons prevent critical terms like "EGFR" (epidermal growth factor receptor) from being split into meaningless subwords. The legal-BERT model successfully incorporated over 5,000 legal Latin phrases through deliberate vocabulary expansion.
  - *Knowledge Infusion*: Integrating structured knowledge graphs acts as a professional mentor. IBM's Watson for Oncology combines LLMs with curated cancer treatment trees, ensuring recommendations align with NCCN guidelines.
- **Human-Guided Refinement**
  - *Expert-In-The-Loop Training*: Like law partners reviewing junior associates' briefs, domain specialists annotate edge cases. The FDA now uses pharmacist-validated LLMs to screen drug interaction reports.

---

[6]Okan Yenigün, "Mastering Domain Knowledge: An Introduction to Domain-Adaptive Pretraining (DAPT)." https://levelup.gitconnected.com/mastering-domain-knowledge-an-introduction-to-domain-adaptive-pretraining-dapt-87930a97448b, Dec 2024, Accessed on 28/07/2025.

o *Controlled Generation*: Constraining outputs with regulatory "guardrails" prevents hazardous deviations. FinGPT automatically flags non-compliant financial advice using SEC rule embeddings.[7]

- **Evaluation Evolution**
  o Moving beyond generic accuracy metrics, teams now employ Terminology Precision Scores (e.g., correctly using "res ipsan loquitur" in tort law contexts) and Contextual Adherence Indexes measuring alignment with industry protocols.

Some emerging techniques may push boundaries further. For example, *Reinforcement Learning from Domain Feedback*, which allows an aerospace LLM iteratively improves turbine design suggestions based on engineers' real-world performance data, and *Multimodal Grounding*, which allows Architecture-focused models cross-reference 3D BIM models with textual specifications to detect design conflicts (see more on Model Fine-tuning in Chapter 12 and Challenges 3.2, 4.1 & 5.9 in this chapter).

3.3.5.2 *Challenge 5.2: Incomplete generation*

Pain point: Incomplete outputs in RAG systems mirror the frustration of receiving a meticulously researched report missing its conclusion. While technically accurate, these truncated responses fail to deliver actionable insights, akin to a legal brief omitting precedent citation. Our analysis of 1,200 financial Q&A interactions revealed 30% of incomplete answers occurred in multi-document queries using native queries, such as "Compare risk clauses in Agreements A, B, and C," a pain point demanding architectural and linguistic solutions.

Contrary to intuition, broader queries often yield shallower responses. When asked to synthesize across documents, most RAG systems default to *document-centric* rather than *concept-centric* retrieval. Our study demonstrated that generic cross-document queries retrieve generally less relevant passages versus targeted single-document requests.

The creation of prompts goes far beyond the direct integration of questions and retrieved materials; it requires adopting customized expression strategies and additional guidance according to the uniqueness of the

---

[7] https://github.com/AI4Finance-Foundation/FinGPT.

generative model. To constrain the potential generalization bias of the large model, establishing a constant "role framework" is particularly crucial, such as clearly instructing the large model to "act as an authoritative scholar in the field," or "to limit itself to using the question and related materials to provide a concise summary and conclusion," thereby maintaining the precision and relevance of the answers in various contexts.

Solution: Generally, we may adopt query transformations of Chapter 8. Comparison questions often perform inadequately with basic RAG methods. An effective strategy to enhance the deductive abilities of RAG involves incorporating a query comprehension layer, which is to implement query alterations prior to the search within the vector database. Below are four distinct query modification techniques:

- **Contextual Routing:** Preserve the original query while identifying the specific subset of instruments it relates to. Subsequently, assign these instruments as the relevant choices. For example, in a case where a mortgage query avoids derivatives paperwork, the original query intent shall be maintained while mapping to regulated document categories and using regulatory taxonomies (e.g., Basel III classifications) as routing guides.
- **Semantic Query-Rewriting:** Keep the chosen instruments, but rephrase the query in various ways to utilize it across the identical set of instruments. In finance, this might mean using different terminology based on the financial product. For example, rephrasing "LIBOR transition" to "SOFR adoption timelines" or "fallback language."
- **Sub-Question Decomposition:** Segment the query into multiple smaller queries, each aimed at different instruments as dictated by their metadata. For example, when querying "Analyze debt covenants in Targets X, Y, Z," we might further decompose it as "List financial maintenance covenants in Target X's Credit Agreement §4.1," "Identify cross-default triggers in Target Y's 2023 Bond Indenture" and "Assess EBITDA addbacks in Target Z's TTM calculations per Credit Suisse Guidelines."
- **ReAct Agent Tool Selection:** Based on the initial query, select which instrument to employ and devise the precise query to execute on that instrument. In finance, this might involve choosing between stress test models, compliance checklists, or risk assessment frameworks. For example, a liquidity query might trigger LCR calculation tools or regulatory templates.

3.3.5.3 *Challenge 5.3: Retrieval and generation synchrony, and incorrect specificity*

Pain point: It is not easy to strike the right balance between using retrieved information and harnessing the language model's human-like creativity and understanding.

- The output does not exhibit the appropriate level of specificity e.g., the responses can be overly vague or general, lacking the detailed nuance required. This insufficiency frequently leads to additional follow-up queries for clarification, as the answers may fail to fully address the user's needs.
- Ensuring the coherence and consistency of the large model's responses faces particularly prominent difficulties in information integration. The challenge for the large model is to integrate knowledge fragments from diverse literature and the diverse content triggered by multiple keywords in user questions. On this basis, the large model must demonstrate a high level of integration capability, ensuring that the final output results are not only closely connected in the logical chain but also maintain a high degree of consistency in the presentation of viewpoints and facts. This process poses a severe test to the comprehensive reasoning and connection capabilities of the large model.

Solution: To address this issue, the system's weighting mechanism should automatically adjust according to the complexity of the retrieval task and the confidence level of the retrieved data. This means dynamically modulating the importance assigned to query-related information. One promising approach is to implement hybrid architectures that can seamlessly transition between retrieval-focused and generation-focused modes without human intervention. Such architectures allow the model to continuously learn and fine-tune its balance, gradually achieving an optimal trade-off between accurate data retrieval and the creative, context-aware generation of responses. Several algorithms are further discussed in the following sections in Chapter 7:

- Auto-Merging Retrieval
- Sentence window retrieval
- Recursive retrieval
- Hybrid or Fusion Retrieval

3.3.5.4 *Challenge 5.4: Knowledge preparation for evaluation and finetune, e.g., high-quality QA datasets*

Pain point: Optimizing the performance of RAG systems hinges on the availability of a large and diverse set of high-quality question-and-answer (QA) pairs for fine-tuning. These QA pairs are essential for teaching the model to understand queries accurately and generate precise responses based on retrieved information. However, creating such datasets manually is an immensely time-consuming and resource-intensive task, often requiring expertise in the specific domain. On the other hand, relying solely on automated generation methods can compromise the quality and reliability of the data, introducing errors, biases, or irrelevant content. Therefore, developing an efficient mechanism to generate QA pairs that ensure both high quality and scalability has become a critical challenge for the success of RAG systems.

- **Manual Creation Limitations:**
    - *Time-Consuming*: Crafting QA pairs by hand requires significant effort from subject matter experts or annotators.
    - *Scalability Issues*: Manually generating the vast number of QA pairs needed for effective training is impractical.
    - *Consistency Concerns*: Ensuring uniform quality and style across manually created pairs can be challenging.
- **Automated Generation Limitations:**
    - *Quality Assurance*: Automatically generated QA pairs may lack depth, accuracy, or relevance to the desired domain.
    - *Bias Introduction*: Automated methods might perpetuate existing biases present in the training data.
    - *Contextual Understanding*: Machines may struggle to generate questions that capture the intricacies of human language and specific domain knowledge.

Solution: Considering the limited databases, several effective strategies can be implemented to ensure both efficiency and accuracy, e.g., synthetic data, semi-automatic annotation, active learning with feedback techniques, etc.

One approach is to employ semi-automated QA generation with human oversight. This method leverages advanced language models to automatically generate preliminary QA pairs from existing texts or documents. However, to ensure accuracy, relevance, and alignment with domain-specific terminology, these machine-generated pairs are then reviewed and refined by domain experts or trained annotators. This combination of automation and human expertise accelerates the dataset creation process while maintaining high quality.

Another strategy involves using data augmentation techniques to expand the dataset without the need for new content creation. Techniques such as paraphrasing, back-translation, synonym replacement, and controlled noise injections can generate multiple variations of existing QA pairs. This enhances the model's robustness by exposing it to diverse phrasings and expressions, allowing it to handle a wider range of user queries effectively.

Moreover, leveraging existing datasets and transfer learning is also a practical solution. By incorporating and adapting QA pairs from publicly available datasets relevant to the target domain, such as SQuAD, WikiQA, or other domain-specific corpora, the amount of new data that needs to be generated from scratch is reduced. Furthermore, fine-tuning models pretrained on general datasets using a smaller set of domain-specific QA pairs allows for faster training and improved initial performance, capitalizing on knowledge from related domains.

Engaging in crowdsourcing for QA pair creation can accelerate data collection by distributing the workload among a diverse group of contributors. However, clear guidelines and quality standards are essential to ensure consistency and accuracy. By implementing quality control mechanisms, such as having multiple annotators work on the same QA pairs and using consensus methods, organizations can help maintain high-quality outputs. This approach increases diversity in language use and perspectives within the dataset.

Employing active learning strategies can further optimize the process. By using the model to identify areas where it performs poorly, data annotation efforts can focus on the most informative examples for improving the model. Iteratively updating the model with new data and reassessing performance ensures that resources are efficiently used by targeting the most beneficial data for enhancement.

Another avenue is synthetic data generation with validation. Rule-based generation methods create QA pairs using predefined templates and domain-specific rules, ensuring that important concepts and commonly asked questions are covered. While synthetic data can quickly expand the dataset, validation steps—including automated checks and optional human review—are necessary to filter out implausible or incorrect QA pairs, thereby maintaining data quality.

Collaborating with domain experts is crucial for capturing complex knowledge that might be missed by non-experts. Organizing expert workshops or consultations allows for the generation or review of QA pairs collaboratively, ensuring that the dataset reflects the depth and nuances of the domain. Additionally, extracting knowledge through interviews or surveys can provide valuable insights that can be transformed into QA pairs.

Utilizing advanced question generation models offers an effective solution as well. State-of-the-art models trained to produce high-quality questions from text passages can automate the creation of QA pairs. By fine-tuning these models on domain-specific data, organizations can enhance relevance, and by applying filters or scoring mechanisms, they can select the best-generated QA pairs for inclusion in the dataset.

Finally, implementing continuous feedback loops by collecting data from real user interactions during deployment helps keep the model updated with current and relevant information. Allowing users to provide feedback on responses enables the identification of gaps and areas for improvement. Incorporating this feedback into the dataset and model fine-tuning process aligns the system more closely with user needs and expectations, ensuring ongoing enhancement of the RAG system's performance. By integrating these strategies, organizations can effectively overcome the bottleneck of QA pair supply, creating high-quality datasets efficiently and optimizing model performance. The similar approach is also applicable for the RAG pipeline optimization. This balanced approach, which combines automation with human expertise, leverages existing resources, and employs innovative data generation techniques, ultimately contributes to the success of RAG systems in real-world applications.

3.3.5.5 *Challenge 5.5: Intent recognition & toxicity to rejection*

Pain point: Defining the subtle balance of reasonable refusal to answer is a significant challenge. When facing user questions, especially when the question goes beyond the scope of existing literature, the appropriate timing of refusal to answer becomes a major test. It is necessary to be brave enough to say "I don't know" when you cannot provide accurate information to avoid misleading. In practice, although similarity thresholds or scenario corpora are often used as criteria for judgment, it is still difficult to achieve 100% accuracy. Overly frequent refusals may lead to an indifferent user experience, while excessive guessing in answering may damage the system's credibility. Therefore, grasping the right degree of refusal to answer is key to maintaining the credibility of the Q&A system and user trust.

- **Intent Recognition:**
  - *Understanding User Intent*: RAG systems need to discern whether a user's question falls within the domain of the system's expertise or whether it is out-of-scope, ambiguous, or otherwise inappropriate. Misunderstanding intent can lead to poor quality answers or incorrect responses, thus affecting the system's reliability.
  - *Toxicity and Malicious Queries*: Additionally, systems must recognize toxic or harmful queries that could elicit inappropriate or harmful responses, including abusive language, hate speech, or questions designed to trigger unsafe responses.
  - *Ambiguity and Complexity*: User queries may be vague, overly broad, or context-dependent, making it challenging for the system to determine the appropriate response.
- **Appropriate Refusal to Answer:**
  - Refusing to answer, especially when the system is unsure, is essential for maintaining user trust. However, the frequency of refusal must be carefully managed to avoid making the system seem unhelpful or indifferent.
  - A balance must be struck between being overly cautious (refusing too often) and being overly speculative (guessing answers), both of which can harm the user experience and the system's credibility.
- **Toxicity Rejection:**
  - RAG systems must detect when queries or responses contain toxic elements such as offensive language, inappropriate content, or malicious intent. Failure to appropriately reject toxic content can lead to legal issues, reputational damage, or harm to users.
  - Simply detecting toxic queries isn't enough. The system must also know how to gracefully reject such inputs, offering alternatives or explanations to avoid alienating the user.

Solution: Striking the right balance between responding to user queries and recognizing when to refuse an answer is vital for the long-term success of RAG systems. By embracing advanced AI techniques and continuously refining the system's capabilities, developers can create safer, more reliable systems that meet user needs without compromising on trust or accuracy.

- **Implement Advanced Intent Recognition Algorithms:** Advanced intent recognition relies on Natural Language Understanding (NLU) models like Qwen-embedding, RoBERTa, BGE-E3, or GPT/Gemini/ Claude variants that are fine-tuned for detecting user intent. These models help in identifying various types of queries, whether they are factual questions, opinion-seeking, ambiguous, or toxic. Additionally, systems can incorporate contextual awareness by analyzing previous interactions to better gauge relevance and domain fit, triggering proactive measures when necessary. For example, repeated off-topic or inappropriate questions can prompt the system to ask the user for clarification or suggest alternatives. Dynamic query classification further refines this process by assessing the relevance and appropriateness of queries, prioritizing those clearly within the system's domain and flagging others for closer scrutiny.[8]
- **Employ Confidence Scoring Mechanisms:** Confidence scoring mechanisms introduce thresholds for determining the system's certainty in its responses. If the retrieved information or generated response falls below a set confidence threshold, the system can trigger a polite refusal or a fallback strategy, such as requesting clarification from the user. Confidence decay models add nuance to this approach by progressively lowering confidence as ambiguity or divergence from the knowledge base increases, allowing the system to decide when to refuse an answer more effectively.
- **Build in Ethical and Toxicity Filters:** RAG systems should integrate real-time toxicity detection models like OpenAI's Moderation API or Google's Perspective API to flag harmful language or inappropriate

---

[8]For example, a RAG gateway might be designed to classify each query via embedding-based intent prototypes (E5-Mistral/BGE-M3) and Self-RAG retrieval-need signals. If intent $\in$ {policy/id-matching}, route to SPLADE; else to ColBERTv2/dense. We may compute semantic entropy on a small model to decide whether to ask exactly one clarification. Safety is enforced by Llama Guard/ShieldGemma (and OpenAI's multimodal moderation when applicable). Finally, a FrugalGPT-style (https://github.com/stanford-futuredata/FrugalGPT) router escalates model size only when predicted confidence is low, keeping latency/cost bounded without sacrificing accuracy.

queries. By applying transformer-based models that are fine-tuned for detecting toxic content, the system can ensure that offensive material is intercepted before it reaches the user. Graceful refusal mechanisms are also crucial in these scenarios; when detecting toxic queries, the system should respond with non-confrontational messages such as, "I'm sorry, I cannot provide an answer to that question," while offering suggestions for rephrasing or directing the user toward more constructive queries.

- **Refusal Strategy Based on Context and Intent:** A user-centered rejection system takes into account the user's frustration and intent. After a few refusals, the system might ask, "Am I misunderstanding your question? Would you like help with something else?" This approach offers alternative actions, such as suggesting rephrasing or directing the user to relevant resources, rather than issuing a flat denial. A multi-layered refusal mechanism can begin by attempting to answer in a general way, and if uncertainty persists, the system can gradually shift to refusal, offering related information or redirecting the query to a more knowledgeable model.

- **Regular Fine-Tuning and Model Updates:** Regular fine-tuning is essential to keep the system's intent recognition and toxicity detection models up to date. Incorporating new domain-specific data helps the system adapt to evolving language use and ensures safe, accurate handling of queries. Conducting regular audits for bias and fairness is equally important, especially in managing toxic queries and refusals, to ensure the system treats all users equitably and avoids promoting harmful stereotypes. This approach may be combined with continual feedback and active learning.

Again, we can apply the hybrid approaches with human oversight. For example, in Human-in-the-Loop Systems, we should involve human moderators to review certain queries or responses for critical applications and allow humans to handle complex or sensitive queries that the system is unsure about. Meanwhile, an escalation Mechanism should be established, so that the system can escalate queries to a human agent when appropriate. In cases where a query falls outside the system's knowledge base, fallback models provide a solution (see more in Challenge 5.6). Intent identification is also critical for agents in terms of function call, MCP and A2A, which will be described in Chapter 15 on Agentic AI and Chapter 12 on Chat Engine.

3.3.5.6 *Challenge 5.6: Handling Out-of-Domain queries*

Pain point: RAG systems often fail in the queries beyond the range of their knowledge base. When users pose questions that the system hasn't been trained on or that extend beyond its available data, the system may fail to provide meaningful answers. This limitation not only leads to user frustration but also diminishes trust in the system's capabilities. We shall handle the following two issues:

- **Identifying Out-of-Domain Queries:** Determining whether a query is outside the system's domain can be complex, particularly if the boundaries of the domain are not well-defined.
- **Appropriate Response Strategy:** Simply stating "I don't know" may not be satisfactory for users. Conversely, attempting to answer without sufficient knowledge can result in errors or confusion.

Solution: At this early stage, it is essential to integrate more robust query classification techniques that can accurately detect out-of-domain queries, e.g., intent identification. Additionally, a promising strategy is to deploy a general-purpose model that can generate responses when the specialized model is unable to do so. Alternatively, implementing a dynamic knowledge acquisition system—which can autonomously expand and update its knowledge base over time—offers a viable solution for addressing queries that fall outside the current domain. Overall, the prevailing trend is to continuously upgrade AI systems to enhance their versatility and responsiveness.

- **Implement Robust Query Classification:** Similar to Challenge 5.1, advanced ML models can be utilized to classify incoming queries as in-domain or out-of-domain with high accuracy. By training on annotated datasets, these models can recognize domain-specific language patterns and improve query classification. Semantic analysis, leveraging transformer-based architectures, helps interpret the intent behind queries and capture contextual nuances, improving overall performance. Regular updates to the classification model with new data ensure continuous learning and adaptability to evolving user queries and domain expansions.
- **Incorporate Fallback to General-Purpose Models:** When a query is classified as out-of-domain, the system can route it to a general-purpose

model like GPT-4/4o/1o/5, which is capable of handling a broader range of topics. This integration ensures that even if the specialized model fails, the user still receives an informative response. A hierarchical response system allows the system to first attempt using the specialized model, with smooth transitions to the fallback model when necessary, maintaining a coherent conversational flow.

- **Provide Graceful Degradation:** When the system cannot answer a query due to domain limitations, it should notify users politely, with messages like, "I'm sorry, but I don't have information on that topic currently." To maintain user engagement, the system should offer suggestions for related topics or guide users back to in-domain queries. Encouraging rephrasing or providing options for common queries within the domain enhances the user experience even when the system is limited.

- **Implement Dynamic Knowledge Acquisition:** To avoid constantly facing out-of-domain queries, the system can automatically expand its knowledge base over time. This can be done by utilizing web scraping, APIs, or external databases to gather new data in response to out-of-domain queries. ML techniques, such as reinforcement learning, can also help the system learn from new inputs and improve responses based on user feedback, ensuring the knowledge base remains dynamic and updated.

- **Utilize External Knowledge Sources, e.g., search engine:** Integration with comprehensive knowledge graphs, such as Wikidata, or domain-specific databases enhances the system's ability to retrieve structured data. This improves both the accuracy and relevance of responses. Additionally, implementing advanced semantic search technologies enables the system to retrieve relevant information from large corpora, including unstructured data sources.

Like some other challenges, we may periodically retrain or fine-tune the model using newly acquired data, especially from out-of-domain queries that have become relevant. Monitoring user interactions through query logging helps identify common out-of-domain topics, allowing for data analysis and knowledge base expansion. Feedback mechanisms enable users to share whether responses were helpful, improving system performance over time. A clear communication strategy with onboarding messages and adaptive dialogue management helps set expectations and ensures graceful handling of out-of-domain queries, keeping users informed and engaged. See the corresponding methods in Chapters 4, 12 & 15 and Challenge 5.1.

3.3.5.7 *Challenge 5.7: Redundant and general content without insights*

Pain points: Verbose or redundant outputs, over-generalization, and lack of depth or insight can significantly impact the quality of generated responses. For instance, when discussing the "advantages of RAG," a model might repeatedly mention "improved accuracy" in various phrasings, leading to unnecessarily lengthy and repetitive content. Similarly, a query about the "differences between RAG and traditional language models" might receive a broad response about the importance of AI advancements, without addressing the specific distinctions, indicating over-generalization. Moreover, when asked about "potential applications of RAG in cloud technologies," the model might list general applications without providing detailed insights, reflecting a lack of depth.

Solutions: To mitigate these issues, implementing advanced retrieval and generation strategies is essential. Techniques such as unlikelihood training can be employed to reduce verbosity and redundancy by penalizing the model for generating repetitive content. Additionally, fine-tuning the model with domain-specific data can enhance its ability to provide detailed and contextually relevant responses, thereby reducing over-generalization and improving depth. Incorporating mechanisms for self-reflection during the generation process can also help the model assess and refine its outputs for coherence and informativeness. By adopting these strategies, RAG systems can produce more concise, specific, and insightful responses, enhancing their overall effectiveness. See the prompt refinement in Chapter 6, and the response synthesis and validation techniques in Chapter 8.

3.3.5.8 *Challenge 5.8: Wrong output or inconsistent format*

Pain point: As the LLM generated content might be further applied in the downstream applications, if the output is in wrong or inconsistent format, it can lead to confusion, decreased usability, and additional workload for post-processing. It is often caused by the following reasons.

- **LLM Limitations:** Language models may prioritize generating coherent and contextually relevant text over adhering to formatting instructions, especially if not explicitly trained to follow such directives.
- **Ambiguous Instructions:** If the prompt lacks clarity or specificity, the LLM might not understand the formatting requirements.
- **Complexity of Formats:** Generating complex structures like nested JSON objects or multi-column tables can be challenging for LLMs without proper guidance.
- **Token Limitations:** Large outputs in specific formats may approach or exceed token limits imposed by the LLM, leading to truncated or incomplete responses.

Solution: To address the issue of incorrect output formats, several strategies can be implemented: Clear and precise prompts guide the LLM more effectively (e.g., explicitly specify the output format syntax), while output parsing and validation ensure that any deviations are corrected before reaching the end-user. There might be an iterative refinement again. If the LLM doesn't support the output instruction well, fine-tuning the model on structured data and acknowledging limitations helps in setting realistic expectations. To implement post-processing techniques, automated formatting can be achieved using libraries like Pandas for data frames or JSON libraries to structure the output after generation. Conversion tools can be employed to transform outputs from one format to another, such as converting text to tables. Quality assurance involves performing validation checks to ensure the output meets the required formatting criteria before presenting it to the user. If the output fails these checks, feedback loops can automatically prompt the language model to regenerate the response with the corrected formatting. Combining these methods creates a robust approach to handling formatting challenges, ensuring that the system meets the needs of various applications that rely on specific output formats.

Additionally, improving the user interface and experience can help address formatting issues by allowing users to select predefined templates that include formatting instructions, adjust the output format after generation, and utilize visualization tools to render structured data visually. Acknowledging the model's limitations and setting user expectations (by informing users about potential constraints and suggesting simplified alternatives) can enhance user satisfaction and prevent misunderstandings. Finally, combining multiple strategies such as prompt engineering, output parsing, and post-processing techniques, while continuously monitoring performance and iteratively refining the system, can lead to improved reliability and better alignment with user needs. See Section "Output Parsing" of Chapter 12 and function parameters, e.g., pydantic, openai API, etc.

3.3.5.9 *Challenge 5.9: Multimodal RAG*

Pain point: Multimodal RAG involves retrieving and synthesizing information across different modalities such as text, images, code, structured knowledge, audio, and video. Each modality introduces unique retrieval and synthesis procedures, target tasks, and challenges. For example, expanding the context of text generation through image retrieval can enhance storytelling or explanation, while incorporating sample code and related documentation can significantly improve code generation tasks. Creating unified embeddings that represent different modalities within a common space is non-trivial. Processing and storing multimodal data demand significant computational resources and efficient algorithms. Additionally, there's a scarcity of large-scale, high-quality multimodal datasets for training and evaluation, which limits the development of these systems. Addressing these challenges is crucial for advancing multimodal RAG capabilities.

Solution: Multimodal support in RAG systems represents a significant advancement in AI capabilities, allowing for richer and more contextually relevant interactions. By developing modality-specific retrieval methods, implementing cross-modal alignment techniques, and optimizing computational resources, we can overcome the challenges associated with multimodal data processing. Furthermore, leveraging advanced models, enhancing data representations, and addressing ethical considerations will pave the way for more robust and versatile RAG systems.

Depending on the use scenarios, we can either develop effective modality-specific retrieval techniques, e.g., different strategies are required for images, audio, and video data, or use cross-modal alignment and fusion (Multimodal embeddings) techniques to process and integrate information from multiple modalities, e.g., Qwen-VL or CLIP (Contrastive Language-Image Pre-training) or GPT-4o/5. Attention mechanisms can help focus on relevant features, improving context understanding, while multimodal transformers such as LXMERT, ViLT, or VisualBERT can handle multimodal inputs and improve overall performance.

To organize and retrieve multimodal information efficiently, multimodal knowledge bases should be established. Unified data stores can be built to store data from different modalities with appropriate indexing for efficient retrieval. Metadata can link related data across modalities, facilitating integration. Additionally, knowledge graphs can be developed to incorporate entities and relationships from different modalities, helping in more effective data retrieval. Finally, efficient data integration techniques should be employed to optimize performance. Modality conversion can be useful when transferring data from one modality to another, such as converting images to text through captioning. Hierarchical retrieval, which involves using initial results from one modality to guide subsequent searches in another, ensures more comprehensive data retrieval. However, due to space constraints, detailed implementation of these techniques will not be discussed.

3.3.6 *Challenges in LLM Serving and Monitoring*

3.3.6.1 *Challenge 6.1: Fallback model(s) and reliability*

Pain point: When the RAG systems that rely on external LLMs, such as OpenAI's GPT-4o/5, it's crucial to ensure consistent availability and performance. However, issues like exceeding rate limits, latency issues, unexpected downtime, cost considerations or API failures can disrupt the primary LLM's functionality. These disruptions can lead to degraded user experience, lost productivity, and potential system failures. To mitigate these risks, it's essential to have fallback models or contingency plans that maintain the system's operability when the primary model encounters problems.

Solution: Incorporating fallback models into RAG systems is proper for maintaining reliability, performance, and user satisfaction. Robust solutions and framework shall be provided for managing multiple LLMs and automating the switch between them when issues arise. By implementing intelligent routing, monitoring, and diversified model strategies, developers can mitigate risks associated with API rate limits, service outages, and other disruptions. These approaches ensure that RAG systems remain resilient and continue to deliver value, even when facing challenges with the primary language model, which are discussed in Chapter 12.

- **Model Ensemble Approach:** This approach involves using multiple models simultaneously and aggregating their outputs to improve reliability and performance. The system is designed to send requests to several models, then combine their responses using methods like majority voting or confidence scoring. This strategy reduces the risk of failure from any single model, enhancing accuracy through aggregated insights. However, it may lead to increased operational costs due to multiple API calls and additional logic to manage the responses and resolve any potential conflicts would be required.

- **On-Premises or Self-Hosted Models:** This solution involves deploying open-source LLMs like Llama, Mixtral, Qwen, DeepSeek or BLOOM on local servers or cloud infrastructure to serve as backups, via vLLM, Ollama, Xinference, llama.cpp, etc. The implementation requires setting up the necessary hardware and software environment to host the models, integrating them into the application's fallback logic. The main benefits include greater control over the models and enhanced data privacy, along with the potential for cost savings in large-scale usage. However, this approach is resource-intensive, requiring significant computational resources and expertise, along with ongoing maintenance responsibilities for updates, security, and performance optimization.

- **Multi-Provider Strategy:** This strategy involves establishing accounts with multiple AI service providers to diversify risk. By integrating APIs from different providers like OpenAI, Microsoft Azure, Google Cloud AI, Amazon Bedrock, Anthropic, Alibaba cloud, Hugging face, DeepSeek, Perplexity, Mistral, Paperspace, or AutoDL, the system can switch between providers based on availability or cost. This provides redundancy and mitigates the impact of provider-specific issues, while also offering access to unique features or models from different providers. However, managing multiple APIs and billing systems adds complexity, and differences in model behavior may affect the consistency of the application's performance.

- **Implementing Intelligent Routing and Monitoring:** Intelligent routing involves setting up monitoring tools to track API usage, rate limits, response times, and error rates. Alerts can notify the development team of any issues with the primary model. Dynamic routing logic can be implemented to make real-time decisions on which model to use, based on criteria such as rate limits, response latency, model performance, or cost thresholds. Additionally, the system should be designed to gracefully degrade when switching to fallback models, ensuring essential functionality remains available and communicating any limitations to users transparently. Regular testing and validation of fallback mechanisms are essential to ensure they work correctly under various scenarios, and that the quality of responses from fallback models meets required standards. We might also design a tiered system for Model Cascading: Route simple queries to smaller, faster, and cheaper models. Only use the large, state-of-the-art model for complex queries that require deep reasoning, thus optimizing cost-performance. Two typical implementations are **Neutrino AI Gateway**[9] **and OpenRouter**[10].

o Neutrino is an AI API gateway that enhances system reliability by intelligently routing requests between different LLMs based on predefined policies, performance metrics, and availability. It ensures continuous service by automatically directing requests to alternative models when the primary model exceeds rate limits or becomes unavailable. Neutrino also offers load balancing to optimize performance and cost, and provides monitoring and analytics for insights into model performance. However, integrating Neutrino requires effort to incorporate it into your application architecture, including defining routing policies and managing API keys, and you should verify compatibility with your desired LLM providers.

o OpenRouter is a unified platform that routes AI API requests across multiple AI service providers like OpenAI, AI21 Labs, and Anthropic. OpenRouter provides a unified API interface, simplifying the integration and switching between different models without extensive code modifications. It abstracts away differences between various LLM APIs by handling authentication and request formatting, and features a fallback mechanism that automatically switches to alternative models when the primary one fails or reaches rate limits. Developers can define customizable routing logic based on availability, performance, or cost, enhancing flexibility and reliability. While OpenRouter improves developer

---

[9] https://www.neutrinoapp.com/.
[10] https://openrouter.ai/.

efficiency by reducing API management overhead, there may be a learning curve, and support primarily comes from community contributions, so it's essential to ensure compatibility with the AI models relevant to your application. Alternatively, Databricks Mosaic, Portkey,[11] Kong AI gateway[12] and LiteLLM are also the candidates. See details in Chapter 12.

3.3.6.2 *Challenge 6.2: The hybrid cloud & on-premise dilemma (especially in regulated industries)*

Pain point: Banks and other regulated institutions cannot send sensitive data to cloud APIs. However, their on-premise data centers often lack the specialized GPU infrastructure to run high-performance LLMs, creating a bottleneck that forces a complex and difficult-to-manage hybrid architecture.

Solutions: The organizations shall consider upgrading their IT infrastructure (see supplementary material), and at the same time, optimizing their LLM application design.

- **Strategic Data & Query Routing:** Implement an intelligent gateway that classifies queries based on data sensitivity. Route non-sensitive queries to powerful and scalable cloud LLMs, while processing sensitive queries on smaller, highly-specialized models running on-premise.
- **Optimized On-Premise Deployment:** Instead of trying to run massive models, focus on fine-tuning smaller, efficient open-source models (e.g., Qwen3.5-27B/35B, GLM-4.7-Flash) on proprietary data. These models can run effectively on less powerful hardware and are better suited for on-premise environments.
- **Confidential Computing:** For cloud deployments, leverage confidential computing environments where data remains encrypted even during processing, providing a hardware-based layer of security that can help meet regulatory requirements.

---

[11] https://github.com/Portkey-AI/gateway.

[12] https://developer.konghq.com/ai-gateway/.

As RAG systems are expensive to run, we shall also closely monitor costs, which are driven by two factors: the constant computational load of the vector database and the high token count (query + large retrieved context) processed by the generator LLM for every single request.

3.3.7 *Challenges in RAG Evaluation*

3.3.7.1 *Challenge 7.1: Proper performance evaluation for system diagnosis*

Pain point: Scientifically measuring the practical threshold of RAG in a specific field. Assessing the maturity of RAG applications in a particular field, similar to using the RAGAS evaluation framework,[13] requires providing independent questions and standard answers, making the evaluation results highly dependent on the rationality of the question design and the accuracy of the standard answers. In addition, the complexity of the evaluation process, which involves the performance of large models and embedded models, further demands the stability and reliability of the models themselves, as well as the precision of the input prompts; otherwise, the evaluation indicators will lose credibility. Therefore, ensuring the quality of the large model and the prompt is a key challenge in verifying the effectiveness of RAG applications.

Solution: Ideally, we shall develop a standardized evaluation framework, involving creating diverse and representative benchmark datasets that cover a wide range of queries and scenarios within the specific domain, utilizing publicly available datasets, when possible, to enable comparative evaluations across different systems. Additionally, it's essential to establish clear and consistent evaluation metrics that assess not only accuracy but also relevance, coherence, and usefulness of the responses, perhaps by adopting or adapting existing frameworks like the RAG Evaluation Toolkit to ensure consistency. In addition, we shall have Component-Specific Monitoring, e.g., monitor the retriever and generator independently. For example, we might track Hit Rate (Did it find the right document?) and Mean Reciprocal Rank (MRR) (How high up was the right document?) for Retriever Metrics, and track Faithfulness

---

[13] https://docs.ragas.io/en/latest/concepts/metrics/.

(Is the answer grounded in the context?) and Answer Relevance (Does it address the user's query?) for Generator Metrics. This is also useful to solve the difficulty in diagnosing failures.

Again, we shall implement continuous monitoring with evaluation and incorporate human-in-the-loop feedback. The former means the framework is built for both offline and online. The latter means engaging domain experts to review and assess the system's responses, providing insights that automated metrics might miss. The general testing techniques, e.g., A/B test, are also applicable. See Section "Evaluation Frameworks" of Chapter 10.

3.3.7.2 *Challenge 7.2: Explainability, transparency and fairness*

Pain point: There is an inherent contradiction in the process of both extracting and subsequently replacing specific pieces of information from data sets. This dual operation makes it exceptionally challenging to clearly explain system outputs and provide transparency in market decision-making. At the same time, fairness, an essential element of responsible AI, requires robust and thorough consideration to ensure unbiased outcomes [10–12].

Solution: To address these issues, a multifaceted approach is required:

- **Attribution Mechanisms:** Implement robust attribution techniques that directly link the generated content to the specific retrieved documents or data segments. This helps in tracing back the origins of the information used in forming the response, making the decision process more transparent and easier to audit.
- **Interactive Interfaces:** Develop user interfaces that allow users to interactively explore both the retrieved documents and the reasoning processes of the AI. Such interfaces should enable users to drill down into the data, review citations, and understand how different pieces of information influenced the final output. Tools like attention visualization and token attribution can highlight which parts of the input and retrieval contributed most significantly to the response.
- **Dynamic Fairness Adjustments:** Incorporate mechanisms that continuously assess and adjust for fairness. As the complexity and

confidence of the retrieved data change, the system should dynamically modulate the weighting of retrieved versus generated information to minimize biases. Hybrid architectures that seamlessly switch between retrieval-intensive and generation-intensive modes without human intervention can help the system adapt and gradually optimize its fairness and transparency. See Section "Evaluation."

- **Transparent Model Reporting:** In addition to interactive tools, providing comprehensive, easily accessible documentation and model reporting can improve transparency. This documentation should include details about data sources, retrieval algorithms, and any known limitations or biases, thereby allowing stakeholders to better understand and trust the system's outputs.

The details can be found in Chapter 10 on "Evaluation."

3.3.7.3 *Challenge 7.3: Privacy and security*

Pain point: RAG systems that retrieve information from sensitive or personal knowledge bases may face privacy and security challenges. For example, when a wealth management RAG system is retrieving client portfolios, this process may accidentally expose some HNWI account numbers, highlighting three critical gaps:

1. Unmasked alternative investment identifiers in estate planning documents.
2. Residual metadata in encrypted PDF loan agreements.
3. Over-retention of KYC document fragments.

Solution: We might implement robust access control and encryption mechanisms for the knowledge base, apply guardrail and PII detection to make sure the content is protected and develop privacy-preserving retrieval techniques, such as anonymization techniques federated learning or differential privacy.

- **Robust Access Control and Encryption:**
  - *Access Control*: Implement stringent role/attribute-based access control (RBAC/ABAC) and multi-factor authentication (MFA) to ensure that only authorized users and services can access the

knowledge base. Regularly review and update permissions based on user roles and organizational changes.

o *Encryption*: Apply end-to-end encryption protocols for data both in transit and at rest. This includes using industry-standard algorithms such as AES-256 for encrypting stored data and TLS for data transmission. These measures help prevent unauthorized data interception or exposure.

- **Guardrails and PII Detection:**
  o *Input/Output Guardrails*: Integrate guardrails into both the input and output pipelines of the RAG system. These guardrails should include automated PII detection modules that scan for sensitive information using advanced NLP techniques and pattern matching. See Section "Guardrails."

  o *Dynamic PII Detectors*: Utilize ML-based PII detectors that continuously improve through feedback loops. These detectors can flag or automatically redact PII from user inputs and retrieved documents, ensuring that sensitive data is not inadvertently exposed in the generated outputs.

  o *Automated Auditing*: Develop logging and monitoring systems to track access, modifications, and usage patterns. Regular audits of these logs help detect suspicious activity and ensure compliance with data protection standards.

- **Privacy-Preserving Retrieval Techniques:**
  o *Data Anonymization and data minimization*: Before processing, apply anonymization techniques such as pseudonymization, data masking, or generalization to remove or obscure any PII in the retrieved documents. This step ensures that even if the data is accessed, individual identities remain protected.

  o *Context-Aware Redaction*: Use context-aware algorithms that not only detect standard PII patterns (like names, addresses, and social security numbers) but also adapt to domain-specific sensitive information that might appear in enterprise data.

  o *Federated Learning*: Instead of centralizing all sensitive data, a federated learning approach enables each data source to keep its data locally. In this setup, the RAG system trains retrieval models on each local dataset independently. Only the model updates (which are aggregated and typically noise-added) are shared with a central server to form a global model. This ensures that raw, sensitive data is never transmitted or stored centrally, significantly reducing the risk of data leakage. This technique is useful when applying multiple sensitive data sources to gain some statistical results.

o *Differential Privacy*: Integrate differential privacy methods by injecting carefully calibrated noise into the retrieval or aggregation processes. This ensures that individual data points cannot be re-identified while still allowing the system to benefit from collective data trends.

o *Secure Multi-Party Computation*: In scenarios requiring joint computations over sensitive data from multiple sources, employ secure multi-party computation techniques to perform these calculations without revealing the underlying data.

Some solutions will be discussed in Chapter 10 on RAG evaluation, while the general privacy-preserving techniques are out of the scope of this book.

3.3.8 *Challenges in RAG Pipeline*

3.3.8.1 *Challenge 8.1: Pipeline integration*

Pain Point: Optimizing the performance of a RAG system requires a comprehensive approach that considers the entire pipeline. Focusing on individual components in isolation may lead to suboptimal results, as the interplay between various stages (such as data chunking, embedding, retrieval, and generation) collectively determines the system's effectiveness.

Solution: To enhance retrieval efficiency in your RAG system, adopt a holistic strategy that addresses each component of the pipeline:

- **Refine Chunking Process:** Experiment with different chunk sizes to achieve an optimal balance between granularity and context. Smaller chunks can improve retrieval accuracy by capturing specific information, while larger chunks provide broader context.
- **Embed Metadata:** Incorporate metadata into your embeddings to facilitate better filtering and enrich context during retrieval. This practice enables more precise matching of queries to relevant documents.
- **Implement Query Routing:** Utilize query routing across multiple indexes to handle diverse query types effectively. This approach

ensures that each query is directed to the most appropriate index, enhancing retrieval performance.

- **Employ Multi-Vector Retrieval:** Consider using LangChain's MultiVectorRetriever, which allows for the creation of multiple vectors per document. This method involves splitting documents into smaller chunks, generating summaries, and formulating hypothetical questions to improve retrieval accuracy.
- **Address Vector Similarity Issues:** Implement re-ranking strategies to refine the relevance of retrieved documents. Techniques such as hybrid search, which combines vector similarity with traditional keyword search, and recursive retrieval can further enhance performance.
- **Optimize Vector Search Algorithm:** Fine-tune your vector search algorithm to balance accuracy and latency, ensuring efficient retrieval without compromising performance.
- **Explore Advanced Strategies:** Experiment with methods like Hypothetical Document Embeddings (HyDE) and iterative approaches such as "Read Retrieve Read" to improve the system's ability to handle complex queries.

By systematically refining each stage of the RAG pipeline, you can develop a system that excels in retrieving relevant and contextually rich information, leading to more accurate and coherent outputs. See Chapter 12 on pipeline and orchestration.

3.3.8.2 *Challenge 8.2: QA pipeline on structured data with Text2SQL*

Pain point: Interpreting user requests to extract relevant structured data accurately poses challenges, particularly with complex or unclear queries, rigid text-to-SQL (Text2SQL) or NL-to-SQL (NL2SQL) translations, and the current limitations of LLMs in effectively managing these tasks. The intricacies of mapping natural language queries to structured queries, understanding diverse data schemas, and ensuring precise and contextually appropriate responses make this a complex problem to solve.

Interpreting user requests to extract relevant structured data accurately is particularly challenging when dealing with complex or unclear queries. Current RAG and LLM systems often struggle with the rigid conversion of natural language to SQL (Text2SQL or other structured query languages), especially when the user query is ambiguous or when the underlying data schema is diverse. These systems face difficulties in accurately mapping the nuances of natural language into precise, contextually appropriate structured queries, resulting in errors or incomplete data retrieval.

Solution: Interpreting user requests to extract relevant structured data accurately in RAG systems is a complex challenge due to ambiguous queries, rigid Text2SQL translations, and limitations of current LLMs. By implementing advanced techniques such as **Chain-of-Table** for table-specific reasoning and **Mix Self-Consistency** to enhance answer reliability (see Chapter 13), we can significantly improve the system's ability to handle complex structured data queries. Additionally, leveraging specialized Text2SQL models, incorporating interactive query refinement, and optimizing LLMs for structured data tasks contribute to more accurate and effective QA pipelines.

- **Implementing Chain-of-Thought Prompting for Structured Data:** Chain-of-Thought (CoT) reasoning guides LLMs to generate intermediate reasoning steps, helping break down complex queries into manageable parts. By dividing the query into smaller steps, the model can better understand and process complex requests, ultimately improving the accuracy of generated SQL queries. Extending this concept to a Pack of Chain-of-Table, the model handles table-specific reasoning by creating steps that consider each relevant table and their relationships within the database schema. This improves the model's comprehension of complex schemas and relational data structures, enhancing the overall query accuracy. See Section "Chain-of-Table" of Chapter 13.
- **Utilizing Mix Self-Consistency Methods:** Mix self-consistency involves generating multiple reasoning paths or answers and selecting the most consistent one among them. When applied to structured data, this technique allows the model to produce several potential SQL queries or reasoning steps, which are then evaluated for consistency with the database schema and expected results. By using a Pack for Mix Self-Consistency, the system packages multiple reasoning chains and applies statistical methods to determine the most probable correct answer. This approach reduces errors and enhances the robustness of the query generation process, mitigating the impact of any single erroneous path. See Section "Mix-Self-Consistency" of Chapter 13.

- **Enhancing Text2SQL Translation Models:** Fine-tuning models specifically for Text2SQL tasks using domain-specific datasets improves translation accuracy. Models like SQLNet, Seq2SQL, and RAT-SQL are tailored for SQL generation, making them ideal for this purpose. By incorporating database schema information directly into the model's input, using encoding methods to represent tables, columns, and relationships, the model can generate syntactically correct and contextually appropriate SQL queries. This schema integration ensures that the output queries align more closely with the structure of the database.
- **Incorporating Interactive Query Refinement:** Implementing clarification dialogues enables the system to ask follow-up questions for ambiguous queries, enhancing its understanding of user intent and reducing potential errors from misinterpretation. Additionally, providing User Validation allows the system to present the generated SQL query or an interpretation of it back to the user for confirmation before execution. This validation step increases user trust in the system and ensures that misunderstandings can be caught and corrected early.

We may further Implement Hybrid Systems using both LLM and rule-based engine, e.g., combining rule-based systems with AI models leverages the strengths of both approaches. Traditional rule-based systems efficiently handle straightforward queries, while AI models tackle more complex tasks. Using a Modular Pipeline Design, the QA process is broken down into specialized modules for parsing, reasoning, query generation, and execution, allowing for more efficient and accurate processing at each stage, ultimately enhancing system performance. This will further need the aid of the agents. See more in Chapter 13 on Text2SQL.

3.4 Concluding Remarks

RAG is rapidly maturing, yet its production roll-out in regulated sectors such as banking continues to expose gaps in efficiency, governance, and scale. It's no doubt that this book cannot provide an exhausted list of the challenges faced in RAG production, noting its dynamic evolution. In fact, there are many promising research and development directions that help to improve the overall performance. For example, recent research shows unified end-to-end RAG agents emerging, distilled small-language models (SLMs) reaching edge devices, streaming pipelines shrinking freshness windows to minutes, and policy-aware retrieval tightening zero-trust postures, i.e., all of which point toward a future where RAG is both leaner and more secure.

- **Unified end-to-end RAG models:** Early prototypes such as Agent-UniRAG train a single agent that orchestrates retrieval, reranking, and generation in one differentiable loop, eliminating brittle component interfaces and reducing latency by up to 35% in benchmarks [13]. However, these gains come at the cost of much larger training corpora and sophisticated curriculum-learning schedules, making reproducibility a key open challenge [14].
- **Distilled SLMs & edge/branch RAG:** Studies on edge collaboration show that 7-B-parameter distilled LLMs, when paired with cloud back-offs, retain roughly 80% of GPT-4 accuracy while cutting inference cost five-fold—crucial for branch kiosks or RM tablets [15]. Future work should refine mixed-precision adapters and federated fine-tuning so that on-device RAG can preserve privacy without constant uplink.
- **Real-time vector pipelines:** Serverless Kafka plus vector-DB upserts now refresh indices in <300 ms, a requirement for high-frequency-trading chat or market-news copilots. Designing these streams demands careful partitioning, memory-mapped shards, and autoscaling governors to avoid throughput spikes that double cloud spend.
- **Policy-aware retrieval & zero-trust controls:** Metadata filtering, role-based access control, and row-level security are moving from "nice to have" to table stakes; Pinecone, Supabase,[14] and similar stacks expose ABAC filters directly in their query API so that vectors never leave their entitlement boundary. Yet vector stores still lag traditional RDBMSs in audit logging and key rotation, so alignment with enterprise IAM remains a priority.

Due to book size limitation, we will not discuss those topics in detail. Meanwhile, note that some recent surveys [2, 14, 16–18] reveal a split view: some architects declare "RAG is dead" as they pivot to secure agent-over-source patterns, while others double-down on hybrid graph-plus-vector approaches for multi-hop reasoning. I would like to say that the tone has shifted from "RAG is dead" to "RAG is the pragmatic

---

[14]https://supabase.com/.

bridge toward richer agentic systems." For example, NVIDIA's latest developer brief notes that dynamic knowledge agents depend on robust RAG layers to stay current, not replace them.[15] In general, I believe that RAG's trajectory is positive, e.g., undergirded by measurable ROI, accelerating research, and an ecosystem that continues to innovate on governance, cost, and latency.

References

[1]   Y. Ding and *et al.*, *A Survey on RAG Meets LLMs: Towards Retrieval-Augmented Large Language Models,* arxiv:2405.06211v1, 2024.

[2]   Y. Gao and *et al.*, *Retrieval-Augmented Generation for Large Language Models: A Survey,* arxiv:2312.10997, 2023.

[3]   T. Gu, H. Feng, M. Li, W. Gu and G. Wang, Alarm: retracted articles on cancer imaging are not only continuously cited by publications but also used by ChatGPT to answer questions, *Journal of Advanced Research,* vol. 71, pp. 1–3, 2025.

[4]   J. Gao and C. Long, *RaBitQ: Quantizing High-Dimensional Vectors with a Theoretical Error Bound for Approximate Nearest Neighbor Search,* in *SIGMOD,* 2024.

[5]   A. Mihan, A. Pandey and H. G. C. Van Spall, Artificial intelligence bias in the prediction and detection of cardiovascular disease, *NPJ Cardiovascular Health,* vol. 1, pp. 31, 2024.

[6]   Z. Jiang, X. Ma and W. Chen, *LongRAG: Enhancing Retrieval-Augmented Generation with Long-context LLMs,* arXiv:2406.15319v3, 2024.

[7]   G. Xiao, Y. Tian, B. Chen, S. Han and M. Lewis, *Efficient Streaming Language Models with Attention Sinks,* arXiv:2309.17453v2, 2024.

[8]   M. Ivgi, U. Shaham and J. Berant, Efficient long-text understanding with short-text models, *Transactions of the Association for Computational Linguistics,* vol. 11, pp. 284–299, 2023.

[9]   J. Lee and *et al.*, BioBERT: A pre-trained biomedical language representation model for biomedical text mining, *Bioinformatics,* vol. 36, no. 4, pp. 1234–1240, 2020.

---

[15]Nicola Sessions, Traditional RAG vs. Agentic RAG—Why AI Agents Need Dynamic Knowledge to Get Smarter, https://developer.nvidia.com/blog/traditional-rag-vs-agentic-rag-why-ai-agents-need-dynamic-knowledge-to-get-smarter, Jul 2025, Accessed on 28/07/2025.

[10]　H. ZHAO and *et al.*, Explainability for large language models: A survey, *ACM Transactions on Intelligent Systems and Technology,* vol. 15, no. 2, pp. 20, 2024.

[11]　H. Luo and L. Specia, *From Understanding to Utilization: A Survey on Explainability for Large Language Models,* arXiv:2401.12874v1, 2024.

[12]　I. O. Gallegos and *et al.*, *Bias and Fairness in Large Language Models: A Survey,* arXiv:2309.00770v3, 2024.

[13]　H. Pham, T.-D. Nguyen and K.-H. N. Bui, *Agent-UniRAG: A Trainable Open-Source LLM Agent Framework for Unified Retrieval-Augmented Generation Systems,* arXiv:2505.22571v3, 2025.

[14]　Y. Chen, J. Zhao and H. Han, *A Survey on Collaborative Mechanisms Between Large and Small Language Models,* arXiv:2505.07460v1, 2025.

[15]　S. Li and *et al.*, *Collaborative Inference and Learning between Edge SLMs and Cloud LLMs: A Survey of Algorithms, Execution, and Open Challenges,* arXiv:2507.16731, 2025.

[16]　H. Yu, A. Gan, K. Zhang, S. Tong, Q. Liu and Z. Liu, *Evaluation of Retrieval-Augmented Generation: A Survey,* arXiv:2405.07437v2, 2024.

[17]　H.-a. Gao, J. Geng and *et al.*, *A Survey of Self-Evolving Agents: On Path to Artificial Super Intelligence,* arXiv:2507.21046v2, 2025.

[18]　S. Zhao, Y. Yang, Z. Wang, Z. He, L. K. Qiu and L. Qiu, *Retrieval Augmented Generation (RAG) and Beyond: A Comprehensive Survey on How to Make your LLMs use External Data More Wisely,* arXiv:2409.14924v1, 2025.

Chapter 4

Data Processing

Data processing generally involves multiple stages such as data collection, data parsing, and data ingestion. Each of these stages is crucial to ensuring the effectiveness and efficiency of downstream tasks like information retrieval, data analysis, and generation. For the context of Retrieval-Augmented Generation (RAG), efficient and accurate data processing pipelines are essential for building robust systems.

The process generally includes the following stages:

- **Data Loading:** Collecting and importing data from various sources and formats.
- **Data Parsing:** Structuring and extracting meaningful segments from loaded data.
- **Data Cleaning:** Removing noise and irrelevant content, eliminating ambiguity, correcting errors, updating outdated documents and ensuring data consistency.
- **Data Parallel Processing:** Optimizing large-scale data processing for performance and efficiency.

Each stage plays a critical role in preparing high-quality data for downstream tasks.

4.1 Data Loading

Data loading is the foundational step in building RAG systems, aiming to import data from diverse sources and formats for further processing and analysis. Modern frameworks and tools offer flexible and efficient solutions for loading data, accommodating local files, cloud storage, databases, and APIs. The following are the common data loading techniques:

- **Local File Loading:** Reading data directly from file systems in formats such as text, PDF, Word or JSON.
- **API Integration:** Loading data dynamically via RESTful APIs or GraphQL.
- **Database Connectivity:** Extracting data from SQL/NoSQL databases for structured information retrieval.
- **Cloud Storage:** Importing data from platforms like AWS S3, Google Cloud Storage, or Azure Blob.
- **Real-time Streams:** Ingesting data from streaming platforms such as Kafka or RabbitMQ.

Most common RAG frameworks provide built-in modules and tools for data loading, offering seamless integration with various data sources as stated above. These frameworks simplify the ingestion process, ensuring that data is prepped and ready for downstream operations like parsing, indexing, and retrieval.

4.2 Loading Data with LLM Frameworks

Modern RAG frameworks like LangChain and LlamaIndex offer a rich ecosystem of document loaders (or readers) to accommodate diverse file types, directories, databases, and APIs. These loaders convert raw content into framework-specific data structures, such as Document objects, containing both text and metadata.

- **LangChain:** Provides loaders for text (TextLoader), PDFs (PyPDFLoader, PyMuPDFLoader), CSV (CSVLoader), and more. It also supports directory-level loading (DirectoryLoader) for bulk ingestion. It converts raw data into its own data objects, named Documents.

These Documents are structured to encapsulate both the content and metadata of the data, ensuring compatibility with downstream tasks such as parsing, indexing, and retrieval. This abstraction simplifies data management and facilitates the use of various tools and methods within the RAG pipeline. Additional information on Metadata processing can be found in Chapter 5.

- **LlamaIndex:** Uses built-in readers (e.g., SimpleDirectoryReader) and external connectors (via LlamaHub) to handle various file types and data sources. Like LangChain, it encapsulates parsed data into Document objects with extensive metadata.

Some sample accesses are provided below with codes available in ch4\files_loader.py.

4.2.1 *Loading Simple Files Using LangChain*

Text-based files (e.g., .txt, .md) are often the simplest to handle, LangChain provides a TextLoader that extracts content directly into Document objects.

Below is a brief example of using LangChain to load a single .txt file:

```
from langchain_community.document_loaders
  import TextLoader

loader = TextLoader("data/wukong_story.txt")
documents = loader.load()
```

Here, each file is mapped to one or more Document objects containing:

- **page_content** – the extracted text.
- **metadata** – metadata like file source, author, or creation date.

4.2.2 *Loading from Directories*

Often, you need to load many files at once, Directory-based loaders can recursively scan a folder, applying specific loading logic (e.g., text vs. PDFs) to each file.

In LangChain, DirectoryLoader defaults to Unstructured for parsing multiple formats, but you can customize a specific loader (e.g., TextLoader) for certain file patterns.

```python
from langchain_community.document_loaders
  import DirectoryLoader, TextLoader

# Load only Markdown files under "data/docs"
  using TextLoader
loader = DirectoryLoader(
    path="data/docs",
    glob="**/*.md",
    loader_cls=TextLoader
)
documents = loader.load()
```

Depending on the chosen loader, the results may have subtle differences in how the text is chunked or how metadata is extracted.

4.2.3 *Loading PDFs and Other Formats*

For PDF documents, LangChain offers multiple loaders (e.g., PyPDFLoader, PyMuPDFLoader, UnstructuredPDFLoader), each with a unique approach:

- **PyPDFLoader** or **PyMuPDFLoader** may provide page-level splitting and handle embedded images.
- **UnstructuredPDFLoader** attempts to preserve document structure (headings, tables) where possible.

```python
from langchain.document_loaders import
  PyPDFLoader

pdf_loader = PyPDFLoader("data/handbook.pdf")
pdf_documents = pdf_loader.load()
```

Though under the hood, LangChain may use tools like PyMuPDF or Unstructured for parsing.

4.2.4 *Data Loading with LlamaIndex*

If you prefer LlamaIndex, its SimpleDirectoryReader can also automatically handle a variety of formats (including text, PDF, and markdown). For specialized needs (e.g., handling a unique file type or database), you can install connectors from LlamaHub. For example:

```python
from llama_index import SimpleDirectoryReader

reader = SimpleDirectoryReader("data/ ")
documents = reader.load_data()
```

Some file types (like CSV or JSON) may be chunked automatically, leading to more documents than expected. Hence, be sure to review how LlamaIndex's Reader implementations handle each file format.

4.2.5 *Loading Data with Data Processing Tools*

While frameworks provide convenient abstractions, using document processing tools directly can offer more control and flexibility. Unstructured is a popular open-source library that excels at converting various document formats into clean, structured text while preserving semantic elements.

Here's an example of using Unstructured directly on a document:

```python
from unstructured.partition.auto import
  partition

# Load and partition a document
elements = partition(filename="data/report.pdf")
```

4.3 Connecting to Third-Party Data Sources

Modern RAG systems often need to integrate with external data sources, such as Database Connections, Cloud Storage and API Integration. While LangChain provides a variety of built-in loaders for common data sources (Table 4.1), many enterprise scenarios require custom integration with specialized APIs or proprietary data stores.

Table 4.1. LangChain's built-in loaders for common data sources.

Document Loader	Description	Partner Package	API Reference
AWS S3 Directory	Load documents from an AWS S3 directory	✗	S3 Directory Loader
AWS S3 File	Load documents from an AWS S3 file	✗	S3 File Loader
Azure AI Data	Load documents from Azure AI services	✗	Azure AI Data Loader
Azure Blob Storage Container	Load documents from an Azure Blob Storage container	✗	Azure Blob Storage Container Loader
Azure Blob Storage File	Load documents from an Azure Blob Storage file	✗	Azure Blob Storage File Loader
Dropbox	Load documents from Dropbox	✗	Dropbox Loader
Google Cloud Storage Directory	Load documents from a GCS bucket	✓	GCS Directory Loader

For data sources not covered by built-in loaders, you'll need to implement custom integration via APIs provided by the data source system.

4.4 Data Parsing

Data parsing is the process of transforming data from its initial raw format into structured and meaningful formats suitable for further processing. This step is critical for extracting valuable information from unstructured data sources such as text documents, PDFs, and web pages. Effective parsing ensures that the extracted data retains its semantic integrity and is compatible with downstream tasks like indexing and querying. Below are some common data parsing techniques:

- **Text Extraction:** Using libraries like MinerU, MarkitDown, LangExtract, Docling or Apache Tika to extract textual content from PDF, DOCX, or HTML files.
- **Metadata Extraction:** Capturing additional context such as timestamps, authorship, or source information from data.
- **Natural Language Processing (NLP):** Employing NLP tools like spaCy or Hugging Face Transformers to tokenize, lemmatize, and extract named entities from text.

- **Format Conversion:** Converting data into standardized formats like JSON, Markdown, HTML, XML, CSV, or Parquet for easier manipulation and storage.
- **Custom Parsing Logic:** Implementing tailored parsers for domain-specific data structures, such as log files, configuration files, or scientific datasets.

Parsing libraries and frameworks often include error handling mechanisms to deal with incomplete or corrupted files. For instance, LangChain and other RAG frameworks provide utilities for parsing and structuring data into meaningful chunks for downstream indexing and retrieval.

4.4.1 *PDF Parsing*

PDF documents, which can be considered the epitome of unstructured data, pose significant challenges in the extraction of information. Unlike HTML or DOCX formats that organize data through tags that define logical structures within the document such as <p>, <w:p>, <table>, and <w:tbl>, PDFs operate differently. A PDF file contains a sequence of printing instructions that directs a PDF reader or printer on the placement and presentation of symbols on a screen or printed page.

The intricacy in parsing PDF documents stems from the need to precisely extract the page layout and convert the contents (e.g., tables, titles, paragraphs, and images) into a textual representation. This process frequently contends with errors in text extraction, image recognition, and the disambiguation of row-column relationships in tables. There are different approaches to parse PDF documents, e.g., rule-based and specific visual model and multimodal approaches.

4.4.1.1 *Rule-based approach*

This method relies on defining the style and content of sections based on the document's structural characteristics. Typical tools include PyMuPDF, PDFMiner, PyPDF, PDFPumber, and Camelot. A sample code is available at ch4\rule_based_parser.py.

However, due to the vast diversity in PDF formats and layouts, this approach struggles with generalizability as it cannot be uniformly applied to all PDF types.

- **Poor Generalization:** Difficult to handle different PDF formats
- **High Maintenance Cost:** Rules need constant updating
- **High Error Rate:** Easily affected by document quality

4.4.1.2 *Specific visual models*

This approach utilizes advanced solutions that integrate object detection with Optical Character Recognition (OCR) models or specific layout recognition models. It is more dynamic and can adapt to the varied and complex structures found within PDF documents. Below are some PDF Parsing Tools.

- **Unstructured**
  The "Unstructured" tool has been incorporated into LangChain, offering effective table recognition capabilities, especially when using the high-resolution (hi_res) strategy with infer_table_structure=True. This setting enhances the accuracy of table detection within PDFs. Conversely, the fast strategy, which lacks object detection models, tends to underperform, often erroneously recognizing images and tables due to its simplified processing approach. A sample code is available at ch4\unstructured_loader.py.
- **PP-StructureV2**[1]
  This tool employs a variety of model combinations for document analysis, generally delivering above-average performance. PP-StructureV2 is tailored for robust document parsing, effectively handling diverse and complex document formats.
- **Layout-Parser**[2]
  Despite a potential decrease in processing speed, for parsing complex structured PDFs, we may try Layout-Parser. It is important to note, however, that the models within Layout-Parser have not been receiving any updates in the past two years, which might affect their competitiveness with newer technologies. There are many others including PDF-Extract-kit[3] and pdf-craft,[4] etc.

---

[1]https://github.com/PaddlePaddle/PaddleOCR.

[2]http://github.com/Layout-Parser/layout-parser.

[3]https://github.com/opendatalab/PDF-Extract-Kit. It efficiently extracts high-quality content from complex and diverse PDF documents.

[4]https://github.com/oomol-lab/pdf-craft. It can convert (scanned) PDF files into various other formats.

Each of these tools offers specific strengths and limitations in the realm of PDF document analysis. Selection of the appropriate tool or strategy should be based on the specific requirements and constraints of the project, such as the need for speed versus accuracy and the complexity of the document structures involved.

4.4.1.3 *Parsing with multimodal large models*

Involves using large-scale multimodal models to parse complex structures or extract crucial information from PDFs. These models leverage their extensive capabilities to interpret and process the multimodal aspects of PDFs more effectively.

For example, you could use pdf2image to convert each page of a PDF into an image and then send these images to GPT-4o/5 or Gemini3, along with a prompt asking it to extract specific information or summarize the document's content. The model can then analyze the images and text to identify key elements like tables, headings, and figures, and extract the relevant information accordingly. This holistic approach can be particularly useful for complex PDFs with intricate layouts or for tasks that require a deeper understanding of the document's content beyond simple text extraction. A sample code is available at ch4\image_converter_extractor.py.

Similarly, you may serve OSS VLMs (Vision Language Models), e.g., InternVL3, Qwen-2.5-VL, DeepSeek-OCR or OlmOCR,[5] using Ollama or vLLM in openAI format. The implementation of these models is similar to what we have gone through above.

4.4.1.4 *Comparison of different PDF parsing tools*

Currently, there are many tools that apply the hybrid approach. Each method offers a unique strategy to tackle the inherent challenges of working with PDF documents, aiming to transform them from rigid

---

[5]https://github.com/allenai/olmocr. olmOCR is a PDF-focused language-model training toolkit built around the olmOCR-7B-0225-preview vision-language model, itself fine-tuned from Qwen2/2.5-VL on roughly 250,000 GPT-4o-annotated PDF pages (scanned and text-born) released as the olmOCR-mix-0225 dataset. It offers an SGLang-optimised inference pipeline for low-cost, high-throughput batch processing; page anchoring that records coordinates of key elements, such as text blocks and images, and injects them alongside the raw text extracted from the PDF binary; and flexible deployment that runs on a single local GPU or scales across multiple nodes using AWS S3 for parallel processing.

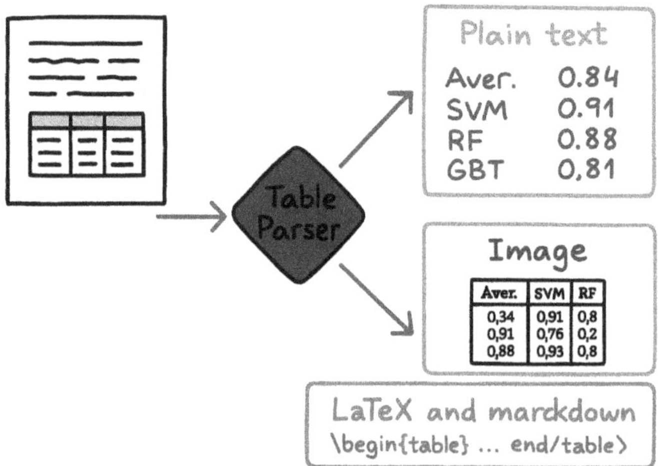

Figure 4.1.   Table parsed into different formats.

instructional sets to accessible and manipulable textual data, as illustrated in Figure 4.1. Table 4.2 compares the capabilities of different tools in overcoming these challenges and effectively extracting information.

Choosing the right PDF parsing tool can significantly impact the efficiency and accuracy of your information extraction pipeline. One ought to explore the key differences between these tools to help make an informed decision.

4.4.2 *Table Parsing*

When optimizing RAG schemes for unstructured documents, particularly tables in image formats such as scanned files, significant challenges arise. The difficulties include:

- **Complexity of Scanned Images:** Scanned images often feature complex structures with non-text elements and a mix of handwritten and printed text. These complexities hinder the accurate and automatic extraction of table information, potentially leading to structural damage that affects the integrity and semantic understanding of the RAG output.
- **Accurate Extraction of Table Titles:** Precisely extracting table titles and effectively linking them to the corresponding tables is crucial for correctly understanding the meaning of the tables. This process ensures

Table 4.2. Comparison of typical LLM extensions.

Tool/Method	Approach	Pros	Cons	Suitable For
OpenAI Vision API + GPT-4o(-mini)	Convert PDF to images, use GPT-4o(-mini) to reconstruct as markdown	– Excels in table parsing – Handles complex layouts well – Preserves formatting and visual elements	– Issues with Chinese characters – Misaligned columns in borderless tables – Potential token limit issues – Requires access to OpenAI API	Complex layouts, accurate table extraction, maintaining visual fidelity
Marker	Rule-based + OCR	– Auto-selects extraction method – Markdown output – Handles images and scanned documents	– Requires license for larger companies – Missing text – Misaligned columns in borderless tables – Issues with Chinese from edited PDFs	Standardized PDFs, mixed text and images
Deepdoc (RAGFlow)	OCR-based with rule-based customization	– End-to-end RAG solution – Manual editing of results – Integration with agents and GoogleRAG – Handles various document formats	– Table parsing performance needs improvement – May require technical expertise for setup and customization	Building RAG applications, diverse PDF formats, integrating with existing workflows

(*Continued*)

Table 4.2.   (*Continued*)

Tool/Method	Approach	Pros	Cons	Suitable For
Open-parse	Rule-based with some OCR and ML for tables	– Markdown output – Well-organized chunks – Open-source and actively maintained	– Limited OCR (tables only) – Issues with Chinese text – May struggle with complex layouts	Simple PDFs, basic table extraction, open-source projects
MegaParse (LlamaIndex)	Rule-based with Tesseract OCR	– Simple to use with LlamaIndex – Open-source	– Poor Chinese support – Messy output chunks – Limited accuracy for complex PDFs	Basic PDF parsing within LlamaIndex
Unstructured	Rule-based	– Handles diverse data formats beyond PDFs – Part of the LangChain ecosystem	– Messy output chunks – Limited accuracy for complex PDFs – May require preprocessing	General-purpose data extraction, integration with LangChain
MinerU	Rule-based with dedicated ML for table parsing. Similar functionality to MarkitDown.	– Open-source – Good performance on both text and tables – Outputs markdown and JSON formats – Active development and community support	– May require fine-tuning for specific PDF types – Relatively new tool, may have undiscovered limitations	Extracting text and tables from various PDFs, research papers, open-source projects

that the semantic intent and data described by the table are fully comprehensible.

- **Development of Efficient Indexing Structures:** Developing advanced indexing structures to store and retrieve deep semantic information of tables is essential. Such structures enable more effective data retrieval processes that are critical for enhancing the functionality and responsiveness of RAG systems.

This section will first discuss the core technologies involved in RAG table processing, then evaluate existing open-source solutions.

4.4.2.1 *Approaches to parsing tables*

The primary role of this module is the high-precision extraction of table structures from unstructured documents and images, with a strong emphasis on the accurate extraction of table titles and their convenient association with the tables themselves. Here's an overview of existing methods:

- **Multimodal LM Applications:** Utilizes VLM tools (e.g., GPT-4v/4o, Qwen3-VL, Gemma3, GLM-4.1V, Xiaomi MiMo) to recognize and extract table information from PDF pages. These LMs/VLMs are adept at understanding and extracting structured data from mixed media formats, leveraging their capacity to interpret complex document layouts.
- **Dedicated Table Detection Models:** Examples include Table Transformer,[6] which is specifically designed to parse table structures. This model focuses on identifying and structuring the components of tables, making it easier to extract and interpret data accurately.
- **Utilization of Open-Source Frameworks:** Frameworks like unstructured, MarkitDown and MinerU are used for comprehensive document analysis to extract table-related content. These frameworks apply a variety of techniques to deconstruct and understand the format and data of tables embedded within larger documents.
- **End-to-End Model Solutions:** Models such as Nougat (Donut)[7] and LlamaParser parse documents and directly extract table information

---

without the need for standalone OCR. These solutions are particularly effective in automatically recognizing and associating table titles. As table titles often summarize the essential content, their correct identification is crucial for understanding the table's content.

These strategies represent a spectrum of approaches, each leveraging different technologies to address the challenge of extracting and processing table data from complex document formats. They illustrate the advancements in document processing technology that facilitate more accurate data retrieval and analysis.

4.4.2.2 *Index structure*

The indexing methods for tables can be categorized based on the format and structure of the data they are designed to handle. Here's a summary of various indexing approaches:

- **Image Table Indexing:** Specifically targets tables in image format. This method involves using image recognition technologies to identify and index tables that are embedded as images in documents.
- **Text/JSON Table Indexing:** Focused on tables formatted as plain text or structured in JSON. This method leverages parsers that can interpret and organize text or JSON data into a structured format suitable for indexing.
- **LaTeX Table Indexing:** Specializes in handling tables formatted in LaTeX, a widely used markup language for scientific documents. This method parses LaTeX to extract and index table data accurately.
- **Table Summary Indexing:** Utilizes LLMs or multimodal models to generate summaries of tables. This approach is particularly useful for creating concise representations of table contents, facilitating quick overviews and easier retrieval.
- **Structured Hierarchical Indexing:** Organizes table content hierarchically from smaller elements (such as rows) to the entire table (including

---

table titles. The tables are parsed into LaTeX format, which is beneficial for academic and scientific documents where precision and formula representation are crucial. An example of Nougat's capability is demonstrated below with a table where the title has been successfully associated and formatted. This illustrates Nougat's effectiveness in processing and summarizing complex table data, proving it a valuable tool in the proposed RAG-based document handling solution.

images, text, and LaTeX tables). This method categorizes data based on content size or document summaries to cover all parts of a table comprehensively.

Additionally, tables can be queried directly by submitting images, text, and other context to multimodal LMs to receive instant answers. It is important to note that not all strategies rely on RAG:

- **Targeted Data Training:** Utilizes BERT-like models, such as TAPAS,[8] which are specially designed for table comprehension. These models are trained on specific datasets to enhance their ability to understand and interpret table data.
- **Utilizing LLMs for Table Comprehension Tasks:** Through pre-training, fine-tuning, or prompt adjustment, LLMs like the method demonstrated by GPT4Table[9] are used to perform table comprehension tasks. These models apply advanced NLP techniques to extract and understand information from tables effectively.

These indexing methods provide a comprehensive framework for managing and retrieving table data across various formats, enhancing the accessibility and usability of structured data in digital documents.

Finally, we must also consider the table representation format in the LLMs. Although markdown is becoming the most popular context format, it has no native support for merged cells in the table (rowspan/colspan). Therefore, we recommend using HTML, XML or Latex for complex cases.

4.4.2.3 *Open-Source integrated solutions*

LlamaIndex Options: LlamaIndex leverages multimodal models in the first three of its four methods:

- **Image Retrieval Followed by VLM Response:** Images are retrieved and processed, with responses generated by VLM.
- **PDF Page to Image Conversion, Indexed by VLM:** PDF pages are converted into images, which are then analyzed and indexed by VLM. Queries are handled through an Image Reasoning Vector Store.

---

[8]https://huggingface.co/docs/transformers/en/model_doc/tapas.
[9]https://github.com/Y-Sui/Table-meets-LLM.

- **Cropping Tables from Images for VLM Processing:** Tables are extracted from images, and GPT-4o/5 processes these cropped tables to generate answers.
- **Table Image OCR to Text, Processed by LLM:** OCR is used to convert table images into text, which is then fed into an LLM for generating answers.

Langchain Solutions: Langchain covers various approaches in semi-structured and multimodal RAG configurations, e.g.,

- **Semi-Structured RAG:** Utilizes the 'Unstructured' component for parsing PDF content, stores it with a multi-vector retriever, and implements Q&A via LCEL (Langchain Customized Entity Linking), without involving multimodal LLMs.
- **Semi-Structured + Multimodal RAG Options:**
  - *Option 1*: Uses embeddings from models like CLIP and Qwen-2.5/3-VL to integrate images and text, directing queries to a multimodal LLM for direct responses.
  - *Option 2*: Multimodal LLMs generate image-text summaries, which are then processed by traditional RAG pathways for Q&A.
  - *Option 3*: Generates summaries and embeddings, uses these to locate the original images, and completes Q&A with a multimodal LLM.

As illustrated in Figure 4.2, these solutions showcase the versatility of current open-source tools in handling complex document formats and content types by blending advanced text and image processing technologies with sophisticated language understanding capabilities. As such, these multimodal models are capable of enhancing data retrieval and analysis tasks.

4.4.3 *Word Parsing*

Regulated sectors (i.e., finance, government) still hold large .doc and .docx corpora. Compared with PDF, Microsoft office documents like Word and PowerPoint files are relatively easier to parse. For example, .docx follows the OOXML standard (a ZIP of XML parts). Table merges (e.g., gridSpan, vMerge) are explicitly represented, so parsing can be reliable. .doc is a legacy, proprietary binary format with sparse public

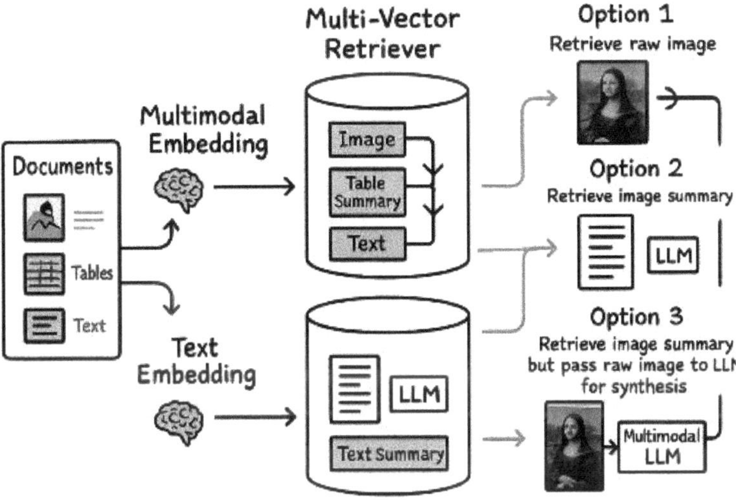

Figure 4.2. Solutions with multimodal models.

documentation. Libraries (e.g., python-docx, POI's HWPF) expose only low-level signals; merged-cell semantics are largely absent. However, we can always convert .doc into .docx before parsing.[10]

We may adopt the following pipeline for word parsing:

- **Bidirectional structure → semantics:** extract headings, hierarchy, lists, emphasis, citations/footnotes into Markdown to preserve semantic cues for chunking and retrieval.
- **Tables as HTML (not Markdown):** to preserve rowspan/colspan faithfully for accurate display and grounding.
- **Images:** keep as URL references with correct document order/anchors and optional alt text; allow OCR/multimodal enrichment when needed.
- **Cost/performance controls:** CPU-first fast parsing with optional GPU models for heavy cases; efficient conversion + embedding throughput.

---

[10]It is also suitable to convert word files into PDFs, and then apply the PDF parsing techniques in some cases, e.g., visual fidelity outweighs semantic tags, documents rely on heavy formatting or field codes, deterministic rendering across platforms are required, compliance/archival needs, etc.

An industrial massive processing approach using "Preprocess + Inject" combines Apache Tika (unified parsing)[11] and Apache POI (Office specifics),[12] then surgically replaces Tika's coarse table output with custom HTML built from higher-fidelity analyzers. This procedure can be roughly divided into 5 steps:

1. Use POI's XWPF APIs to enumerate all tables.
2. Apply a CellMergeAnalyzer to reconstruct true merges.
3. Build canonical HTML tables via HtmlTableBuilder.
4. Let Tika parse the rest (body text, headings, lists, etc.).
5. Inject/replace Tika's table fragments with the custom HTML, yielding Markdown + embedded HTML output.

4.4.4 *Integration of Multiple Techniques*

We can also combine the abovementioned techniques into a unified document preprocessing framework that integrates structure recognition, OCR, and multimodal analysis to process documents efficiently with high precision, as shown in Figure 4.3. The pipeline begins with segmenting documents into distinct components, such as paragraphs, tables, and figures, using structural recognition models. Each component is then handled by specialized tools: OCR is applied to images (especially scanned content), table structure models extract relational data, and multimodal models interpret visual elements like flowcharts or charts.

This integrated method excels in both depth and breadth, capturing nearly every meaningful unit within a document. It is particularly valuable for domains requiring complete and accurate data extraction, such as legal, compliance, or enterprise knowledge systems. For instance, it can accurately parse contractual clauses, financial tables, or procedural diagrams, maintaining semantic context throughout. For embedded metadata, such as footnotes, cross-references, or annotations, custom rule-based or learning-based extractors can be employed to preserve referential integrity in downstream use cases.

However, the approach is still resource intensive. It involves multiple specialized models, which can limit throughput and automation when

---

[11]Apache Tika, https://tika.apache.org/.

[12]Apache POI™ — the Java API for Microsoft Documents, https://poi.apache.org/.

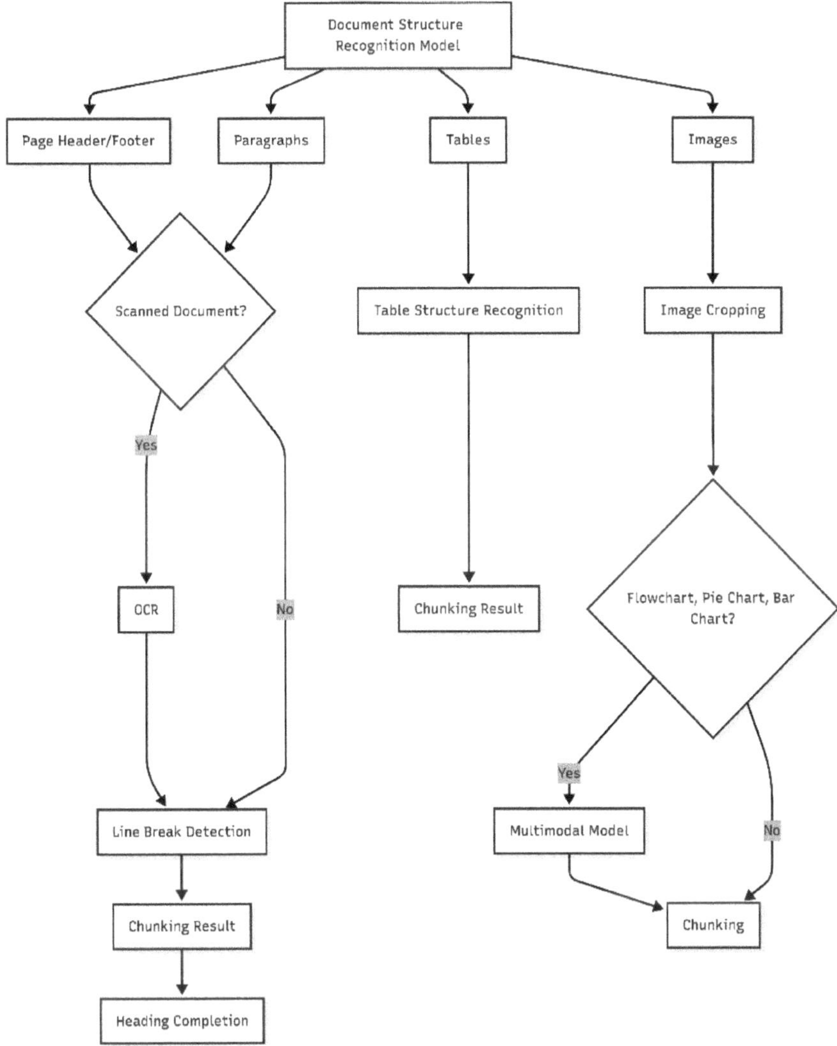

Figure 4.3.   Unified approach for document preprocessing.

applied to large-scale datasets. While it offers unmatched accuracy, it may require complementary strategies, like parallel processing or active learning, to improve efficiency and scalability for enterprise-wide deployment.

4.5 Data Cleaning

Data cleaning is a crucial step that ensures the quality and consistency of the data used in the RAG pipeline. Poor-quality data can lead to inaccurate retrieval results and negatively impact the performance of the system. This stage involves removing noise, correcting errors, and handling missing or inconsistent data.

The best practices for data cleaning include:

- **To improve content quality:**
  - Remove duplicate or near-duplicate content
  - Filter out boilerplate text and irrelevant sections
  - Standardize formatting and encodings
- **To improve structural integrity:**
  - Preserve document hierarchy and relationships
  - Maintain metadata consistency
  - Handle missing or incomplete data
- **To improve language processing:**
  - Implement language detection and validation
  - Correct spelling and grammatical errors
  - Normalize text representations

Frameworks like Pandas, Dask, and PySpark offer powerful tools for large-scale data cleaning. Additionally, domain-specific cleaning techniques can be integrated into the pipeline to address unique challenges. Listed below are some general steps:

1. **Remove noise and irrelevant information:** Eliminate special or unwanted characters, stop words (common words like "the," "an," "I" and "a"), and markup (e.g., HTML tags).
2. **Identify and correct spelling, grammatical and typographical errors:** Correct those errors to maintain context consistency and improve the information quality using tools like spell checkers and language models.
3. **Deduplicate content:** Eliminating duplicates helps to preserve the diversity of the knowledge base, ensuring a balanced representation of data. Use advanced similarity detection algorithms to identify and

remove records that are overly similar, which might otherwise skew the relevance ranking and lead to redundant or biased outputs.

A sample code is available at ch4\data_cleaning.py. For more data cleaning functionalities, you can refer to Unstructured.io and their core library offering cleaning capabilities.

4.6 Data Parallel Processing

As the size and complexity of data grows, parallel processing becomes essential for ensuring timely and efficient data handling. Parallel processing involves dividing the workload across multiple threads, cores, or distributed systems to speed up data processing tasks.

Techniques for parallel processing include:

1. **Multithreading and Multiprocessing:** Using Python's threading and multiprocessing modules to parallelize tasks on a single machine.
2. **Distributed Computing:** Leveraging frameworks like Apache Spark, Dask, Modin, Polars or Vaex for processing large datasets across clusters.
3. **Batch Processing:** Breaking data into smaller batches for concurrent processing and aggregation.
4. **Vectorized Operations:** Employing libraries like NumPy (Numba) or Pandas to perform efficient, low-level operations on large arrays or dataframes.

RAG frameworks often include optimizations for parallel processing. For example, LangChain's document loaders and retrievers can process data in parallel to improve performance when handling large corpora. LlamaIndex's implementation achieves up to 15x faster processing through optimized pipeline orchestration and workload distribution. LlamaIndex's parallel processing pipeline handles:

1. Document loading and parsing
2. Text splitting and chunking
3. Embedding generation
4. Vector storage operations
5. Metadata extraction

A sample code is available at ch4\data_parallel_processing.py. Check out LlamaIndex's full notebook for more details.[13] A similar idea can be applied to index building. Another popular tool for parallism is Ray, whose sample code is available at Github.[14]

4.7 Data Version and Update

Data versioning and updating are critical components in maintaining a robust RAG system. As knowledge bases evolve and expand, managing data versions becomes essential for tracking changes, ensuring consistency, and enabling rollback capabilities. Additionally, efficient update mechanisms allow the system to incorporate new information while maintaining historical context.

Different update strategies can be implemented depending on the use case and requirements:

1. **Full Refresh:** Complete replacement of existing data
2. **Incremental Updates:** Adding or modifying specific portions
3. **Differential Updates:** Updating only changed content
4. **Schedule-based Updates:** Regular updates at predetermined intervals

The versioning and update system ensures that the RAG pipeline maintains data quality and reliability while accommodating evolving information needs. By maintaining version histories, a RAG system can reference the correct version of data based on the temporal context, enabling it to provide accurate information that reflects changes over time. For instance, if a user inquiry about the CEO of a company as of a specific date, the system can retrieve the correct information from the version of the data corresponding to that date. A simple example of implementing data versioning using a database with time-stamped entries is available at ch4\data_version.py.

In addition to versioning, maintaining historical data access is important for handling queries that pertain to past events or time periods.

---

[13]https://github.com/run-llama/llama_index/blob/main/docs/examples/ingestion/parallel_execution_ingestion_pipeline.ipynb.

[14]https://github.com/benman1/generative_ai_with_langchain/blob/main/search_engine/indexing.py.

By archiving historical data, the system ensures that even as the knowledge base evolves, information from previous versions remains accessible when needed.

Regularly updating and maintaining the knowledge base is crucial for the system to remain current. Automated updates can be set up using data pipelines that ingest new data as it becomes available. This can be achieved using tools like Apache Airflow, Apache NiFi, or custom scripts that fetch and process data at scheduled intervals. A mockup example using Python's schedule library to automate data ingestion is available at ch4\data_scheduled_ingestion.py.

To ensure that the data remains fresh and accurate, implementing data freshness monitoring is essential. Monitoring systems can track the age of data entries and flag outdated information for review or update. This can involve setting thresholds for data validity and generating alerts when data exceeds these thresholds. See Chapter 11 for more details.

Integrating real-time data sources enhances the system's ability to provide the most current information. Connecting to APIs and feeds, such as news feeds, stock market data, or weather updates, allows the RAG system to access live data streams. An example using the requests library to fetch stock price data from a real-time API is available at ch4\data_real-time_fetch.py.

To handle streaming data and update information in near real-time, streaming data processing technologies like Apache Kafka, Apache Flink, or Apache Spark Streaming can be utilized. These platforms allow the system to process incoming data streams efficiently and update the knowledge base accordingly. Here's a conceptual example using Kafka with Python's confluent_kafka library: ch4\data_kafka.py.

By combining data versioning, automated updates, data freshness monitoring, and integration with real-time data sources, RAG systems can maintain a robust and up-to-date knowledge base. This ensures that the LLM provides accurate and timely responses, enhancing the overall effectiveness of the system.

Implementing these strategies not only improves the system's performance but also builds user trust by consistently providing relevant and current information. As data continues to evolve rapidly, having a solid version control and data management framework becomes indispensable for any RAG system aiming to deliver high-quality results.

4.8 Concluding Remarks

Industry scaling RAG agrees that data processing (e.g., how content is ingested, secured, refreshed, logged and cost optimized) is now the main gating factor for production success. Freshness gaps, duplicate stores that bypass access controls, soaring vector footprints and audit expectations all collide with strict regulatory norms. This chapter discussed some common issues faced by developers and their corresponding (partial) solutions using existing techniques. Table 4.3 concludes this chapter with a summary of main issues and their corresponding solutions.

Table 4.3. Summary of pain points and solutions.

Problem Description	Cause Analysis	Solution
Missing blocks	Layout analysis model has detection blind spots, causing some areas to be missed.	1. Use a domain-specific layout analysis model for targeted handling. 2. Use a general-purpose layout analysis model for optimization. 3. Use OCR-Detect to fill missing blocks.
Overall speed is too slow	Especially for densely packed text documents, the OCR rec model is very slow.	1. Split large documents into non-editing and editing paths; only the necessary parts are sent to OCR. 2. Deploy multiple nodes, use multi-threading and asynchronous processing. 3. Use different models based on document types. For example, handwritten text requires heavy models but others may not, enabling lightweight processing. 4. Use batch inference and optimize through Onnx acceleration. 5. For tools like Flowchart, Datachart, consider using small models or 1.5B models, and directly connect to larger models (balancing precision and speed).
Table parsing is inaccurate	Model generalization is insufficient, especially for cross-page table merging.	1. Add sample data, refine categories, and differentiate between wired and wireless tables. 2. Design rule-based or depth parsing strategies. 3. For cross-page tables, construct segmentation models; for complex single tables, split header rows and information areas separately.

Some new techniques further close the gaps of issues, e.g., VLM for documents processing, streaming upserts, policy aware filters, lineage first pipelines, model distillation for edge devices and hybrid graph plus vector retrieval. As mentioned earlier, modern RAG pipelines may start with more document intelligence preflight checks instead of simple text extraction. VLMs such as OpenAI GPT-4o, InternVL3 and Qwen-3-VL, restore complex table layouts in scanned statements with transformer based split merge decoding, sharply reducing structural OCR errors and boosting downstream retrieval precision in financial tabular data sets. Once the text is clean, more smart semantic chunking techniques may be applied, e.g., adaptive windows align to section headers, code blocks or semantic breaks.

The second pillar is event driven freshness at ingest. Changing data capture streams push inserts and updates from core ledgers through Kafka topics to embedding workers, allowing new facts to surface in the vector store within minutes rather than waiting for the next batch window. Prototype pipelines implemented with Upstash Kafka show end-to-end latencies of <300 ms between a database row change and an upserted embedding later, eliminating the overnight ETL blind spot that plagued early pilots.[15] To keep that real time flow affordable, banks hash every incoming blob and apply SimHash or MinHash filters.

Third, fine grained security and lineage are moving on with the various techniques. In fact, whether firms adopt on-chain ledgers (e.g., blockchain anchored hash commits) or conventional off-chain logs, the current gold standard is to fuse access control with immutable lineage metadata at the preprocessing layer; vector databases such as Pinecone now let engineers embed attribute or role based filters directly in similarity queries so that only chunks authorized for the caller can ever leave storage, closing long standing "shadow repository" loopholes.[16] Qdrant offers comparable RBAC and metadata filtering primitives to enforce row level or tag level segregation inside the index without costly duplication; upstream, data engineering teams wire their extract–embed pipelines

---

[15]Prabhakar Chandrasekaran, "Stream ingest data from Kafka to Amazon Bedrock Knowledge Bases using custom connectors" https://aws.amazon.com/blogs/machine-learning/stream-ingest-data-from-kafka-to-amazon-bedrock-knowledge-bases-using-custom-connectors/, Apr 2025, Accessed on 27/07/2025.

[16]Roie Schwaber-Cohen, "RAG with Access Control," https://www.pinecone.io/learn/rag-access-control/, Apr 2024, Accessed on 27/07/2025.

through OpenLineage instrumented DAGs[17] (or, in on-chain variants, hash each transform onto a permissioned ledger) so every step (from raw PDF ingestion through chunk hashing and upsert) emits a tamper evident event. This gives auditors a single provenance graph that satisfies forthcoming Digital Operational Resilience Act (DORA) expectations for end-to-end traceability.[18] More details on metadata will be discussed in Chapter 5.

---

[17]https://github.com/openlineage/, Accessed on 27/07/2025.

[18]Sabrina Aquino, David Myriel, "A Complete Guide to Filtering in Vector Search" https://qdrant.tech/articles/vector-search-filtering, Sep 2024, Accessed on 27/07/2025.

Chapter 5

Embedding and Vector Database

In the context of RAG systems, the integration of embedding techniques and vector databases with proper indexing is pivotal for enhancing information retrieval processes. Embedding usually transforms high-dimensional data into low-dimensional, dense vector spaces, where semantic relationships are preserved. These embeddings are then stored in vector databases, which are optimized for fast and efficient similarity searches crucial for RAG systems. Indexing these embeddings allows for rapid retrieval of contextually relevant information from large datasets. In practice, high-dimensional embeddings are served with approximate k-NN (nearest-neighbor) indexes, notably graph-based HNSW (Hierarchical Navigable Small World), quantization-based IVF/PQ (Inverted File with Product Quantization), or LSH (Local Sensitive Hashing). Tree-based structures like KD-trees/Ball Trees are generally effective only in low- to moderate-dimensional settings and degrade in very high dimensions. Together, these components form a robust backbone for RAG systems, enabling them to efficiently process and retrieve necessary information to augment the language model's response generation capabilities. This architecture not only speeds up the retrieval process but also ensures that the responses generated by the language model are contextually accurate and informative, leading to improvements in both the speed and quality of generated content.

However, the sequential processes inherently introduce information loss at multiple stages:

- **Chunking and Embedding:** Dividing text into chunks and generating embeddings for each segment can lead to information loss. The choice of chunk size is critical; overly large chunks may exceed model input limits, while excessively small chunks might disrupt semantic coherence. Additionally, embedding models may not capture all nuances of the original text, leading to potential omissions.
- **Retrieval:** The retrieval phase, which involves selecting top-k chunks based on semantic similarity, is also susceptible to information loss. The similarity function may not perfectly align with human judgment, and the top-k constraint can result in the exclusion of relevant information, especially in cases where pertinent data is ranked just outside the selection threshold.
- **Response Generation:** During response generation, the LLM synthesizes information from the retrieved chunks. However, limitations such as maximum token counts and the inherent capabilities of the LLMs can lead to incomplete or imprecise outputs. The models may omit less prominent details or fail to integrate all relevant information effectively.

These cumulative losses underscore the importance of optimizing each component of the RAG pipeline to enhance overall system performance and fidelity. In this chapter, we will discuss how to alleviate this issue in the chunking and embedding phase first.

5.1 Chunking Techniques

Effective chunking is essential due to the fixed and limited input sequence length of Transformer-based models. Even with increasingly large context windows, the semantic meaning of a sentence or a small group of sentences is more accurately captured when represented by dedicated vectors, rather than by averaging over long passages of text, such as several pages. Consequently, it is critical to split documents into coherent chunks that preserve their full meaning (typically at sentence or paragraph boundaries) while avoiding fragmentation of individual sentences. Many text splitting algorithms and tools are available to perform this task efficiently.

However, in practice, many developers inadvertently compromise system performance by adopting overly simplistic approaches, such as splitting text strictly by a fixed number of characters or lines. While this may appear to be an efficient shortcut, it is often the underlying cause of poor retrieval results and degraded system behavior. Consider the consequences: when a coherent narrative, an essential code example, or a critical definition is arbitrarily cut in half, the resulting fragment loses its integrity. The retriever may return incomplete or context-deficient chunks, leaving the language model with only partial information and severely limiting its ability to generate accurate answers.

Additionally, improper chunking can cause key information to be scattered across chunk boundaries. This dispersion makes retrieval akin to searching for a needle in a haystack, either partial fragments are retrieved, or crucial content is missed entirely. On the other hand, excessively large chunks introduce unnecessary noise, diluting semantic relevance and impairing precise matching. Conversely, chunks that are too small may lack sufficient context, also reducing retrieval quality. Both extremes contribute to inefficiencies in retrieval and increase the likelihood of returning superficially relevant but ultimately useless results.

Therefore, the size of the chunks is an important parameter to consider. It depends on the embedding model and its token capacity. For instance, a BERT-based Sentence Transformers can handle up to 512 tokens, whereas OpenAI's GPT-4 Turbo model can handle sequences up to 128K tokens (see more in Chapter 9 on the latest model context size). The trade-off lies in providing sufficient context for the language model to reason effectively, while ensuring that the generated text embeddings remain detailed enough for efficient and accurate search execution (see Table 5.1 for more details). Research on chunk size selection can

Table 5.1. Trade-off between smaller and larger chunks.

Chunk Size	Advantages	Disadvantages
Smaller Chunks	– More focused information and concentrated semantics. – Easier for query vectors to precisely match relevant chunks containing specific keywords or closely related concepts. – Increases retrieval precision by improving "target accuracy."	– May lose important contextual information. – Individual chunks may contain only isolated facts without sufficient background, preconditions, or follow-up explanations. – Can lead to poor understanding of complex queries requiring multi-faceted reasoning, like "seeing only part of the elephant."

(*Continued*)

Table 5.1.   (*Continued*)

Chunk Size	Advantages	Disadvantages
Larger Chunks	– Retain richer contextual information. – Preserve longer logical chains or comprehensive background descriptions. – Help LLMs better understand complex concepts, causal relationships, and interrelated information points.	– May introduce excessive irrelevant information (noise), weakening semantic relevance. – Risk of retrieving large chunks with partially relevant terms but incorrect overall topic. – Increase LLM processing load and raise the potential for hallucinations. – Reduce retrieval precision.

offer insights into these concerns. In LlamaIndex, the NodeParser class addresses this by allowing the definition of custom text splitters, metadata, and relationships between nodes and chunks, among other advanced options. A pipeline to find the (sub-)optimal size are discussed in the section of "Hyperparamter Tuning" of Chapter 12.

5.1.1 *General Methods*

The core objectives of text chunking in RAG systems can be summarized into three key principles:

- **Overcoming Context Window Limitations:** This is the most fundamental motivation for chunking. All large language models (LLMs), whether from the GPT family, Gemini, Claude, LLaMA, Deepseek, or others, operate within strict context window limitations. That is, the maximum number of tokens that the model can process in a single pass. In contrast, source documents often exceed these limits, ranging from thousands to hundreds of thousands of tokens. Without chunking, it would be impossible to deliver the full content of lengthy documents to the model in a usable form. Chunking ensures that each segment passed to the model remains within the boundaries of what the model can effectively process, making long documents manageable for LLMs.
- **Improving Retrieval Accuracy and Efficiency:** Well-designed chunking significantly enhances both the precision and performance of the retrieval process. To illustrate, imagine a user querying, "What are the key safety features of the Boeing 787?" A small chunk that exclusively

describes these safety features will be far easier for a similarity-based retriever to locate than a large document chunk that contains the entire aircraft design manual, including unrelated details on aerodynamics, materials, and production schedules. Large, unfocused chunks introduce unnecessary noise that interferes with similarity calculations. In contrast, smaller, topic-centered chunks improve matching precision. Additionally, breaking documents into smaller units streamlines the process of converting text to embeddings and speeds up indexing. A well-structured vector database can quickly scan smaller, targeted chunks and return relevant results with lower computational overhead, drastically reducing retrieval latency.

- **Maintaining Contextual Integrity:** The art of chunking is not just about size but about preserving the meaning and completeness of the content. Effective chunking avoids splitting sentences, breaking tables in technical specifications, or severing numbered steps in a process guideline. For example, when segmenting a product troubleshooting manual, each chunk should contain an entire diagnostic procedure rather than fragmenting it midway through a step. The objective is to ensure that each chunk remains a coherent and self-contained unit of information, capable of delivering useful context to the language model without requiring external references from adjacent chunks.

Traditional chunking methods include rule-based techniques, such as those utilizing fixed chunk sizes or overlapping adjacent chunks to define multi-level document separators, or document structure-based chunking, which maintains the heading information (e.g., title-chapter-section-subsection). Tools like Langchain's RecursiveCharacterTextSplitter enable multi-level chunking, and grobid[1] or minerU can extract the structure information.

However, rule-based chunking can sometimes lead to issues like incomplete retrieval contexts or overly large chunks that include irrelevant information due to its rigid predefined rules. Thus, semantic chunking, which aims to ensure that each chunk contains as much semantically independent information as possible, emerges as a superior method.

The content below delves into the various approaches to semantic chunking, highlighting three main types: embedding-based, model-based, and LLM-based methods. More details can be found in Table 5.2.

---

[1]https://github.com/kermitt2/grobid.

1. **Embedding-based Methods:**
    o   Implemented by platforms like LlamaIndex and Langchain, these methods use embeddings to segment text, focusing on semantic continuity within chunks.
2. **Model-based Methods:**
    o   **Naive BERT/GPT/T5-like Approach:** Utilizes BERT's Next Sentence Prediction (NSP) capability to assess the semantic link between adjacent sentences. If the link is weak (below a threshold), it identifies a potential segmentation point.
    o   **Cross Segment Attention:** Introduces more sophisticated models such as the BERT+Bi-LSTM and hierarchical BERT, which use enhanced local and extended contexts to determine text segmentation boundaries more accurately.
    o   **SeqModel:** employs BERT to encode several sentences simultaneously, modeling dependencies within longer contexts before computing sentence vectors. It then predicts if text segmentation takes place after each sentence. Furthermore, this model utilizes the self-adaptive sliding window method to boost inference speed without compromising accuracy.
3. **LLM-based Methods:**
    o   Discusses the innovative use of LLMs to generate propositions, small semantic units within text that each represent a distinct factoid. This method, while resource-intensive, offers a refined approach to semantic chunking by leveraging LLMs to analyze and segment text based on deeper linguistic structures. A similar idea is also applied to the Dense X Retrieval method in Chapter 7.

While each method offers unique advantages, they also present specific challenges such as computational efficiency and the need for extensive training data. The choice of chunking strategy often depends on the specific requirements of the RAG system and the complexity of the documents being processed. Other methods include atomic units, i.e., [1] indicates that decomposing chunks into atomic statements and generating synthetic questions based on these atoms significantly improves the retrieval step in enterprise RAG.

Figure 5.1 illustrates the differences between some typical methods. Note that once the method is confirmed, its retrieval parameters such as top_k (top k relevant chunks to be selected) also matter. Table 5.2 provides a brief comparison of various methods and implementation tips, and

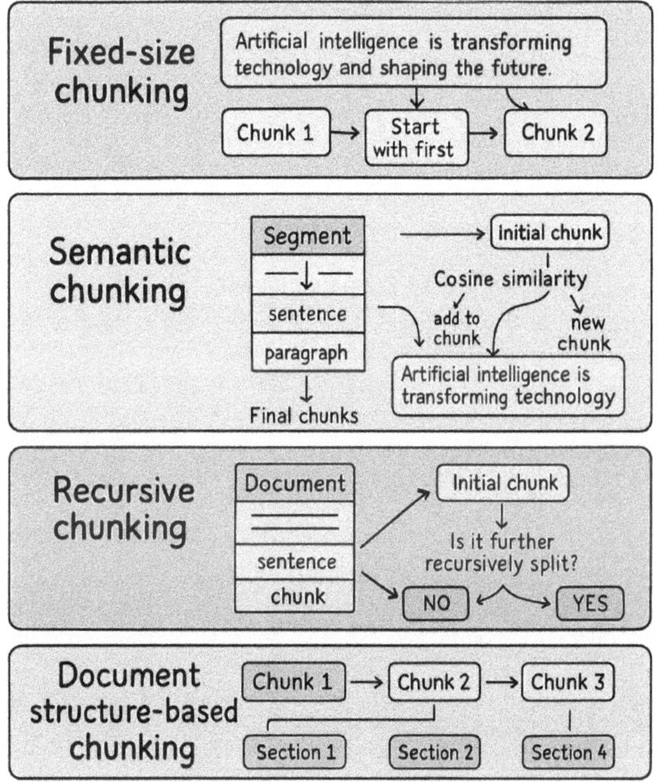

Figure 5.1.   Chunking illustration.

Table 5.3 suggests the general selection criteria and their suitable scenarios from different perspectives. More relevant details can be found in Chapter 7.

In practice, we usually implement a multi-layered hybrid optimization strategy, applying differentiated chunking methods tailored to various document types. This approach enhances both retrieval accuracy and response quality in downstream applications such as question answering.

For instance, employee handbooks are segmented by chapters, with each chunk containing approximately 500 tokens. This ensures that each unit of information is well-contained, neither too sparse nor too dense, facilitating efficient retrieval and precise response generation.

In the case of structured documents like contracts, we adopt a clause-based chunking strategy. Each chunk is organized around specific clause

Table 5.2.   Comparison of various chunking techniques.

Strategy	Description	Advantages	Disadvantages	Implementation Tips
Fixed-length Chunking	Rule-based. Split the text by a predefined length (e.g., tokens or characters). Best for simple documents or very fast processing.	Simple, uniform, efficient.	Context loss, weaker relevance, potential information loss.	Choose an appropriate chunk size; use overlapping windows to preserve context when needed.
Sentence-based Chunking	Rule-based. Segment at sentence boundaries so each chunk is a complete thought, which is useful for short answers such as customer queries.	Preserves context, easy to implement, good readability.	Inconsistent chunk sizes; unsuitable for very long sentences; limited control.	Detect sentences with an NLP library; merge very short sentences when necessary.
Paragraph Chunking	Rule-based. Divide by paragraphs; each paragraph usually contains a full idea or topic, which suits well-structured documents.	Richer context, logical segmentation.	Uneven chunk sizes; may exceed token limits.	Monitor chunk size and split long paragraphs if required to maintain context.
Document-level Chunking	Rule-based. Treat the whole document as one chunk (or split only minimally) to keep the full structure and context, common for legal or medical texts.	Complete context retained; simple for highly structured text.	Poor scalability, low efficiency, hard to extract fine-grained information.	Use only when splitting would disrupt the workflow.
Sliding-window Chunking	Rule-based. Create overlapping chunks with a sliding window so neighboring chunks share content which maintains cross-segment context.	Continuous context, better retrieval quality.	Redundant data, higher computational cost.	Tune window size and overlap; use deduplication techniques to handle redundancy.
Recursive Chunking	Rule-based or hybrid. Iteratively split with hierarchical or multiple delimiters to produce small chunks, ideal for large, highly structured documents.	Layered context, scalable, meaningful chunks, fine-grained control.	Implementation complexity, possible context loss, compute-heavy.	Use explicit structural markers; store chunk-position metadata for reconstruction.

Embedding-based Semantic Chunking	Semantic-based. Use embeddings or ML models to cut the text where semantic coherence changes, ensuring each chunk stays on a single theme.	Higher contextual relevance, flexible, improved retrieval precision.	Complex, slower processing, compute-intensive, threshold tuning required.	Leverage pre-trained models; balance cost against desired granularity.
Proposition Chunking	Semantic-based. Divide the content into smaller and atomic propositions, which is usually processed by LLM or specific NLP models.	Generate fine-granularity and condensed knowledge unit for precise retrieval.	Complex & costly; heavily dependent on model extraction capability; potential loss of relation & small differences.	Careful consideration on the connection of the discrete propositions and provision of an overview.
Context-augmented Chunking	Semantic-based. Attach a summary or metadata of neighboring chunks to each chunk to keep inter-chunk context, suited to long documents.	Strengthens context, enhances coherence.	Complex; large storage overhead.	Generate concise summaries; consider key terms or concepts as metadata.
Modality-specific Chunking	Semantic-based. Process each content type (text, tables, images, etc.) separately and chunk according to that type's characteristics.	Tailored approach, higher accuracy.	Implementation is complex; integration is difficult.	Apply OCR to images, convert tables to structured data, and keep index alignment consistent.
Agent-based Chunking	Semantic-based or hybrid. Use AI (e.g., LLMs) to suggest boundaries based on content structure and semantics, or organise chunks around specific agent tasks.	Task-oriented efficiency, focused data, flexible, intelligent segmentation.	Complex setup, compute-intensive, costly; risk of over-specialization and losing the global view.	Apply selectively; craft effective LLM prompts; explicitly define agent roles and task rules.

Table 5.3.   Chunking strategy.

Factor	Preferred Strategy	Rationale
Document Type	*Semi-structured documents* (e.g., specific reports and scientific papers) → Paragraph Chunking *Unstructured documents* (e.g., chat logs) → Semantic Chunking	Structured texts already contain logical breaks, whereas unstructured texts need semantic detection to preserve meaning.
Query Complexity	*Complex, multi-facet queries* → Semantic or Agent-based Chunking *Simple look-ups* → Fixed-length Chunking with sliding windows	Sophisticated queries benefit from deeper contextual cues; simple queries value speed over context richness.
Resource Availability	*Limited compute or memory* → Fixed-length or Sentence-based Chunking (possibly with sliding windows)	These methods require minimal preprocessing and incur the lowest runtime overhead.
Desired Outcome	*Speed-first* → Fixed-length Chunking *Accuracy-first* → Semantic Chunking *Context retention-first* → Sliding-window Chunking	Each strategy optimizes a different metric, aligning the choice with the primary business goal.

types (e.g., termination conditions, payment terms, liability provisions), preserving the logical coherence of the content. This not only avoids redundancy or fragmentation of meaning but also makes it easier to extract contextually relevant information when queried.

For technical manuals, especially those related to IT systems or machinery, we segment by functional modules or step-by-step procedures. This allows engineers or operators to retrieve only the relevant portion, such as "System Setup" or "Error Code Troubleshooting," without having to sift through entire documents.

In compliance guidelines or regulatory policies, chunking is performed by thematic sections (e.g., "Data Privacy," "Customer Due Diligence"), making it more effective for compliance officers or auditors to locate relevant content during reviews or internal assessments.

Additionally, meeting transcripts or interview records are chunked based on speaker turns or topic transitions. This preserves the dialogue context and improves the quality of response when users ask for specific opinions, decisions, or key statements.

The adoption of these different chunking strategies enables the system to precisely identify relevant vector segments during retrieval, significantly improving both the efficiency and accuracy of the answer generation process.

An example with three different chunking methods using LangChain and LlamaIndex, i.e., Fixed-Length Chunking, Semantic/Sliding Window Chunking and Structure-Based Chunking, is available at ch5\chunking_methods.py.

5.2 Embedding Methods

Once the text is chunked, a proper embedding model shall be select to embed the chunks. The term "dimensionality" in embedding models refers to the number of dimensions used to represent textual data as vectors in a certain dimensional space. In natural language processing (NLP), we often use sparse and dense vectors to differentiate between two categories.

The traditional models like TF-IDF, Query likelihood or BM25,[2] which are usually keyword-based, are often considered as sparse methods. Another sparse method is SPLADE (SParse Lexical AnD Expansion model) [2].[3] The traditional dense models include Word2Vec/GloVe/Fasttext for word embeddings, and Universal Sentence Encoder or RoBERTa/ALBERT/T5 for sentence embeddings; these transform words, phrases, or sentences into numerical vectors. This dimensionality, which can range from a few tens to several thousands of dimensions, directly influences the model's ability to capture both semantic and syntactic nuances. While higher-dimensional embeddings tend to capture more detailed and subtle linguistic features, they also require more computational resources and may increase the risk of overfitting in downstream tasks.

---

[2]It often works with inverted index for full-text search, e.g., inverted index helps to locate the documents, while BM25 find the most relevant ones.

[3]SPLADE combines the strengths of lexical and semantic search by representing queries and documents as sparse vectors over the vocabulary. Instead of dense embeddings, it outputs term-weighted sparse vectors using a transformer, enabling compatibility with traditional inverted indexes like BM25. It learns to expand queries and documents with semantically related terms while preserving interpretability and fast retrieval. This makes it highly efficient and scalable, bridging the gap between neural and symbolic retrieval.

In the context of LLMs, embedding dimensionality plays a crucial role in representing intricate relationships within the data. Generally, higher dimensionality can lead to improved performance by encoding richer semantic details, albeit at the cost of increased processing time and resource consumption. Below is a list of popular (dense) text embedding models along with their typical dimensionalities:

1. **BAAI/bge-large-en-v1.5:** With a dimensionality of 1,024, this model is highly performant and excels at embedding entire sentences and paragraphs, capturing more detailed semantic information. It is search optimized, like E5-large (1024; E5-base 768). BGE-large-zh-v1.5 is dedicated for Chinese texts.

2. **Sentence-transformers/all-MiniLM-L6-v2:** This model has a dimensionality of 384 and is suitable for general-purpose use cases of embedding sentences and paragraphs that require lower dimensionality efficiently.

3. **OpenAI text-embedding-3-large and text-embedding-ada-002:** The most recently announced text-embedding-3-large features an embedding size of 3,072 dimensions, while text-embedding-ada-002 has a dimension size of 1,536.

4. **Cohere Embed v3:** Cohere's latest embedding model offers versions with either 1,024 or 384 dimensions. The providers claim it is the most efficient and cost-effective embeddings model available, balancing performance with computational efficiency.

5. **Qwen3-embedding:** It is specifically designed for text embedding and ranking tasks, with various sizes (0.6B, 4B, and 8B). This embedding model ranks No. 1 in MTEB (Massive Text Embedding Benchmark)[4] as of June 2025.

6. **Tencent's Youtu-embedding:** It uses a three-stage recipe, LLM-based pretraining, weakly supervised alignment, and a Collaborative-Discriminative fine-tuning framework with a unified data format, task-specific losses, dynamic single-task sampling, and hard-negative mining. It learns robust, general-purpose text representations for easy enterprise adoption of RAG/IR/STS via open-sourced weights, self-host or cloud API options, and plug-and-play integrations with LangChain and LlamaIndex.

---

[4]https://huggingface.co/spaces/mteb/leaderboard. Accessed on 10/06/2025.

7. There are quite some other options, e.g., NV-Embed-v2 (based on Mistral-7B), M3E-base (social media), nomic-embed-text (fast with 768 dimension & suitable for edge device), multilingual-e5 (multilingual models). For more information, just check the MTEB multilingual leaderboard for the latest updates.

Balancing the trade-off between performance and computational efficiency is an important step, as higher-dimensional embeddings often provide better performance but require higher computational resources.[5] Ongoing research focuses on finding the optimal dimensionality that maximizes model performance while minimizing resource usage, aiming to achieve the best possible results without unnecessary computational expense. Table 5.4 lists some general considerations when choosing a proper model. Some of the typical approaches are proposed as follows.

- Quantized embedding compresses dense vectors using lower-precision data types, e.g., from fp32 to int8. An extreme case is binary embeddings.
- Variable dimension approach, e.g., Matryshka embeddings [3], provides flexible embedding size, which allows to encode information hierarchically and adapt to different tasks or constraints with preserved semantic meaning.
- Multi-vector embedding, e.g., ColBERT,[6] uses multiple vectors (i.e., more space) instead of one pooled vector to represent a complex text to

---

[5]It's a common misconception that higher vector dimensionality directly equates to greater precision in retrieval; however, these are distinct concepts. Dimension refers to the length of the vector, such as 384, 768, or 1,024, which primarily dictates the upper limit of expressive capability, allowing the model more room to represent complex semantic relationships, though it can also introduce noise. Precision, on the other hand, typically measures the model's retrieval accuracy on a specific task, often quantified by metrics like Top-1 hit rate, Mean Reciprocal Rank (MRR), or Recall@K. This accuracy is largely determined by how well the model's training data aligns with the target task, rather than the vector's dimension alone; for instance, a 384-dimensional model fine-tuned on legal texts might outperform a 1024-dimensional general-purpose model in the legal domain due to its specialized training.

[6]ColBERT (Contextualized Late Interaction over BERT) generates contextualized vectors for each token. At retrieval time, ColBERT uses a "late interaction" mechanism that compares each query token with all document tokens via a MaxSim operation, aggregating the highest similarities. This allows precise matching between query intent and specific

Table 5.4.   Factors to consider when choosing an embedding model.

Factor	Description
Task type	Match the model to the requirement (e.g., QA, search, classification, clustering).
Domain specificity	General-purpose vs. domain-specific (finance, manufacturing, medicine, law, etc.).
Multilingual support	Important when the content spans multiple languages.
Dimensionality	Balance information richness against computational cost. After a certain value, high dimensionality may introduce redundancy and reduce the performance.
Licensing	Open-source models (commercial usage) vs. proprietary services.
Maximum tokens	Choose an embedding model whose context window is large enough for your use case.
Infrastructure	Match the model performance and efficiency with the available infrastructure.

capture more details of context, addressing the limitations of single-vector representations for long documents by preserving fine-grained information often lost during compression.

- Hybrid embedding, e.g., BGE-M3,[7] is designed as an all-in-one solution for retrieval tasks across languages, granularities, and formats.

Note that it is also possible to use dense passage retrieval (DPR) with two independent encoders instead of one, i.e., Query Encoder processes a short and often informal query string (e.g., "What is the capital of Japan?") and learns to extract intent and semantic cues; while Passage Encoder processes longer, information-rich text (e.g., Wikipedia

---

document segments, making it especially effective for capturing nuanced relevance in lengthy, detail-rich texts. As one of tensor rerankers, we will discuss it in the next chapter.

[7]ColBERT (Contextualized Late Interaction over BERT) generates contextualized vectors for each token. At retrieval time, ColBERT uses a "late interaction" mechanism that compares each query token with all document tokens via a MaxSim operation, aggregating the highest similarities. This allows precise matching between query intent and specific document segments, making it especially effective for capturing nuanced relevance in lengthy, detail-rich texts. As one of tensor rerankers, we will discuss it in the next chapter.

paragraphs) and learns to summarize and represent *relevant knowledge*. Both encoders are jointly trained to maximize the dot-product similarity between a query and its relevant passage, as well as to minimize similarity with irrelevant passages. A typical example from Facebook is their facebook/dpr-ctx_encoder-single-nq-base and dpr-ctx_encoder-single-nq-base models in Huggingface. An end2end implementation of the chunking & vectorization step inside a full data ingestion pipeline by LlamaIndex are illustrated in the example of ch5\ingestion_pipeline.py.

If an open-source embedding model is employed, fine-tuning it is an excellent strategy to improve retrieval accuracy. LlamaIndex, for example, offers a detailed step-by-step guide on fine-tuning open-source embedding models,[8] and their experiments show that fine-tuning consistently boosts performance across a variety of evaluation metrics. A sample code snippet using SentenceTransformersFinetuneEngine in ch5\finetune.py demonstrates how to create a fine-tune engine, run the fine-tuning process, and obtain the fine-tuned model. Fine-tuning allows the model to adapt its embeddings to the nuances of your specific data, which results in more precise and relevant retrievals. More information can be found in Chapter 12.

5.3 Vector Database

A vector database is specifically designed to store data and metadata, index, and retrieve high-dimensional data points, also known as vectors. This contrasts with traditional relational databases that organize data like numbers and strings in tables. Vector databases are tailored for handling unstructured data within a multi-dimensional vector space, making them particularly well-suited for AI and machine learning tasks, where data often exists in the form of an image, text, or other types of embedding. Due to their specialization, vector databases are sometimes referred to as one of AI databases. These databases are adept at conducting similarity searches, rapidly locating vectors within a dataset that closely matches a given query. This functionality is critical for applications inherent to RAG systems, recommendation engines, and natural language processing tasks, all of which require efficient processing of complex, high-dimensional data. In sum, its primary functions, as stated, are:

---

[8]https://docs.llamaindex.ai/en/stable/examples/finetuning/embeddings/finetune_embedding.html.

- **Fast Storage and Retrieval of Vectors:** They are optimized for storing high-dimensional vectors, which are numerical representations of data (like text, images, or audio) in a continuous vector space. This optimization allows for quick insertion and retrieval of these vectors.
- **Support for Fast Search, e.g., Approximate Nearest Neighbor (ANN):** This is arguably their most crucial feature. Unlike traditional databases that rely on exact matches or precise range queries, vector databases excel at finding vectors that are "similar" to a given query vector. Because an exact nearest neighbor search in high-dimensional spaces can be computationally expensive, vector databases use ANN algorithms to find approximate matches very quickly, which is sufficient for most real-world applications like semantic search, recommendation systems, and anomaly detection.

There are many open source or proprietary tools available on the market. Table 5.5 summarizes the key attributes of each vector database, providing a clear comparison to help in understanding their capabilities and suitability for different use cases. For local vector management in small applications (or "apps") and dev experiments, we can choose Chroma and Faiss with simple python installation. Weaviate, Vespa, and Qdrant can be tailored for single node apps. pgvector, as an extension of PostgreSQL, can have much better scalability. Note that although we

Table 5.5. Vector databases.

Vector Database	Open Source	Programming Language	Main Features
Chroma	Yes	Python, JavaScript	Embeddings, vector search, document storage, full-text search, metadata filtering, multi-modal support. Suitable for light-weight applications with locality and limited functionalities.
Faiss	Yes	Python	High-speed search, handle large datasets with minimal memory usage. Supports user-defined distance metrics and indexing for performance tuning. Suitable for local deployment with limited requirements on metadata.

Table 5.5.   (*Continued*)

Vector Database	Open Source	Programming Language	Main Features
Milvus	Yes	Go, C++, Python	High-speed vector similarity search, scalable to billions of vectors, supports hybrid search for combining structured and unstructured data, multi-vector search, and metadata filtering. Suitable for large-scale enterprise cases.
pgvector pgvector-rs	Yes	C/Rust	Vector similarity search extension for PostgreSQL, supports exact and approximate nearest neighbor search, integrates with SQL queries.
Qdrant	Yes	Rust	Fast and scalable (both dense and sparse) vector similarity search, cloud-native scalability, supports multimodal data, built-in compression, and quantization. Suitable for large-scale applications.
Pinecone	No	Python	Serverless architecture, low-latency vector search, real-time updates, hybrid search capabilities, metadata filtering, and integration with various AI frameworks.
txtai	Yes	Python	Embeddings, semantic search, document indexing, and retrieval; support various data types. claimed to be an all-in-one embeddings database.
Vellum	Yes	Python	Vector search, embeddings management, and integration with machine learning workflows.
Weaviate	Yes	Go/Python	AI native; built-in RAG, support native vector search, embeddings, GraphQL and integration with data pipelines for AI applications; support multimodal models.
Vespa	Yes	Python	Built to scale out and ensure high availability, allows multiple applications to operate on a single instance.
Elastic	Partial	Java	Matured commercial search engine for enterprise; expensive but easy to use.

suggest building a "global" vector database for easy use, it might not be necessary to access it for all applications. For some small-scale and non-critical-non-sensitive internal-client facing applications, local vector database might be more suitable, considering the limited bandwidth of the global one and the low security level.

5.3.1 *Indexing*

The search index stores the vectorized content produced during the embedding step. In its simplest form, a flat index is used (e.g., IndexFlatIP of Faiss) this involves performing brute-force distance calculations between the query vector and each of the chunk vectors. While this method is straightforward (the advantage of indexing is that all vectors are stored "as it is" without compression or quantization. Searches use brute-force cracking for higher accuracy), it becomes computationally expensive when dealing with large-scale data (i.e., 10,000+ elements).

For efficient retrieval at such scales, more sophisticated methods are necessary. Advanced vector indices leverage Approximate Nearest Neighbors (ANN) algorithms. Libraries like faiss, nmslib or annoy implement techniques such as clustering, tree-based methods, and the HNSW algorithm to significantly improve search performance. In addition, other methods (e.g., IVF systems, product quantization (PQ), and hybrid approaches) offer further optimization for high-dimensional vector search.

Managed solutions and vector databases also play a crucial role. Platforms like OpenSearch and ElasticSearch now support vector search capabilities, while dedicated vector databases such as Pinecone, Weaviate or Chroma and Milvus automatically handle the entire data ingestion pipeline, indexing, and retrieval processes. By combining these sophisticated techniques, RAG systems can achieve both high accuracy and efficient performance, even when operating over large and complex datasets.

Depending on your index choice and specific data and search requirements, you can also store metadata alongside vectors. This allows for advanced search functionalities, such as filtering information by date or source. Table 5.6 compares the different types of indexes and provides guidance in optimizing the vector database performance.

Faiss and LlamaIndex support a variety of vector store indices, but also includes simpler index implementations like list index, tree index,

Table 5.6.   Index types.

Index Type	Best For	Example Use Cases
Flat	Small datasets, high precision.	Audio fingerprinting, similarity in <10k images.
HNSW	Low latency, high recall, medium-large datasets.	Real-time recommendations, e-commerce search.
IVF	Large datasets with moderate precision needs.	Document retrieval, basic NLP search.
IVF+PQ	Massive datasets, memory-efficient with trade-offs.	Video metadata search, embedding-based analytics.

and keyword table index, which will be discussed further in the "Fusion Retrieval" section of Chapter 7. The sample code ch5\index.py demonstrates the general procedure.

5.4 Metadata

Metadata refers to structured information attached to documents or content chunks, such as timestamps, authorship, category labels, data source, data security level, language, or file type. In RAG systems, effectively leveraging metadata plays a critical role in improving the precision and contextual relevance of retrieval outcomes. It is usually recorded during chunking phase.

While vector-based semantic search excels at identifying content with similar meaning, it often returns results that are semantically aligned but contextually irrelevant. This limitation becomes particularly evident when retrieved documents are outdated, from untrusted sources, or written in the wrong language. Metadata filtering offers a robust solution to this problem by introducing explicit constraints during retrieval. For instance, a user query can be restricted to documents published after a certain year, or to content originating from a specific source or authored by a particular individual. Even if a text segment ranks high in semantic similarity, it will be excluded from the results if it fails to meet the predefined metadata criteria.

Another valuable application of metadata is the incorporation of temporal awareness. By indexing documents with timestamps, a RAG system can prioritize newer information, helping to ensure that generated answers are not only accurate but also up to date. Similarly, metadata enables multilingual filtering (i.e., retrieving only documents in the desired language), or domain-specific control, such as limiting retrieval to research papers, legal filings, or internal reports. Developers can also embed data attributes and role characters to define the essence for role-based access control (RBAC) and attribute-based access control (ABAC) in entitlement.

However, the design and selection of metadata fields must be carefully considered. The chosen attributes should meaningfully contribute to retrieval relevance without introducing unnecessary complexity or overhead in the indexing and filtering process. Integrating metadata into the RAG pipeline significantly enhances search accuracy and contextual alignment. By narrowing the retrieval scope and enforcing relevant constraints, metadata-driven filtering ensures that the content surfaced by the system is more aligned with user intent, thereby improving the overall quality of LLM-generated responses.

Below is an example to show how to embed document attributes for user entitlement purposes. In the code segment below, each document is annotated with an "entitlement" key to define access levels for Document Metadata. The Indexing is handled by LlamaIndex, where metadata is added using extra_info, and FAISS, where metadata is stored alongside embeddings. Finally, metadata filtering ensures users only access documents they are entitled to. This can be done dynamically in both LlamaIndex and FAISS retrieval steps, as demonstrated in ch5\metadata.py.

A more complex scenario where the user extracts both the vector database and relational database is illustrated in Figure 5.2 (see more in Chapter 13). The process begins when a user logs in via the user interface, triggering the authentication service. If authentication fails, access is denied. If successful, the user submits a query, which the LLM interprets to determine the appropriate processing agents. Two parallel paths are then initiated: the first involves database agents generating or retrieving structured data through SQL queries on a relational database, applying role-based filtering; the second path embeds the query as a vector to search for a vector database, also filtered by metadata and user roles. Both structured and unstructured query results are independently filtered and then merged into a unified dataset. A prompt template is applied to the merged data, and the LLM generates the final response, which is presented to the user. This architecture balances secure access control with effective multi-modal data retrieval and response generation.

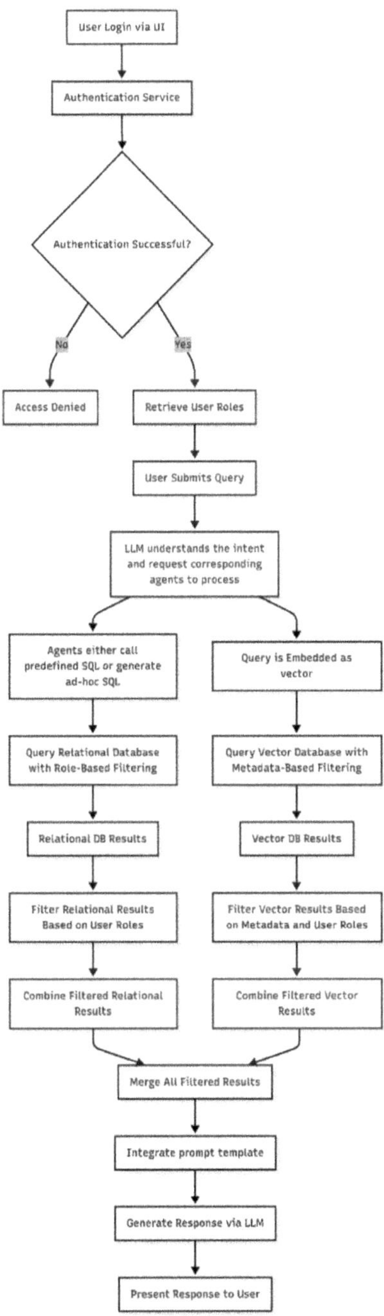

Figure 5.2.   RBAC entitlement.

5.5 References

References or citations are important for ensuring transparency and credibility when generating answers using multiple sources. When an answer is derived from various documents (either due to the complexity of the initial query requiring multiple subqueries or because relevant context was found across different documents) accurate back-referencing of sources becomes essential. The corresponding metadata can be also used for filtering purposes, which sometimes is used for user entitlement. Here are a couple of ways to achieve this, e.g.,

1. **Embedding Source Information in Prompts:**
    o   Incorporate the referencing task into the LLM prompt, asking the model to mention the IDs of the sources used. This approach requires that citation information accompanies the retrieval context. Therefore, when building the embedding and index, the vector database should also record the corresponding metadata, such as document ID, page number, and chunk ID.
2. **Matching with Original Text Chunks:**
    o   Match parts of the generated response to the original text chunks in the index. Tools like LlamaIndex offer efficient fuzzy matching solutions for this purpose. Fuzzy matching is a powerful string-matching technique that identifies approximate matches between strings, ensuring that the generated response accurately reflects the original sources.

By using those methods, it is possible to maintain accurate citations and ensure that all sources are properly referenced, thereby enhancing the reliability and traceability of the generated answers.

5.6 Vector Operations

Effectively managing vectors when documents are **inserted, deleted, or updated** is crucial for maintaining data integrity and ensuring accurate similarity searches.

Inserting data into a database involves adding new rows, or records, to a table. When you perform an insert operation, you specify the table where the data will be stored and provide the values for each of the columns in the new row. For example, if you have a table labeled "Customers," an insert operation might add a new customer's name, address, and contact information. This process expands the dataset, making more information available for queries and analysis.

Deleting data, conversely, removes existing rows from a table. When you execute a delete operation, you typically specify a condition that determines which rows should be removed. For instance, you might delete all records for customers who haven't made a purchase in over five years, or remove a specific product from an inventory table once it's discontinued. This action permanently removes the selected data, so careful consideration of the "delete" condition is crucial to avoid unintended data loss.

Updating data modifies existing rows within a table without adding or removing them entirely. An update operation allows you to change the values in one or more columns for specific rows that meet a defined condition. For example, you might update a customer's address if they move, or change the price of a product in your inventory. This operation is essential for maintaining the accuracy and currency of the information stored in your database, ensuring that your data always reflects the latest state.

Below are some of the best practices and tips when executing these functions:

- **chunk_id and doc_id:** Always include a chunk_id (unique for each chunk) and a doc_id (unique for the entire document) in your metadata. These IDs are invaluable for tracking, updating, and debugging your vector data.
- **Vector Model Updates:** If you ever update your embedding model, you must re-vectorize all your existing documents. Old vectors generated by a different model will reside in a different vector space, making them incompatible with new queries and leading to poor search results.
- **Index Management:** For large vector databases, consider strategies for index management. As data changes, your vector index might become fragmented or less optimal. Periodically rebuilding or optimizing your index can maintain search performance.

By diligently following these practices, you can ensure your vector database remains a robust and accurate source for your document knowledge base, providing relevant answers and insights.

5.7 Concluding Remarks

We can expect more intelligent methods for preprocessing knowledge bases, including adaptive chunking strategies that maintain semantic coherence and improve retrieval efficiency.

Vector embedding (and retrieval) for RAG is now served by a compact but robust toolsets, such as Milvus/Zilliz, Pinecone, Qdrant, Weaviate, pgvector and lakehouse native services (e.g., Databricks Mosaic AI Vector Search), all of which wrap proven ANN indexes (HNSW, IVF, PQ) and optional hybrid dense plus sparse scoring in production APIs. This convergence lets financial institutions move from hand rolled Chroma/FAISS clusters to governed, SLA backed services without surrendering search quality.

Cloud offerings have caused costs to decline through serverless separation of storage, and compute and tiered storage that parks cold embeddings on cheaper media while still allowing relevance boosted recall when they are needed. Some service providers' serverless and "Bring Your Own Cloud" (BYOC) editions add private networking and customer managed keys so that regulated data never leaves the bank's region yet still benefits from automatic scaling.

To satisfy strict data residency and zero trust mandates, some financial institutions and government agencies are adopting an "SaaS grade, self-hosted" pattern: run the vector engine inside the bank's fire walled Kubernetes cluster or private VPC, hard enforce role or attribute based access control (RBAC/ABAC) at query time, encrypt every path (in transit, at rest, and inside memory where supported), and accelerate index build/search with GPUs to meet sub second SLAs. This architecture keeps sensitive embeddings on prem while preserving cloud like elasticity: Helm or Operator charts automate sharding, high availability and snapshots; customer managed keys plug into existing HSM/KMS; OpenLineage style hooks emit immutable provenance for each ingest step; and GPU back ends such as FAISS CUDA or RAPIDS cuVS give HNSW/IVF indexes the throughput needed for trading desk workloads. The result is a deployment that mirrors the governance posture of core databases yet scales to billions of vectors without exposing data to the public cloud.

The next wave of enhancements centers on lakehouse convergence, confidential compute (security) and GPU-accelerated/compressed GPU indices. Databricks is embedding vector search directly into Unity

Catalog so structured rows, embeddings and lineage share one policy plane.[9] Snowflake is taking a parallel path: Cortex Search exposes hybrid vector + keyword + semantic rerank retrieval through a single SQL call, and the recent AISQL preview lets teams build RAG pipelines inside the data cloud without standing up separate infrastructure.[10] Underpinning both ecosystems, Apache Iceberg's roadmap shall add native vector data types and optimized layouts, positioning Iceberg tables as the open standard for unified OLAP/vector workloads across Spark, Trino and Flink.[11]

On the performance and security fronts, postgres (pgvector) is adding HNSW and scalar quantization for larger on prem Postgres clusters; Milvus is experimenting with learned product quantization and encrypted similarity via Intel SGX [4]; and NVIDIA cuVS (CUDA Vector Search via RAPIDS RAFT) offers production ready GPU kernels for HNSW, IVF PQ and K means clustering.[12] Collectively, these innovations position on prem vector stores to deliver sub second retrieval, zero trust security and audit ready provenance even as regulated document volumes grow by double digits each quarter.

References

[1] V. Raina and M. Gales, *Question-Based Retrieval using Atomic Units for Enterprise RAG,* https://www.doi.org/10.48550/arxiv.2405.12363, 2024.

[2] T. Formal, B. Piwowarski and S. Clinchant, *SPLADE: Sparse Lexical and Expansion Model for First Stage Ranking,* arXiv:2107.05720, 2021.

[3] A. Kusupati and *et al., Matryoshka Representation Learning,* arXiv: 2205.13147v4, 2024.

---

[9]Databricks manual, https://docs.databricks.com/aws/en/generative-ai/vector-search, Accessed on 30/08/2025.

[10]Snowflake manual, https://docs.snowflake.com/en/user-guide/snowflake-cortex/cortex-search/cortex-search-overview, Accessed on 30/08/2025.

[11]As of August 2025, Iceberg v3 introduces deletion vectors, row lineage, geospatial and semi-structured types. The community is exploring vector/full-text indexing, but native vector data types are not yet in the spec.

[12]Corey Nolet, Isabel Hulseman and Nathan Stephens, "Optimizing Vector Search for Indexing and Real-Time Retrieval with NVIDIA cuVS," https://developer.nvidia.com/blog/optimizing-vector-search-for-indexing-and-real-time-retrieval-with-nvidia-cuvs/, Jul 2025, Accessed 27/07/2025.

[4] Z. Fan, Y. Zeng, Z. Zheng and Y. Tong, FedVSE: A privacy-preserving and efficient vector search engine for federated databases, in *Proceedings of the 31st ACM SIGKDD Conference on Knowledge Discovery and Data Mining (SIGKDD 2025)*, Toronto, ON, Canada, 2025.

Chapter 6

Query Transformation and Prompt Engineering

A user query can undergo various transformations and decompositions before being executed within a Retrieval-Augmented Generation (RAG) query engine, agent, or other processing pipelines. The idea behind query transformations is that there are cases where the retriever may not consider the user's initial prompt particularly similar to the relevant documents in the database. This calls for the modification of the query to increase its relevance to our database's sources before retrieving and feeding them to the language model. Similar situations may also arise during augmentation and generation phases.

This chapter details several methods for transforming and decomposing queries, and identifying the set of relevant tools for such processes. Each technique is tailored to specific use cases, optimizing the processing workflow.

For clarity, we refer to the underlying processing system as a "tool." Below are some typical query transformation techniques:

- **Query Rewriting:** While the set of tools remains constant, the query is rewritten in multiple ways to optimize execution against these tools. This may involve paraphrasing, expanding, or rephrasing the query to enhance compatibility and retrieval effectiveness.
- **Query Decomposition (Sub-Questions):** The original query is decomposed into several smaller, more manageable sub-questions, each

directed towards different tools based on their metadata and specialization. This decomposition allows for more granular and precise information retrieval from specialized sources. We can also view it as a sub-category of rewriting.

- **Routing:** The system's goal is to analyze the query and identify the most appropriate subset of tools that are capable of processing it, while the original query remains unchanged. Essentially, the system "routes" the query to those tools whose specialties match the query's intent, ensuring that only the relevant modules are engaged. The advanced version becomes agent-based.

- **ReAct Agent Tool Picking:** This technique involves a two-step process. First, given the initial query, the system identifies the most appropriate tool to use. Second, the system formulates a specific query tailored for execution on the chosen tool, ensuring optimized retrieval and response generation. This tool will be further discussed in Chapter 14.

These transformation techniques enhance the flexibility and precision of query processing, allowing for intelligent decision-making and improved response quality across various information retrieval scenarios.

6.1 Query Rewriting

In RAG systems, addressing the discrepancies between user queries and the semantic content of target documents is critical. A common challenge arises when user queries are phrased inaccurately or lack necessary semantic details, potentially leading to irrelevant or incorrect responses from the language models. To mitigate this issue, query rewriting technology serves as a vital tool within RAG, functioning as a pre-retrieval method to refine and enhance the queries before they are used to fetch documents.

The role of query rewriting is to bridge the semantic gap between the user's input and the information stored in documents, ensuring that the retrieval process is both accurate and effective. For example:

- Multi-query uses a prompt template to generate multiple search queries based on a single input query, while the subQuery technique uses a divide-and-conquer approach to handle complex questions. i.e., break down the query into several smaller questions, each targeting different tools as determined by their metadata.

- Hypothetical Document Embeddings (HyDE) helps align the semantic spaces of the query and the documents by generating hypothetical documents that better match the query's intent.
- Rewrite-Retrieve-Read framework alters the traditional sequence of retrieval and reading by prioritizing query rewriting, thereby refining the query prior to document retrieval.
- Step-Back Prompting enables the LLM to perform abstract reasoning and retrieval based on higher-level concepts derived from the query.
- Query2Doc technique involves creating pseudo-documents from prompts provided by LLMs, which are then combined with the original query to form a new, more aligned query.
- ITER-RETGEN method iteratively combines the outcomes of previous generations with the initial query, retrieves relevant documents, and generates new content based on the refined query. This cycle is repeated to progressively refine the response.

By implementing these advanced query rewriting techniques, RAG systems can significantly improve the alignment between user queries and the semantic structure of the data they aim to retrieve, enhancing the overall accuracy and relevance of the generated responses.

6.1.1 *Multi-Query and SubQuery*

This method generates alternative phrasings through paraphrasing, expanding, or rewording, which is designed to better align with the input expectations of the tools. By doing so, the system can run these different query formulations across the same set of tools to maximize the chance of retrieving the most relevant information, ensuring that nuances in language or ambiguity are addressed. An extension method proposed is diverse multi-query rewriting, which generate sets of strategically varied rewrites to capture different information granularities. This method improves recall and downstream answer quality [1]. We can customize the rewriting procedure as shown in ch6\multiquery.py. where the script uses LlamaIndex's PromptTemplate to create a formatted prompt that instructs an LLM to generate multiple search queries from a single input query, demonstrating how to leverage prompt templating for query expansion in RAG applications. When it is asked "How to build a machine learning model efficiently?," it returns the following possible result:

```
Generated queries:
1. Best practices for optimizing machine
   learning model performance
2. Tools and techniques for streamlining
   machine learning model development
3. Strategies for reducing training time in
   machine learning models
4. Tips for improving the accuracy of machine
   learning models
5. Resources for learning about efficient
   machine learning model building techniques
```

The sub-query method uses a divide-and-conquer approach to handle complex questions, as shown in Figure 6.1. Essentially, the query is

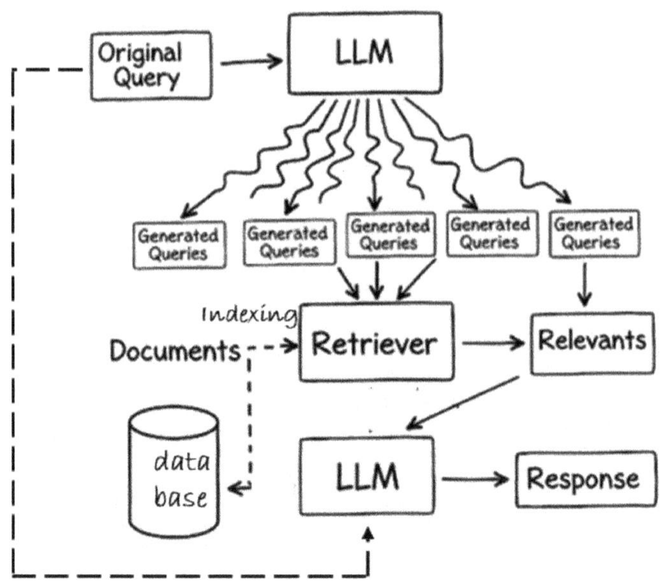

Figure 6.1.   Sub-query.

broken down into several smaller questions, each targeting different tools as determined by their metadata. Below are the three steps of the process:

1. It first analyzes the questions and breaks them down into simpler sub-questions.
2. Each sub-question targets different relevant documents that can provide part of the answer.
3. The engine then gathers the intermediate responses and synthesizes all the partial results into a final response.

The code ch6\queryengine.py demonstrates how to use Sub-QuestionQueryEngine with QueryEngineTool of LlamaIndex to decompose complex queries into simpler sub-questions that are routed to specialized query engines for more accurate and targeted responses. It generates the following possible result for the same question as above:

```
[ml] Q: What are the best practices for data
 preprocessing in machine learning?
[ml] Q: How to select the most suitable
 machine learning algorithm for a specific
 task?
[ml] Q: What techniques can be used for
 feature selection in machine learning
 models?
[ml] A: Feature engineering plays a critical
 role in extracting relevant characteristics
 from raw data in machine learning models.
[ml] A: High-quality data collection,
 cleaning, handling missing values, and
 normalizing features are essential steps in
 data preprocessing for machine learning.
 Feature engineering is crucial for
 extracting relevant characteristics from raw
 data, improving model performance, and
 reducing the risk of bias and overfitting.
[ml] A: Choosing the right algorithm for a
 specific task in machine learning involves
```

> considering the nature of the task itself,
> whether it is classification, regression, or
> clustering. Depending on the task
> requirements, practitioners typically select
> from a range of algorithms such as linear
> regression, decision trees, support vector
> machines, neural networks, and ensemble
> methods. The selection process is crucial to
> ensure that the chosen algorithm is well-
> suited to the data and can effectively
> address the objectives of the task at hand.

A similar function is implemented as a Multi Query Retriever in LangChain.[1]

6.1.2 *HyDE*

HyDE (Hypothetical Document Embeddings) [2] is a query-rewriting technique for generating document embeddings to retrieve relevant documents without needing actual training data, as shown in Figure 6.2. First, an LLM creates a hypothetical answer in response to a query. While reflecting patterns of relevance to the query, this answer includes information that might not be factually accurate.

Next, both the query and the generated answer are transformed into embeddings. The system then identifies and retrieves actual documents from a predefined database that are closest to these embeddings in the vector space.

LlamaIndex provides HyDEQueryTransform as its implementation. See ch6\hyde.py for the details.

Newer variants (e.g., SL-HyDE [3], HyQE [4]) extend the idea into domain adaptation and ranking. These methods are simple to add ahead of dense or hybrid retrievers and remain strong unsupervised baselines.

---

[1]https://python.langchain.com/docs/modules/data_connection/retrievers/MultiQuery Retriever/?ref=blog.langchain.dev.

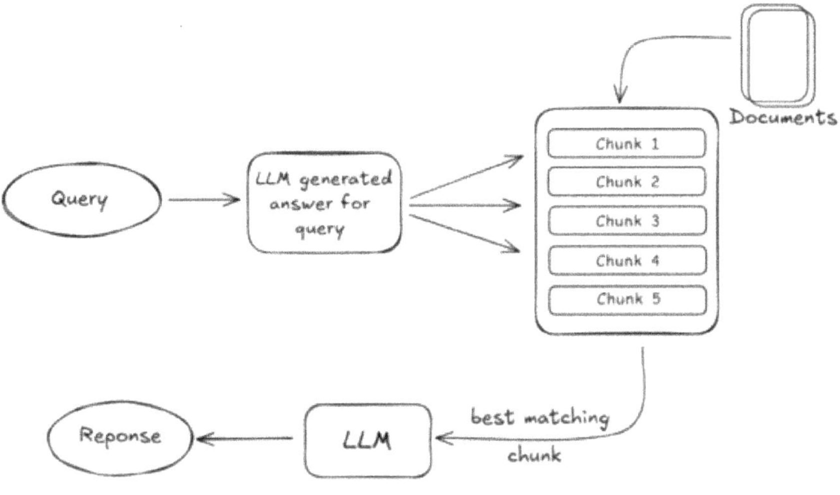

Figure 6.2. HyBE.

6.1.3 *Query2doc*

Query2doc [6] leverages LLMs to enhance query understanding and retrieval effectiveness in Dense Retrieval systems. This approach involves generating pseudo-documents from a few LLM-generated prompts, which are then concatenated with the original query to form a new, expanded query. This process is visually depicted in an illustration where, due to space constraints, some in-context examples are omitted.

In practical terms, the expanded query, denoted as (q^+), is constructed by simply concatenating the original query $((q))$ with the generated pseudo-documents $((d'))$, separated by a "[SEP]" token: $q^+ = \text{concat}(q, [SEP], d')$. This method aims to enrich the query with additional context that mirrors the semantics of potential answers or related content, thereby enhancing the retrieval process by providing more detailed cues to the retrieval system.

Query2doc critically evaluates the assumptions of another model, HyDE, which posits that the groundtruth document and the generated pseudo-documents convey identical semantics in different formulations. Query2doc points out that this assumption may not always hold true, as the nuances of certain queries might lead to variations in the semantic alignment between the original and pseudo-documents.

A significant feature of Query2doc is its use of a supervised dense retriever that is trained explicitly, differing from HyDE's approach. This training involves learning to effectively use the enriched context provided by the pseudo-documents to improve the precision of retrieving relevant documents. In the sample code below, we implement the algorithm in four steps:

1. **Pseudo-Document Generation:**
   - The LLM generates contextually rich pseudo-documents that align semantically with the user query.
   - This step enriches the retrieval process by simulating potential answers or related content.
2. **Query Expansion:**
   - The original query is concatenated with the generated pseudo-documents, separated by a [SEP] token, to form an enriched query.
3. **Dense Retrieval:**
   - The enriched query guides the retrieval process by providing more semantic cues, improving precision in retrieving relevant documents.
4. **Dense Vector Search:**
   - FAISS is used as the dense vector search engine to fetch the most relevant documents.

The sample code in ch6\query2doc.py illustrates the execution of these steps using both LlamaIndex and LangChain. In this example, some texts on LangChain are mixed with other contents. When the query of "What is LangChain framework?" is asked, the following pseudo-documents are generated:

```
Generated Pseudo-Documents: ["The LangChain
  framework is a technological framework
  designed to revolutionize the language
  industry. This framework is based on
  blockchain technology and artificial
  intelligence (AI), … <skip some sentences>…
  leading to reduced costs and increased
  efficiency."]
```

```
Expanded Query: What is LangChain framework?
  [SEP] … <skip some sentences>…
Overall, the LangChain framework aims to
  democratize the language industry by
  providing a decentralized platform where
  anyone can offer or access language
  services. It seeks to eliminate the need for
  intermediaries in the language industry,
  leading to reduced costs and increased
  efficiency.
```

And the correct documents are retrieved:

```
Retrieved Documents:
LangChain Development: Build your applications
  using LangChain's open-source components and
  third-party integrations. Use LangGraph to
  build stateful agents with first-class
  streaming and human-in-the-loop support.
LangChain is a framework for building
  applications with LLMs.
```

6.1.4 *Step-Back Prompting*

Step-Back Prompting is a technique that directs LLMs by identifying high-level concepts and fundamental principles from user inquiries and leveraging these elements to guide the reasoning process. This method can significantly enhance the LLM's ability to follow the appropriate reasoning path to solve problems effectively. The idea of generating more queries is similar to the mutli-step query or subquery techniques.

For example, consider a query on the pressure of a system, given a specific temperature and volume. In the initial response of this hypothetical scenario, both the direct answer and the chain-of-thought reasoning are incorrect. However, by employing step-back prompting, a more general question is first formulated based on the original query, such as identifying the underlying physical formula relevant to the problem. The answer to this broader question is then obtained and provided to the LLM alongside the original question, resulting in the generation of the correct and accurate answer.

The code in ch6\step_back.py shows the simplified workflow:

1. The system generates a more general step-back version of the user query.
2. Retrieves relevant context for both the original and step-back questions.
3. Combines the context and generates a final answer using the LLM.

When the query of "Can LangChain be used for building document-based retrieval pipelines?" is asked, a step-back question is generated

```
Is LangChain suitable for creating pipelines
  that involve retrieving documents?
```

Which further generates the final Answer:

```
Yes, LangChain can be used for building
  document-based retrieval pipelines, as it is
  a framework for building applications with
  LLMs, including prompt management and
  chains. Additionally, LlamaIndex provides
  tools for data retrieval and processing in
  LLM workflows, which can complement
  LangChain's capabilities in building AI
  applications.
```

6.1.5 *Multi-Step Query*

The multi-step query transformation strategy utilizes the self-ask method, where the language model internally formulates and answers follow-up questions before addressing the primary query. This technique enables the model to integrate distinct facts and insights acquired during pretraining. An advanced variation of this approach might incorporate a search engine to enhance its capabilities.

Research has revealed that LLMs often struggle to combine two separate facts, even when they understand each one individually. For example, a model may recognize Fact A and Fact B but fail to infer the combined implication of both.

The self-ask method is designed to address this limitation. During testing, the prompt and question are provided to the model, which then autonomously generates the necessary follow-up questions to link facts, develop reasoning steps, and determine when to conclude the process.

We can call the function MultiStepQueryEngine directly from llamaIndex, which is implemented in ch6\multistep.py with a simple customized prompt. When we ask "How can we design and implement an end-to-end, production-oriented machine learning system covering all aspects of the machine learning lifecycle?," the following new queries are generated (with default 3 steps):

```
> New query: What are the best practices for
  monitoring and maintaining the performance
  of a deployed machine learning model in a
  production environment?
> New query: What strategies can be employed
  to ensure data quality and integrity
  throughout the machine learning lifecycle,
  from data collection to model deployment?
> New query: What are the key considerations
  for selecting appropriate evaluation metrics
  for different types of machine learning
  models and tasks?
```

6.2 Rewrite-Retrieve-Read

As discussed earlier, a fundamental challenge in utilizing LLM for information retrieval is the inefficacy of original user queries, which may not always be optimized for effective retrieval. In real-world scenarios, user queries can often be ambiguous or misaligned with the underlying data, which can impede the retrieval process and affect the quality of the generated responses.

Thus, to overcome this, we implement different methods of rewriting or improving the queries that are better suited for retrieval. This process enhances the relevance and specificity of the query, aligning it more closely with the available data, which is crucial for accurate retrieval and effective answer generation. This section summarizes the rewrite-retrieve-read pipeline [5], as illustrated in Figure 6.3 below. The pipeline contrasts

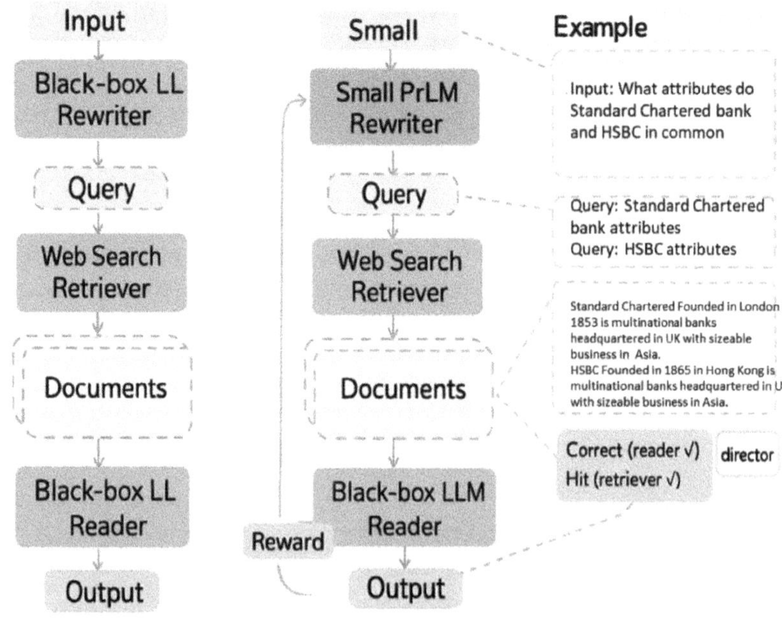

Figure 6.3.   Rewrite-retrieval-read scheme.

against the standard retrieve-then-read method by incorporating an intermediate step of query rewriting before the retrieval and reading phases.

For example, consider the incoherent query "The NBA champion of 2020 is the Los Angeles Lakers! Tell me, what is the LangChain framework?" In the rewrite-retrieve-read model, this query would first be rewritten to clarify and focus the request, perhaps simplifying it to "Explain the LangChain framework." This rewritten query would then guide the retrieval process to focus on the most relevant documents, thereby enhancing the accuracy and relevance of the information retrieved, which in turn improves the prediction performance of the LLM in generating the final answer.

Ch6\rewrite_retrieve_read.py is a sample code with the three steps:

1. **Rewrite Step:**
   - The rewrite_query function uses an LLM (e.g., GPT-4) to simplify and focus the user query.
   - The rewritten query aligns closely with the retrieval requirements.

2. **Retrieve Step:**
   - The retrieve_documents function fetches the most relevant documents from the index using the rewritten query.
   - A retrieval index (e.g., SimpleKeywordTableIndex) optimizes the process.
3. **Read Step:**
   - The read_documents function uses an LLM (e.g., GPT-3.5) to synthesize a coherent and accurate response from the retrieved documents.
   - It provides the final answer in the context of the original user query.

Its sample document contains mixture content of langChain, LlamaIndex and NBA. When the user asks "The NBA champion of 2020 is the Los Angeles Lakers! Tell me what is Langchain framework?," which is purposely added some confusions, the program correctly generates the rewritten query:

```
"Explain the Langchain framework."
```

With the final Response:

```
"LangChain is a framework for building
 applications with large language models
 (LLMs). It includes tools for managing
 prompts and creating chains to connect
 multiple tasks in a pipeline."
```

6.3　Query Routing

Routing retains the original query while identifying the specific subset of tools that are most relevant to it, designating those as the appropriate options for processing. Each query transformation proved useful for different cases. The sub-question decomposition works best for questions that can be broken into simpler sub-questions, like comparing LangChain and Llamaindex's best use scenarios.

Multi-step transformation works best for queries that require exploring context iteratively, like linking multiple facets of information. Simple queries might not need any transformation, so applying them would be a waste of resources.

In the example of ch6\routing2.py, we show how a query can be used to select the set of relevant tool choices. We use our selector abstraction to pick the relevant tools (query_routing, query_rewriting, query_react) - it can be a single tool, or a multiple tool depending on the abstraction. We have the following selectors: combination of (LLM or function calling) x (single selection or multi-selection).

When we input "Tell me more about rewriting methods", only single tool is selected:

```
Single Selector - Selected Tool:
  Name: query_rewriting
  Description: This tool contains the
  Wikipedia page about rewriting method
  Reason: The Wikipedia page about rewriting
  method is directly related to the question
  and provides more information on the topic.
```

When we further ask "Compare routing and rewriting methods," two tools are selected:

```
Multi Selector - Selected Tools:
  Name: query_routing
  Description: This tool contains a paper
  about routing method
  Reason: Routing method is directly related
  to the topic of routing, which is relevant
  to the question
  ---
  Name: query_rewriting
  Description: This tool contains the
  Wikipedia page about rewriting method
  Reason: Rewriting method may also be
  relevant as it involves changing or
  transforming information, similar to routing
  methods
```

The LangChain version is also available.[2]

---

[2]https://python.langchain.com/v0.1/docs/expression_language/how_to/routing/?ref=blog.langchain.dev.

An extended version of this scenario might be one where we incorporate a classifier for the routing. For example, the RAG system should *profile* the user's queries, then "serve the right dish for the right guest." We distinguish four pragmatic intent classes in Table 6.1.

Table 6.1.   Query classes.

Class	User's Primary Need	Typical Output	Implementation
Factual	Exact answers, data points, formal definitions	A concise, source-backed statement	*Strategy* Optimize the query with the LLM to isolate **core entities and relations**, then retrieve narrowly. *Execution* Vector similarity → top-k → **LLM reranking** for factuality confidence.
Analytical	Causal explanations, mechanisms, holistic viewpoints	A structured analysis or synthesis	*Strategy* Ask the LLM to break a complex question into ~3 focused sub-questions. *Execution* Retrieve two passages per sub-question, deduplicate, then merge to guarantee breadth.
Opinion	Positions, debates, alternative viewpoints	A balanced set of arguments	*Strategy* Prompt the LLM to hypothesize three contrasting stances. *Execution* Issue a separate search for each stance; prefer diversity *before* relevance, then fill with high-score passages.
Contextual	Advice personalized to the asker's background	A situation-specific recommendation	*Strategy* Infer implicit user attributes (domain, role, constraints). Append them to the original query to craft a **custom search phrase**. *Execution* After retrieval, rerank with a context-aware LLM scorer that privileges passages matching the inferred scenario.

A lightweight LLM such as **Qwen3-1.7b** can act as an intent classifier. A pseudo python code will be as such:

```python
def classify_query(query: str) -> str:
    """

    Return one of {"Factual", "Analytical",
    "Opinion", "Contextual"}.
    """
return llm("Classify the intent of this
  question:", query).strip()
def adaptive_retrieval(query, vstore,
  user_ctx=None):
    intent = classify_query(query)

    if intent == "Factual":
        return factual_strategy(query, vstore)
    elif intent == "Analytical":
        return analytical_strategy(query, vstore)
    elif intent == "Opinion":
        return opinion_strategy(query, vstore)
    elif intent == "Contextual":
        return contextual_strategy(query,
        vstore, user_ctx)
    else:                        # Fallback
        return factual_strategy(query, vstore)
```

For analytical, opinionated and contextual strategies, they often require dynamic reasoning before retrievals. Therefore, the abovementioned query rewriting, sub-query and ReAct approaches can be applied only for some specific query classes. Therefore, we can keep a balance between efficiency and accuracy.

6.4 Concluding Remarks

This chapter discusses query transformation, which is essentially within the domain of prompt engineering. Beyond the core strategies above, several innovative extensions on query transformation can further enhance the RAG pipeline in the literature [7–9], e.g.,

- Recognizing that some queries may contain mixed intentions, such as seeking both explanations and factual data, dynamic intent blending can be employed to assign soft labels and orchestrate multiple retrieval strategies in parallel.
- Rather than entangling plan and context inside the model window, we may externalize a DAG-style plan of sub-queries and execute them in parallel or sequence; or use LLM-based retrieval plans to cut latency versus purely agentic loops. These methods make decomposition explicit and controllable.
- To manage resource constraints, a cost-adaptive routing mechanism can be implemented, enabling the system to switch between high-precision and low-cost variants of embedding models or rerankers based on real-time performance or budget requirements. It can be extended to performance-adaptive/aware routing. Further, we can let the model decide *when* to retrieve (and how much), while Self-Route/ Selective routing [10] decides *whether* to use RAG at all or rely on long-context reading, which trades cost and accuracy dynamically. This line cleanly connects transformation policies to compute governance.
- Incorporating a reinforcement feedback loop based on user interactions, such as click-through rates or satisfaction scores, allows for continuous fine-tuning of both the intent classifier and reranker, improving overall answer quality and reducing hallucinations.
- For use in regulated industries or privacy-sensitive contexts, on-device inference can be used to generate anonymized context vectors for personalization, ensuring user data remains secure while still allowing for customized retrieval. Sometimes, we should also apply temporal awareness and time-sensitive queries.
- Emerging benchmarks and methods explicitly encode recency and temporal relevance in rewriting and retrieval, improving answers where facts evolve.
- Automatic prompt optimization might be also useful in some cases. For example, DSPy[3] optimizes the correctness and performance of the LLM's output by automating prompt engineering, while Parlant[4] provides fine-grained, rule-based control over an agent's behavior to ensure safety and alignment with business rules ("Alignment Engine").

---

[3]https://dspy.ai; https://github.com/stanfordnlp/dspy.
[4]https://parlant.io; https://github.com/emcie-co/parlant.

Directed by the insights mentioned above, a next-generation "query transformation policy" should possess these traits:

Unified, learned policy: Query transformation is converging on a single policy that treats rewriting, decomposition, routing, and tool-specific synthesis as coordinated actions optimized for multiple objectives such as faithfulness, latency, cost, and privacy. Test-time planners externalize multi-hop plans while selective-retrieval methods (e.g., Self-RAG [10]) decide *when* to retrieve; coupled with cost-aware routers (e.g., CARROT) and standardized routing benchmarks (RouterBench[5]), teams can explicitly trade off accuracy vs. spend and latency for each deployment.

Context, intent, and time awareness: Beyond surface paraphrasing, systems increasingly infer user intent, task constraints, and conversational state to generate purposeful rewrites; hypothetical expansion methods (e.g., HyDE) remain strong, low-overhead baselines. For evolving corpora, time-aware RAG is maturing; new frameworks and benchmarks add temporal triggers, timeline retrieval, and evaluation tailored to time-sensitive questions, which is key for domains like news and finance. Memory- and profile-driven personalization (e.g., Mem0[6] and recent personalization surveys [11]) will make transformation proactive, anticipating the next turn and pre-fetching context when appropriate.

RAG-Aware Routing: Instead of just routing based on the query's text, emerging models like RAGRouter [12] make routing decisions based on the retrieved documents themselves. This allows the system to dynamically select the best LLM or pipeline to generate an answer given the specific context that was found, acknowledging that different models have different strengths.

Structure-first and small-model copilots: Graph- and table-aware transformation is moving into the mainstream so sub-queries target entities, relations, and intervals directly; exemplars include GraphRAG and LightRAG/MiniRAG [13], with new structure-aware planning over evolving knowledge graphs. In parallel, compact planners/routers are emerging as cost-efficient "copilots" that decompose tasks and steer

---

[5]https://github.com/withmartian/routerbench.
[6]https://github.com/mem0ai/mem0.

retrieval or model choice, with practical guidance to route routine work to smaller models and escalate only when necessary.

Governance and explainability by design: Transformation policies are beginning to embed compliance controls—PII masking, DP-based safeguards, and region-aware routing for data residency—alongside evaluation that checks not only *answers* but also *attribution faithfulness*. Expect wider adoption of differential-privacy RAG and hybrid/edge deployments for residency, plus routine use of RAG evaluation suites (e.g., RAGAS) and emerging work showing why "correct" answers can still lack faithful citations.

References

[1] Z. Li and *et al.*, *DMQR-RAG: Diverse Multi-Query Rewriting for Retrieval-Augmented Generation,* arXiv:2411.13154v1, 2024.

[2] L. Gao, X. Ma, J. Lin and J. Callan, *Precise Zero-Shot Dense Retrieval without Relevance Labels,* arXiv:2212.10496, 2022.

[3] L. Li, X. Zhang, X. Zhou and Z. Liu, *AutoMIR: Effective Zero-Shot Medical Information Retrieval without Relevance Labels,* arXiv:2410. 20050v1, 2024.

[4] W. Zhou, J. Zhang, H. Hasson, A. Singh and W. Li, *HyQE: Ranking Contexts with Hypothetical Query Embeddings,* arXiv:2410.15262v1, 2024.

[5] X. Ma, Y. Gong, P. He, H. Zhao and N. Duan, *Query Rewriting for Retrieval-Augmented Large Language Models,* http://arxiv.org/ abs/2305.14283 arXiv:2305.14283, 2023.

[6] L. Wang, N. Yang and F. Wei, *Query2doc: Query Expansion with Large Language Models,* http://arxiv.org/abs/2303.07678 arXiv:2303.07678, 2023.

[7] Y. Gao and *et al.*, *Retrieval-Augmented Generation for Large Language Models: A Survey,* arxiv:2312.10997, 2023.

[8] Y. Ding and *et al.*, *A Survey on RAG Meets LLMs: Towards Retrieval-Augmented Large Language Models,* arxiv:2405.06211v1, 2024.

[9] B. J. Chan, C.-T. Chen, J.-H. Cheng and H.-H. Huang, *Don't Do RAG: When Cache-Augmented Generation is All You Need for Knowledge Tasks,* arXiv:2412.15605v2, 2025.

[10] D. Wu, J.-C. Gu, K.-W. Chang and N. Peng, *Self-Routing RAG: Binding Selective Retrieval with Knowledge Verbalization,* arXiv:2504.01018v1, 2025.

[11] Z. Zhang and *et al.*, *Personalization of Large Language Models: A Survey,* arXiv:2411.00027v2, 2025.
[12] J. Zhang, X. Liu, Y. Hu, C. N. Niu, F. Wu and G. Chen, *Query Routing for Retrieval-Augmented Language Models,* arXiv:2505.23052, 2025.
[13] Z. Guo, L. Xia, Y. Yu, T. Ao and C. Huang, *LightRAG: Simple and Fast Retrieval-Augmented Generation,* arXiv:2410.05779v3, 2025.

Chapter 7

Retrieval Techniques

7.1 Introduction

In RAG systems, the retriever module is a critical component that significantly influences the performance and effectiveness of the system. For a given query, retrieval is to identify and extract the relevant information. In some literatures [1–3], it divides into three stages, pre-retrieval, in-retrieval and post-retrieval. Figure 7.1 provides a brief taxonomy of the retrieval techniques.

- Pre-retrieval techniques are applied before the actual retrieval step to improve the query's effectiveness or prepare the data quality. Chunking, indexing and query optimization falls in the pre-retrieval stage (discussed in Chapters 5 and 6). Do not forget the simple pre-filtering based on metadata information.
- In-retrieval techniques directly improve the core retrieval mechanism. Based on the search mechanism difference, we further divide them into sparse retrieval, dense retrieval, advanced retrieval, hybrid retrieval, as compared in Table 7.1. This chapter will focus on this part.
- Post-retrieval techniques are applied after the initial retrieval to refine the retrieved documents or their scores. The augmentation is part of pos-retrieval stage (will discuss in Chapter 8).[1]

---

[1]Note that in some literatures [1], reranking and retrieved content enhancement are also considered as part of in-retrieval process. Here, we will put them into augmentation process. However, it will not impact the overall process. In fact, it is always better to consider them within a unified pipeline.

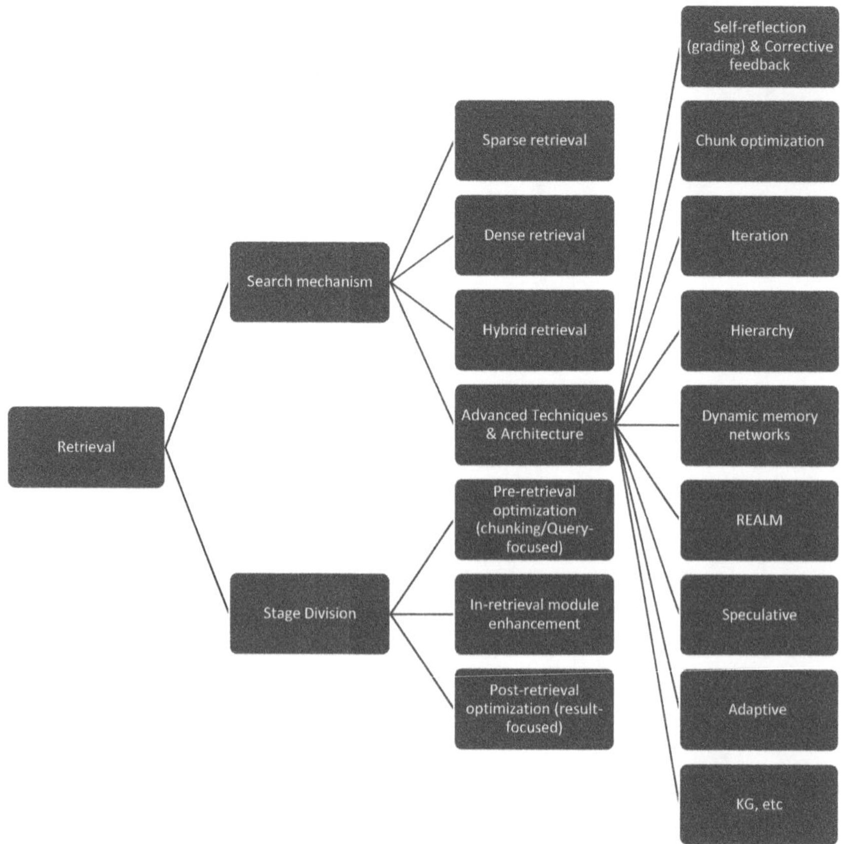

Figure 7.1. Taxonomy of retrieval techniques.

One primary challenge within this module is managing the trade-off between retrieval efficiency and retrieval quality. Retrieval efficiency focuses on the speed at which relevant information can be accessed, optimizing processes such as encoding, indexing, and batch querying within the datastore. On the other hand, retrieval quality aims to ensure the relevance of the information retrieved. This involves refining chunk representation learning and employing advanced ANN algorithms to improve the accuracy and relevance of the search results. Balancing these two aspects of efficiency and quality is crucial for enhancing the overall performance of RAG systems, ensuring they provide timely and contextually appropriate responses.

Table 7.1.   Comparison of sparse, dense, and hybrid retrieval.

Feature/Type	Sparse Vector Retrieval	Dense Vector Retrieval	Hybrid Search (simple)
Mechanism/ Foundation	Inverted Index, Term Expansion	Neural Embeddings, Vector Databases	Combination of Sparse & Dense
Primary Focus	Keyword/Lexical Matching	Semantic/Contextual Similarity	Both Keyword & Semantic
Vector Characteristics	High-dimensional, few non-zero values	Lower-dimensional, most/all non-zero values	Combines both types of vectors
Strengths	Interpretability, handles specific jargon/terms, effective for short queries	Captures semantic nuance, understands complex queries, good for similarity matching	High accuracy, robustness, comprehensive coverage, mitigates individual weaknesses
Weaknesses	Limited semantic understanding, struggles with synonyms/ paraphrases	May struggle with specialized jargon or exact keyword matches	Increased implementation complexity, requires careful tuning of combination strategies
Use Cases/Best For	Exact phrase matching, highly specialized domains, short inputs	Open-domain QA, semantic search, finding conceptually similar content	General purpose RAG, enterprise search, complex queries requiring both precision and recall
Example Technologies	BM25, Neural Sparse Search (OpenSearch)	OpenAI Embeddings, Cohere Embed, Pinecone, Weaviate, Milvus	Vertex AI Search, Amazon OpenSearch Service Hybrid Search, Elastic

7.2 In-Retrieval Techniques

7.2.1 *Sparse Retrieval Techniques*

Sparse retrieval techniques rely on matching indexes and/or keywords between the query and the documents, such as TF-IDF, Query likelihood or BM25 discussed in Chapter 6. Sometimes, they are also called keyword-based approach or lexical technique. Some other techniques, such as edit distance between natural language words or abstract syntax tress (AST) of code snippets [1] or name entity recognition (NER) based approach, can be viewed in this category too, although they do not require a calculating representation. Those techniques are generally effective for exact matches but struggle with semantic understanding.

7.2.2 *Dense Retrieval Technique*

These techniques understand the meaning or intent behind a query and documents, even if exact keywords aren't present. They achieve this by converting text into high-dimensional numerical representations called embeddings. Sometimes, it is called as semantic search.

A dense retrieval module typically consists of three key elements: an encoder (embedding generator), an indexing system, and a datastore (vector database). The encoder transforms input data into embeddings, which are compact vector representations that capture the semantic essence of the data. These embeddings are then indexed in a system designed to support efficient approximate nearest neighbor (ANN) searches, facilitating quick retrieval of the most semantically similar entries from the datastore. The datastore itself is structured to hold external knowledge in a key-value format, allowing for rapid access and retrieval. The encoder model plays the key roles as discussed in Section "Embedding Models" of Chapter 5.

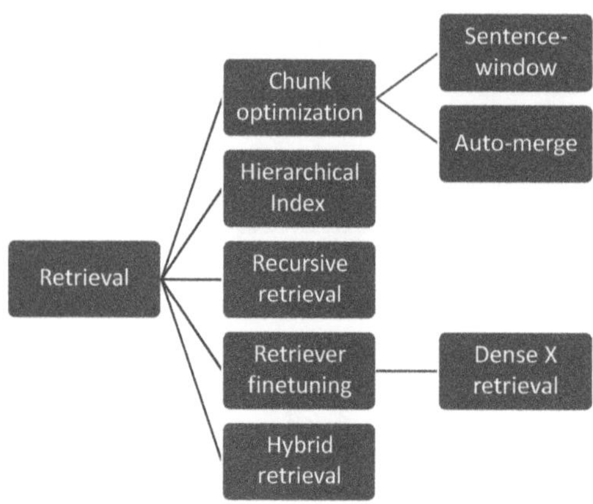

Figure 7.2. Retrieval techniques.

7.2.3 *Advanced Techniques*

The advanced techniques usually go through multiple stages of a RAG pipeline. However, we only discuss their key ideas related to the retrieval parts. Below, let's have a glance at some typical techniques with a summary illustrated in Figure 7.2 [1–3].

1. **Chunk Optimization:** Chunk optimization focuses on improving retrieval effectiveness by carefully designing and adjusting the size and structure of text chunks. This has been partially discussed in Chapter 5. Approaches like sentence-window retrieval retrieve small, highly relevant snippets of text along with adjacent sentences, providing the language model with sufficient surrounding context for better reasoning. Another sophisticated method is auto-merge retrieval, exemplified by LlamaIndex, which organizes documents into hierarchical tree structures where parent nodes encapsulate the content of their child nodes (e.g., articles → paragraphs → sentences). This hierarchical strategy allows fine-grained retrieval at the child level, while ultimately returning parent nodes to enrich context. Furthermore, we can also enhance chunk optimization by applying recursive embedding, clustering, and summarization techniques, building multi-level trees of text chunks that facilitate efficient and contextually rich retrieval.

2. **Recursive Retrieval:** Recursive retrieval improves the quality and depth of retrieved information by performing multiple rounds of search. Instead of relying on a single query, the system breaks down complex information needs into smaller, manageable sub-queries. Each sub-query explores different aspects of the content space to gather more comprehensive and diverse results. The information retrieved from these multiple steps is then consolidated and delivered to the language model, enhancing the accuracy and richness of the final output. The main idea was discussed in Chapter 6, such as the query rewriting and sub-questions. Some implementation includes LlamaIndex RecursiveRetriever that explores linked nodes/ query engines; hierarchical summary trees (e.g., RAPTOR [4]) enabling top-down drill-down and graph/community expansions in GraphRAG. Chapter 12 will further discuss the iterative RAG in the pipeline.

3. **Iterative Retrieval:** This technique generally refers to a step-by-step, looping process that continuously refines the query or information. It suggests a sequence of independent retrieval steps that build upon each other. For example, a system might ask "What are the causes of X?," retrieve some information, then generate a new query like "What are the effects of X based on the causes?," and so on, until the information is complete. It iterates through a series of increasingly specific searches. The typical mechanisms include Self-RAG's on-demand "retrieve/skip and reflect" during generation [5]; and iterative conversational query reformulation (IterCQR) [6] and multi-hop/step methods that fetch evidence step-by-step.

   Both recursive and iterative retrieval are often used interchangeably to describe a multi-step, self-refining retrieval process. Both approaches move beyond a single, static search to progressively refine the information gathered for a more accurate and comprehensive response. While the terms are largely synonymous in practice, they can sometimes have slightly different connotations depending on the specific implementation, as shown in Table 7.2. If we consider their

Table 7.2. Iterative vs. recursive retrieval.

Aspect	Iterative Retrieval	Recursive Retrieval
Core idea	Re-run retrieval in time with evolving queries or plans.	Traverse corpus structure (hierarchy/graph) top-down or along links.
Typical loop	Query → retrieve → read/rerank → reformulate/plan → retrieve again until stop.	Fetch parent/summary → descend to children/follow edges → fetch finer evidence until depth/ budget met.
Query evolution	Actively changes (rewrites, sub-queries, plan updates).	Usually fixed; scope narrows by following structure, not by rewriting.
Corpus dependence	Works on any index; no structure required.	Requires hierarchical or graph indices (summary trees, communities, KG links).
Examples	Self-RAG retrieve/skip/reflect; iterative CQR; multi-hop stepwise retrieval.	RAPTOR (summary trees), GraphRAG (community/edge traversal), LlamaIndex RecursiveRetriever.

Table 7.2. (*Continued*)

Aspect	Iterative Retrieval	Recursive Retrieval
Control signals	Confidence/uncertainty, answerability, coverage, disagreement.	Parent-child scores, edge weights, summary relevance, depth limits.
Stop criteria	Max rounds, marginal-gain threshold, confidence $\geq \tau$, token/latency budget.	Max depth/breadth, relevance falloff, budget (tokens/time).
Strengths	Adapts to ambiguity; good for multi-step or underspecified questions.	Scales long docs; preserves topical locality; avoids scattered chunks.
Risks	Query drift; latency from extra rounds; cost escalation.	Over-expansion on noisy links; missing off-path evidence.
Retriever choices	Hybrid BM25+dense; rerank after each round; dynamic k.	Starts with summaries; descend with tighter k; local rerank per node.
Budgeting	Round caps, adaptive k, early-exit on high confidence.	Depth caps, per-level token budgets, breadth throttling.
Evaluation focus	Round-wise recall/precision, stability (no drift), latency per round.	Depth coverage, on-path recall, final answer grounding to leaf citations.
Implementation fit	Easy to add to existing RAG stacks; planner/rewriter required.	Needs prebuilt hierarchies/graphs (summary trees, KGs, TOCs).
Best combined pattern	Iterative controller decides whether to retrieve again.	Chooses how deep to recurse on each step/branch.

core idea in different axes: time vs. structure; they can be combined, i.e., powerful together for complex corpora.

4. **Hierarchical Index Retrieval:** Hierarchical index retrieval addresses the challenges of searching within large document collections by applying a structured two-stage retrieval process. In the first stage, the system searches through a summary index composed of document abstracts or concise summaries to quickly narrow down the pool of potentially relevant documents. Once a smaller, more manageable set is identified, the second stage involves a detailed search within a chunk index that contains fine-grained segments of the selected documents. This hierarchical approach significantly reduces retrieval time

and computational resources while maintaining high relevance and accuracy of the results. By effectively balancing coarse filtering with detailed examination, hierarchical index retrieval provides an efficient pathway for synthesizing information into high-quality answers.

5. **Retriever Finetuning:** Retriever finetuning enhances the capability of the system to accurately identify and group semantically similar pieces of information. While pre-trained models can provide a strong starting point, further adaptation using high-quality or task-specific data allows the retriever to align more closely with the unique requirements of the application. By improving the representation of content in vector space, the retriever can more effectively select relevant information for the language model, resulting in a more efficient and accurate generation process. Various tuning strategies can be applied to optimize the relationship between retrieved content and downstream generation quality. More information can be found in Chapter 12.

6. **Graph-base Retrieval:** Integrating knowledge graphs to leverage structured information (entities, relationships) can be used for more nuanced and connected retrieval, moving beyond simple factual recall to more sophisticated relational reasoning [7]. This approach is especially useful in contexts where data structure is crucial for understanding, such as social networks, or semantic web applications. By leveraging graphs, the model can retrieve not only isolated pieces of information but also their intricate connections. More details can be found in Chapter 14.

7.2.4 *Hybrid Techniques*

Hybrid retrieval combines multiple retrieval strategies or integrates information from diverse sources to increase robustness and precision. For example, one approach may focus on capturing the overall semantic meaning of a query, while another complements it by retrieving content that matches exact terms or keywords. Hybrid systems may also leverage data from different modalities, such as combining textual and visual information, or apply decision logic to dynamically select or merge results based on confidence levels. This multi-faceted approach improves both the completeness and reliability of the retrieved information, supporting more informed and accurate language model outputs. A fusion algorithm is usually applied to the multiple results, e.g., Reciprocal Rank Fusion (RRF) and weighted fusion (see Rerank in Chapter 8 for more details).

In the remaining content, we will discuss some typical implementations in detail as shown in Figure 7.2.

7.3 Sentence Window Retrieval

In this scheme, each sentence in a document is embedded separately, offering high accuracy for query-to-context cosine distance searches, as illustrated in Figure 7.3. To enhance the reasoning ability of the LLM upon retrieving the most relevant single sentence, the context window is extended by including k sentences before and after the retrieved sentence, assuming the context continuity in the documents. This extended context is then sent to the LLM, providing a broader and more informative context for better reasoning and understanding.

We employ the SentenceWindowNodeParser from llamaIndex to divide documents into individual sentence nodes, ensuring that each node represents a single sentence. Additionally, each node is augmented with a "window" that contains the sentences immediately preceding and following the primary sentence. This window provides essential context that enhances understanding and retrieval accuracy. After the retrieval phase, before the sentences are passed to the LLM, the Metadata-ReplacementNodePostProcessor replaces the isolated sentence with the

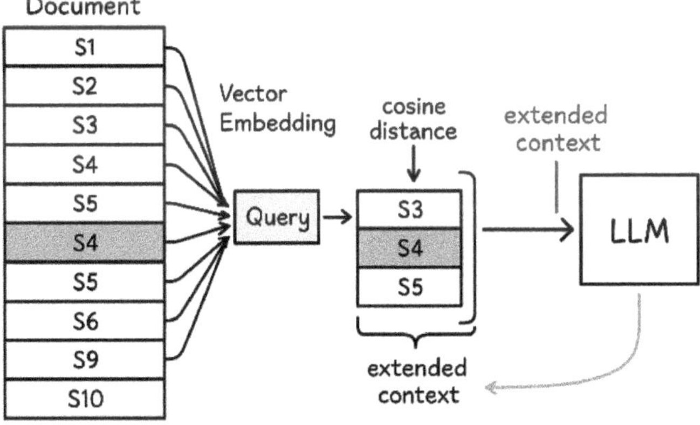

Figure 7.3.   Sentence window retrieval.

complete window. Sometimes, this method is called relevant segment extraction (RSE).

This method is particularly beneficial for large documents or extensive indexes, as it enables the system to capture more fine-grained details by preserving contextual information. The functionalities are implemented in ch7\sentence_windows.py. By default, the window includes 5 sentences on either side of the original sentence, offering a balanced view that helps the LLM generate more precise and context-aware responses. This approach not only improves retrieval quality but also enhances the interpretability of the results by ensuring that the surrounding context is readily available for analysis. In the code, the build_ sentence_window_index function creates a vector index with sentence window parsing to capture contextual information, while get_sentence_ window_query_engine sets up a query engine that uses metadata replacement and reranking to improve retrieval accuracy by expanding context windows and ranking results.

7.4 Auto-Merging Retrieval

The Auto-Merging Retriever (aka Parent Document Retriever; Small-to-big Retriever) operates similarly to the Sentence Window Retriever, focusing on retrieving more granular pieces of information and then expanding the context window before feeding it to a LLM for reasoning. This method involves splitting documents into smaller child chunks that refer back to larger parent chunks, as shown in Figure 7.4.

Here's how it works:

1. **Document Splitting:** The documents are divided into smaller child chunks. Each child chunk is associated with a larger parent chunk, maintaining a hierarchical structure.
2. **Granular Search:** The retriever first searches through these smaller child chunks to find the most relevant pieces of information.
3. **Context Expansion:** Once the relevant child chunks are identified, the context window is extended to include the larger parent chunks. This provides the LLM with a broader context for better reasoning and understanding.

This approach allows for a more detailed and precise search, ensuring that the LLM receives comprehensive context while maintaining efficient retrieval of specific information.

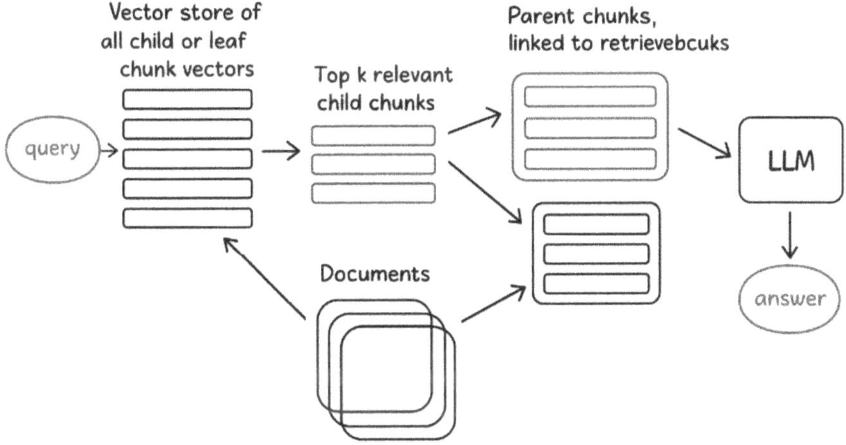

Figure 7.4.   Auto-merging retrieval.

A sample implementation is available at ch7\utils.py. The build_ automerging_index function creates a hierarchical index structure using HierarchicalNodeParser to generate nodes at multiple chunk sizes, then builds a vector index from the leaf nodes while storing all hierarchical nodes in the storage context. The get_automerging_query_engine function sets up a query engine that uses AutoMergingRetriever to automatically merge smaller chunks into larger parent chunks during retrieval, improving context coherence. Both functions incorporate reranking with SentenceTransformerRerank to enhance the final result quality by reordering retrieved nodes based on relevance scores.

7.5 Dense X Retrieval

Dense retrieval plays a pivotal role in providing relevant context for RAG tasks, with the performance significantly influenced by the choice of retrieval unit—whether a document, passage, or sentence. The "Dense X Retrieval," introduces propositions as an effective retrieval unit. This approach enhances our understanding of dense retrieval's impact in NLP by introducing the Propositionizer, a text generation model fine-tuned using a two-step distillation process with two LLMs, e.g., GPT-4 and Flan-T5-large, as illustrated in Figure 7.5. This method not only leverages the strengths of pre-trained language models but also enhances them with task-specific fine-tuning, utilizing 1-shot demonstrations and distillation

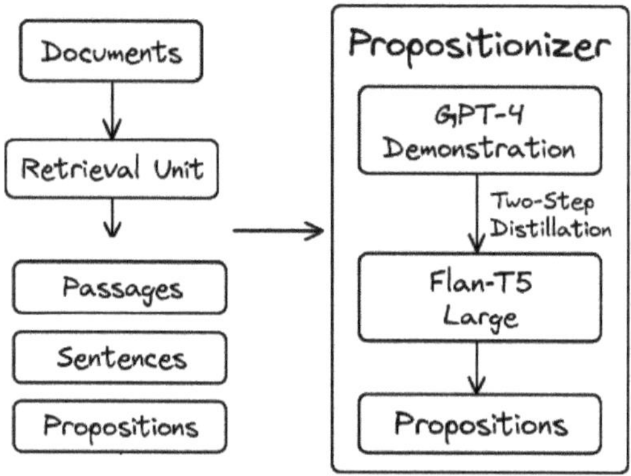

Figure 7.5.   Dense X retrieval.

techniques to refine the model's ability to parse passages into propositions effectively.

A sample code is implemented in ch7\dense_x_retrieval.py. It uses a specialized retrieval pack named DenseXRetrievalPack (pip install llama-index-packs-dense-x-retrieval), comparing its performance against a standard vector store index. It loads documents, creates both DenseX and baseline retrieval engines, then executes queries to demonstrate the improved retrieval capabilities of the DenseX approach.

7.6 Hierarchical Index Retrieval

When dealing with a large number of documents, long complex documents, and multi-topic repositories, efficient search and retrieval are crucial for finding relevant information and synthesizing it into a single answer with references to the sources. An effective approach in the context of a large database involves creating two indices:

1. **Summary Index:** This index consists of summaries of the documents.
2. **Chunk Index:** This index is composed of the document chunks.

The retrieval process is conducted in two steps:

1. **First Step – Search in Summary Index:** Filter out the relevant documents by searching through the summaries. This step narrows down the number of documents to a manageable subset.

2. **Second Step – Search in Chunk Index:** Conduct a detailed search within the chunks of the relevant documents identified in the first step.

This two-step approach, as shown in Figure 7.6, ensures efficient retrieval by quickly filtering out irrelevant documents and focusing the detailed search only on the relevant subset, thereby optimizing the search process in a large database.

Now we look into the sample code available at ch7\hierachical2.py with roughly 6 steps:

1. **Initialization:**
   - o An LLMPredictor is initialized with the OpenAI model gpt-3.5-turbo.
   - o An OpenAIEmbedding model is initialized for embedding generation.
   - o A ServiceContext is created using the LLMPredictor and OpenAIEmbedding.

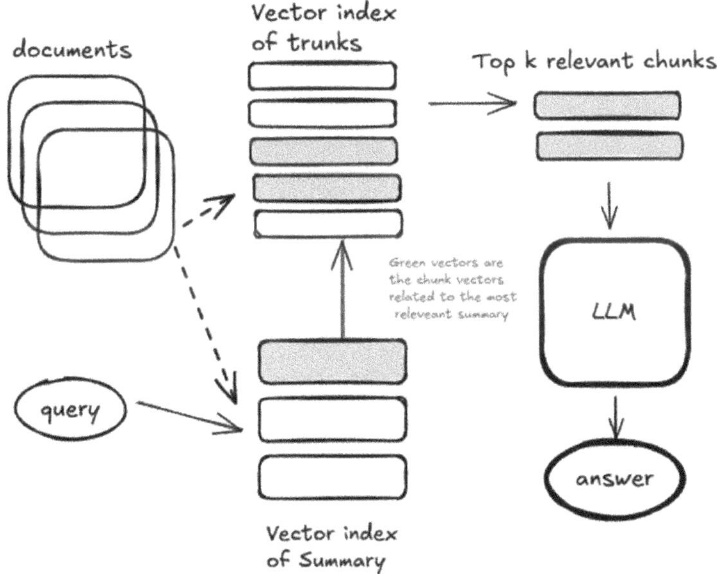

Figure 7.6. Hierarchical index retrieval.

2. **Document Loading and Chunking:**
   o Documents are loaded from a specified directory using SimpleDirectoryReader.
   o Each document is split into chunks of 1024 tokens using SentenceSplitter.
3. **Index Creation:**
   o A VectorStoreIndex (chunk_index) is created for the document chunks, facilitating efficient vector-based retrieval.
   o Summaries for each document are generated using the LLMPredictor.
   o A VectorStoreIndex (summary_index) is created for the document summaries.
4. **Retriever Setup:**
   o Retrievers are created for both the summary_index and chunk_index with a similarity_top_k parameter to limit the number of retrieved nodes.
5. **Hierarchical Retrieval Function:**
   o The hierarchical_retrieval function performs a two-step retrieval process: 1) Retrieves relevant summaries from the summary_index based on the query. 2) Retrieves relevant chunks from the chunk_index corresponding to the documents identified in the first step.
6. **Usage:**
   o The hierarchical_retrieval function is called with a user query, and the relevant chunks are processed as needed.

In general, this retrieval approach excels in resource-constrained environments and high-precision domains like legal or medical fields where targeted retrieval is crucial. The two-stage filtering process provides better performance by first identifying relevant documents through summaries, then retrieving detailed chunks only from those pre-selected documents.

7.7 Recursive Retrieval

The Recursive Retriever [4] in Figure 7.7 is designed to enhance information retrieval from hierarchical or structured data sources. It operates by fetching directly relevant nodes and exploring associated nodes through predefined relationships. For instance, a node might represent a summary of a structured table and link to a query engine that can perform detailed

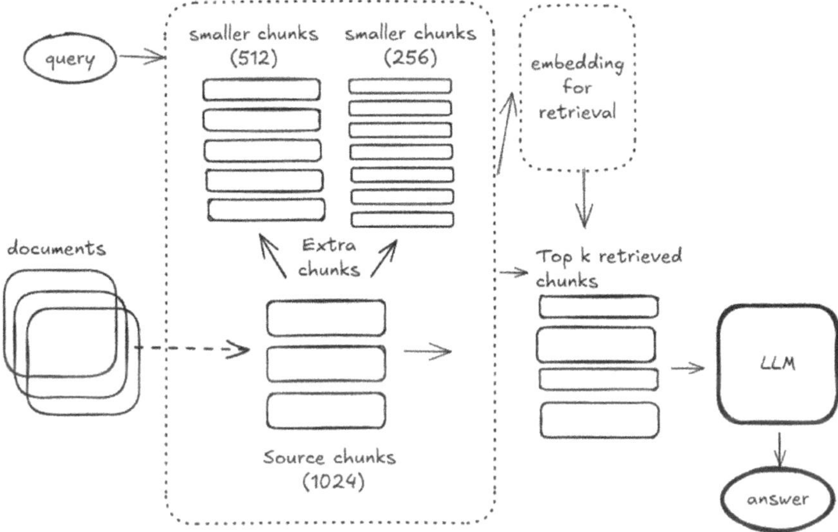

Figure 7.7.   Recursive retrieval.

searches within that table. When such a node is retrieved, the Recursive Retriever delves deeper by querying the linked engine, thereby providing a more comprehensive and contextually relevant response. This method is particularly effective for documents that contain both textual content and embedded structured data, such as PDFs with tables. By leveraging the Recursive Retriever, one can efficiently navigate and extract pertinent information from complex document structures. Note that hierarchical retrieval could be its special case in a sense.

A sample code implements the method with 6 steps in ch7\ recurisive_retrieval.py:

1. **Initialization:** The code begins by setting up the language model predictor and embedding model using OpenAI's model. A service context is then created to manage these components.
2. **Document Loading and Chunking:** Documents are loaded from a specified directory and split into manageable chunks of 1,024 tokens each using the SentenceSplitter.
3. **Index Creation:** A VectorStoreIndex is built from the document chunks to facilitate efficient retrieval.

4. **IndexNode Creation:** For chunks that represent tables, IndexNode instances are created, linking these chunks to specific query engines capable of handling structured data queries. It is also possible to use **camelot** library to extract the table data from PDF, and then apply **PandasQueryEngine.**
5. **RecursiveRetriever Setup:** A RecursiveRetriever is initialized with the vector index retriever and a dictionary of query engines associated with the IndexNodes.
6. **Query Execution:** A RetrieverQueryEngine is created using the RecursiveRetriever, and a sample query is executed to demonstrate the retrieval process.

Above is a type of block-based reference recursive retrieval methods, which involves navigating structured documents by following explicit blocks (e.g., page references), making it ideal for technical manuals or documents with clear citations but less effective in contexts without them. Information-Centric Recursive Retrieval focuses on a central entity, or "seed node," and recursively retrieves related information to build a knowledge graph, useful for legal or financial scenarios requiring a comprehensive view of an entity's relationships. Concept-Centric Recursive Retrieval goes beyond fixed entities by autonomously identifying and exploring related concepts linked to the query, constructing a dynamic web of interconnected information. It is essentially a metadata reference method. This method is valuable for uncovering indirect or second-order relationships. Together, these recursive approaches improve multi-hop retrieval, accuracy, and knowledge storage, allowing systems to develop a memory base for future queries. They also help bridge fragmented or unstructured data across various contexts and refine retrieval efficiency over time.

7.8 Hybrid or Fusion Retrieval

In practice, we can observe that one technique may only cover some retrieval scenarios. In order to fit for more complex situations, it is possible to combine multiple retrieval techniques into a single one. A straightforward way is to create multiple retrievers, and get the corresponding indexes from each retriever. After that, we may further process the multiple indexes.

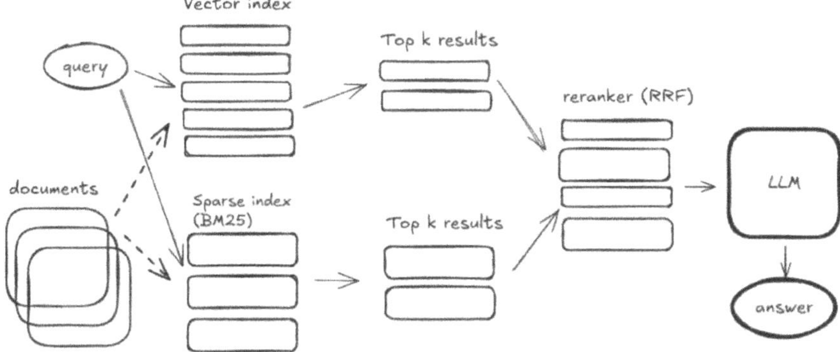

Figure 7.8. Hybrid or fusion retrieval.

Llama_index provides a function "QueryFusionRetriever," which can make the combination simpler. An example using the simplest "VectorStoreIndex" and "BM25"[2] is illustrated in Figure 7.8 with the code in ch7\hybrid.py, which take the advantage of the semantic search via vector cosine similarity and the keyword search via TF-IDF, via applying a fusion step/function to select the best results from each technique.[3]

Another solution is to use "VectorStoreQueryMode,"[4] whose enum provides different retrieval strategies for querying vector stores, allowing users to choose between pure vector search, hybrid search (combining keyword and semantic search), or semantic hybrid search with advanced

---

[2]BM25 (Best Matching 25) is a ranking function that extends TF-IDF by considering term frequency saturation and document length. BM25 effectively ranks documents based on query term occurrence and rarity across the corpus.

[3]In modern retrieval and recommendation systems, the ranking process is typically divided into two distinct stages: Pre-ranking (recall) and Fine-grained (re)Ranking. These stages serve complementary purposes, striking a balance between computational efficiency and ranking precision. Pre-ranking is primarily responsible for rapidly narrowing down a large pool of candidate items, often numbering in the millions, into a much smaller subset that can be further evaluated. Fine-grained (re)Ranking then takes this subset and performs a more detailed, high-fidelity evaluation to determine which items are most relevant to the user's query or preferences.

[4]Although it has the following modes: "DEFAULT," "HYBRID," "LINEAR_REGRESSION," "LOGISTIC_REGRESSION," "MMR," "SEMANTIC_HYBRID," "SPARSE," "SVM," "TEXT_SEARCH," it still needs the native support from the vector databases.

ranking. We can find the sample codes in ch7\hybrid_mode_qdrant.py and cht\hybrid_mode_faiss.py.

The readers can also customize their own fusion function to mix several retreivals. For example, we can consider to build an index first, e.g., "build_sentence_window_index" and "build_automerging_index" as discussed earlier, then call the indexes with the mixed_retrieval function. See ch7\hybrid_custom_index.py for the details.

Note that sometimes, the keyword-based retrieval is very efficient and effective when the domain knowledge is sufficient. The keywords may be automatically generated by the tool "keyLLM," as shown in ch7\keyword.py.

7.9 Concluding Remarks

This chapter examines the fundamental retrieval methodologies, which serve as the pivotal component of the comprehensive RAG pipeline. While we're diving into some of the most effective in-retrieval techniques here, which certainly boost performance, it's crucial to remember they don't operate in a vacuum. Their true power is unlocked only through seamless integration with the pre-retrieval and post-retrieval stages.

Think of it like a symphony: the retrieval module is the lead soloist, but the pre-processing tunes the instruments and sets the stage, while post-processing refines the acoustics for the perfect performance. Just like a well-tuned orchestra, each part of the system must play in harmony to ensure the LLM's output is as impactful as possible.

Therefore, to achieve peak RAG performance, we must always consider the entire pipeline holistically, rather than fixating on individual components in isolation. While optimizing each piece is beneficial, it's their orchestrated effort that truly makes a difference. We will discuss pipeline and orchestration further in Chapter 12.

Below are some future research directions aiming to push retrieval capabilities further:

- **Self-Improving Retrieval Models, Self-Supervised and Adaptive Algorithms:** This involves developing models via meta-learning that can dynamically evaluate and refine their retrieval performance. This includes self-evaluation pipelines that detect inconsistent or incomplete knowledge and trigger additional retrieval cycles. In addition, continuous adaptation to evolving knowledge bases is crucial, particularly for dynamic domains. What's more, developing adaptive retrieval policies that reduce reliance on manual prompt engineering and continuously learn from data. Adaptive retrieval algorithms will iteratively refine search based on real-time feedback, learning to dynamically adjust query strategies.

- **Advanced Hybridization Algorithms:** Systems will blend graph, vector, and keyword retrieval with adaptive rerankers and contextual expansion to optimize for precision, recall, and domain specifics.
- **Human-AI Collaboration for Retrieval Validation:** Integrating human feedback loops (Human-in-the-Loop) will be essential for continuous improvement and refining retrieval accuracy. Native support for user/ranking feedback will incrementally improve retrieval recall and relevance over time, supporting enterprise and mission-critical applications.
- **Scalable Architectures for Cross-Modal/Multimodal Retrieval Fusion:** Developing architectures that can efficiently combine and retrieve information from diverse modalities (text, image, audio, video) at scale.
- **RL-Optimized Retrieval Agents and Multi-Agent Retrieval:** Capsulating retrieval module as an agent can help improving the modularization and efficiency in large RAG applications. RL-driven retrieval optimizers will learn from historical queries to improve performance, potentially through reinforcement learning. Multi-agent approach will enable collaborative agent systems where specialized retrieval agents handle different domains (e.g., legal, finance, healthcare), working together to synthesize comprehensive responses.
- **Causal Reasoning in Retrieval Context:** Future AI will aim to understand causal relationships between retrieved facts, moving beyond mere correlation to deeper reasoning. Think about Chain-of-Retrieval (CoRAG) [8], which Develops more robust methods for iterative retrieval and reasoning chains, potentially with validation and rejection sampling at each step.
- **Scalable, Efficient Retrieval Pipelines:** Sparse and hierarchical retrieval strategies will optimize speed and cost for larger corpora, while modular architectures will make enterprise deployments easier.
- **Parametric RAG Advancements:** Continued research into integrating external knowledge directly into LLM parameters for enhanced efficiency and effectiveness, potentially combining with in-context RAG for even better performance.

In conclusion, the trajectory of retrieval technology in RAG is towards more intelligent, adaptive, and comprehensive systems. The focus is shifting from simple retrieval to sophisticated knowledge management, which encompasses self-correction, multi-modality, personalization, and real-time capabilities. Confronting the enduring challenges of scalability, bias, and factual accuracy will be imperative as RAG evolves, thereby facilitating the development of more dependable, transparent, and potent AI applications across diverse sectors.

References

[1] P. Zhao and *et al.*, *Retrieval-Augmented Generation for AI-Generated Content: A Survey,* arXiv:2402.19473v3, 2024.

[2] X. Ma, Y. Gong, P. He, H. Zhao and N. Duan, *Query Rewriting for Retrieval-Augmented Large Language Models,* http://arxiv.org/abs/2305.14283 arXiv: 2305.14283, 2023.

[3] Y. Gao and *et al.*, *Retrieval-Augmented Generation for Large Language Models: A Survey,* arxiv:2312.10997, 2023.

[4] P. Sarthi, S. Abdullah, A. Tuli, S. Khanna, A. Goldie and C. D. Manning, *RAPTOR: Recursive Abstractive Processing for Tree-Organized Retrieval,* arXiv:2401.18059, 2024.

[5] A. Asai, Z. Wu, Y. Wang, A. Sil and H. Hajishirzi, *Self-RAG: Learning to Retrieve, Generate, and Critique through Self-Reflection,* arXiv:2310.11511, 2023.

[6] Y. Jang, K.-i. Lee, H. Bae, H. Lee and K. J. Jung, "IterCQR: Iterative Conversational Query Reformulation with Retrieval Guidance," in *Proceedings of the 2024 Conference of the North American Chapter of the Association for Computational Linguistics: Human Language Technologies,* 2024.

[7] D. Edge and *et al.*, *From Local to Global: A GraphRAG Approach to Query-Focused Summarization,* arXiv:2404.16130v2, 2025.

[8] L. Wang, H. Chen, N. Yang, X. Huang, Z. Dou and F. Wei, *Chain-of-Retrieval Augmented Generation,* arXiv:2501.14342v2, 2025.

Chapter 8

Augmentation and Refinement Techniques

8.1 Introduction

In RAG systems, the augmentation component, often referred to as post-retrieval refinement or retrieval enhancement or context fusion, plays a crucial role in improving the quality of the output generated by leveraging the retrieved information. These methods refine the retrieved content before it's processed by LLMs, ensuring more accurate and coherent outputs.

Fusion techniques within these systems can be broadly classified into three types: query-based fusion, latent fusion, and logits-based fusion, as discussed in Chapter 1. Query-based fusion involves augmenting the input with retrieved information before it is fed into the generators. This method directly incorporates the context from the retrieved data into the input query, aiming to enrich the generator's context for a more informed output. Logits-based fusion, on the other hand, concentrates on the output logits from the generators. It integrates the logits from the retrieved data to refine the final logits, aiming to enhance the robustness and accuracy of the predictions. Lastly, latent fusion introduces retrieval representations directly into the latent spaces of the generators. This technique subtly improves the models' performance by enriching the internal representations with external knowledge, thus enhancing the depth and relevance of

Table 8.1. Comparison of three fusion techniques.

Fusion Type	Key Feature	Typical Algorithms/ Methods	Applicable Scenarios
Query-based Fusion	Augments or rewrites user queries before retrieval or prior to generation.	HyDE, Self-RAG, Query2Doc, Step-Back Prompting, Rewrite-Retrieve-Read	Useful for improving alignment between user intent and retrieval corpus semantics.
Latent Fusion	Injects retrieved information into the model's hidden (latent) representations.	DenseRetriever+Fusion-in-decoder; latent retrieval integration	Ideal for implicit knowledge injection without modifying visible prompt structure.
Logits-based Fusion	Combines output logits from multiple retrieval paths to stabilize generation.	FiD (Fusion-in-Decoder), RAG Sequence Fusion	Effective in ensemble-style generation or improving robustness across noisy retrievals.

the generated content. Each of these fusion techniques provides a unique pathway to optimize the synthesis of information, ensuring that the final output is not only contextually enriched but also precisely aligned with the input query's intent. Table 8.1 compares their unique features, algorithms and applicable scenarios. In this chapter, we focus on the query-based fusion, where chunk rerank and filtering, context enhancement and content compression are some typical refinement techniques in this phase. Logits-based fusion will be touched on in Chapter 9, and latent fusion will be further discussed in Chapter 12.

8.2 Rerank

Reranking in RAG systems functions as a sophisticated filtering mechanism, critical for refining search results retrieved from vast indexed collections. This process assesses the relevance of each retrieved context to the user's query, sorting them so that the most pertinent ones are prioritized, similar to selecting the best study materials during an open-book exam. The aim is to ensure that the LLM focuses on these top-ranked contexts to generate higher quality and more accurate responses.

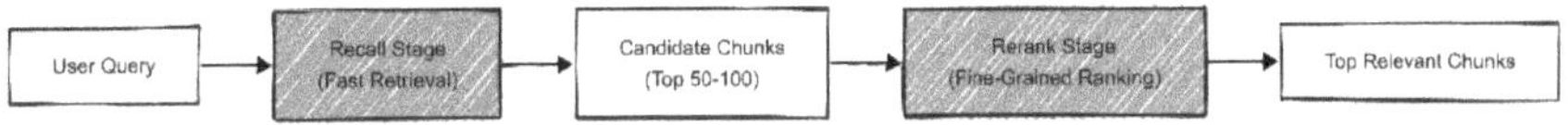

Figure 8.1.   Two stages of ranking.

The necessity of reranking arises because not all contexts retrieved by the initial search are equally relevant; some may be directly relevant, others marginally related, or even irrelevant. By effectively evaluating and prioritizing these contexts, re-ranking helps optimize the efficiency of the answer generation process, enhancing both the speed and accuracy of the LLM's outputs.

The overall idea is similar to the traditional recommendation system that uses a two-stage approach, i.e., recall stage and reranking stage,[1] as shown in Figure 8.1. In Stage 1 of Recall Retrieval, the process begins with one or few fast approaches, e.g., BM25 and approximate nearest neighbor search using vector similarity metrics, to retrieve a broad set of candidate documents (e.g., top 50–100). This over-retrieval strategy prioritizes recall, ensuring that potentially relevant information is not missed. In Stage 2 of Reranking, it involves applying a more precise, albeit computationally intensive, reranking model to the initial set of candidates. This model reassesses the relevance of each document concerning the query, often using advanced techniques such as cross-encoders[2] or LLM-based evaluators. The reranker assigns new relevance scores and reorders the documents accordingly, selecting the most pertinent ones (e.g., top 5–10) to provide as context to the LLM. The reranking stage optionally considers the computational cost and response constraints. Often, the results in the first stage using the hybrid/fusion approach can be pre-ranked directly.

---

[1]The recall stage is also named as pre-ranking, coarse-ranking, fast-retrieval or initial-retrieval stage, and the recall stage is also called fine-grained ranking.

[2]In a cross-encoder setup, the model takes a pair of texts—a query and a candidate document—as a single input. The self-attention mechanism of the Transformer architecture can then analyze the tokens of the query and the document together, allowing for a deep, fine-grained understanding of their semantic relationship. The model's final layer then outputs a single relevance score, indicating how well the document answers the query.

Typical (re)ranking methods use either statistical methods like Reciprocal Rank Fusion (RRF)[3] and weighted fusion[4] to combine multiple retrieval scores (usually applied in the pre-filtering stage) or ML-based methods like cohere cross-encoders to link close between queries and documents. From engineering point of view, we can further look at the following two types:

- **Specific Re-Ranking Models:** These models utilize interaction features between the documents and the query to gauge relevance with greater precision. By analyzing the dynamics of this interaction, re-ranking models can discern subtleties in relevance that might be missed by more straightforward retrieval algorithms. This method is usually applied to vectors.
- **LLM Integration:** The integration of LLMs into re-ranking introduces a profound ability to understand the content at a semantic level. LLMs can process the entire content of both the document and the query, capturing deep semantic information that extends beyond keyword matching. This comprehensive understanding allows LLMs to re-rank the results based on a nuanced interpretation of relevance, thereby improving the contextual alignment between the query and the retrieved documents. This method is often directly applied to retrieved texts.

---

[3]RRF is a rank-based aggregation technique designed to merge the results from multiple retrieval sources into a unified ranked list. This is particularly valuable in hybrid retrieval systems, where results from different modalities, such as sparse retrieval (e.g., BM25) and dense retrieval (e.g., BGE, BCE embeddings), are inherently incomparable due to different scoring distributions. Instead of relying on absolute relevance scores, which vary across retrieval models, RRF operates purely on relative rankings. This makes it robust and model-agnostic, allowing the combination of multiple ranked lists without any need for score normalization or retraining. The detailed formulation can be found in the supplementary material.

[4]**Weighted fusion** combines the scores from different retrieval systems, such as sparse (e.g., BM25) and dense (e.g., embeddings), by assigning them different weights (e.g., 0.3 for BM25 and 0.7 for embeddings) and summing them to compute a final relevance score, usually after normalizing the individual scores. These weights can be set heuristically, tuned experimentally, or learned via machine learning (similar to an assemble approach). This approach effectively balances the precision of keyword matching with the recall strength of semantic retrieval, often yielding more robust and accurate results, especially for diverse queries that vary in lexical and semantic focus.

Pre-ranking (Recall) and rerank models differ primarily in their trade-off between efficiency and precision. Recall models, such as BM25 or Bi-Encoder models like SBERT, encode the query and documents independently, then compute similarity scores to quickly retrieve relevant candidates from large corpora. This process is fast and scalable, akin to two judges independently scoring an essay. In contrast, rerankers (typically Cross-Encoders) concatenate the query and each candidate document as a single input to a deep model (e.g., BERT or RoBERTa), enabling fine-grained token-level interaction through attention mechanisms. This allows for more precise semantic matching, like a single judge carefully comparing the essay to the prompt word by word. While rerankers offer significantly higher-ranking accuracy, their computational cost is much higher, so they are usually applied only to the top-k results from the recall stage. Some common reranker models include Cohere Rerank, BAAI's bge-reranker series, Qwen reranker series, and Jina AI's Reranker, all of which are optimized for deep re-ranking tasks. Table 8.2 lists some commonly used open source reranker models and Table 8.3 compares their efficiency and suitable scenarios.

Re-ranking, therefore, plays a pivotal role in enhancing the functionality of RAG systems, acting as an intelligent layer that refines the initial retrieval results to ensure that the subsequent generation phase is based on the most relevant and contextually appropriate information available. This step is crucial for maintaining the efficacy of RAG systems, especially in complex queries where precision and accuracy are paramount.

8.2.1 *Implementation*

A direct and simple approach is to use a prompt as a reranker for LLM. For example, a simplified prompt might be:

```
You are a RAG (Retrieval-Augmented Generation)
  retrievals ranker. Given a query and a
  retrieved text block, evaluate how relevant
  the block is to the query.
## Instructions:

### Relevance Reasoning:
  Briefly explain how the block relates to the
  query. Focus on whether it provides a direct
```

Table 8.2. Overview of common re-ranking models.

Model Name	Developer	Supported Languages	Core Technology	Application Scenario	Description
ColBERT	Open-source community (based on BERT)	English, multilingual	Fine-grained contextual late interaction	Fast and accurate retrieval	Enables scalable BERT-based search over large text collections in tens of milliseconds.
MonoBERT or Mono-T5	Open-source community	English	Pointwise scoring (BERT/T5)	Medium-scale document retrieval tasks	Provides stable performance with reasonable resource consumption. Well-suited for scenarios where budget and computing resources are limited but accuracy remains important.
BGE-Reranker	Beijing Academy of Artificial Intelligence (BAAI)	English, Chinese, multilingual	Optimized cross-attention mechanism	High-precision document reranking	Optimized for question answering and search result refinement. Open-source and high-performing, often recommended for general reranking tasks.

Table 8.3.   Rankers' comparison.

Ranking Model	Characteristics	Effectiveness	Performance
RRF	Simple score aggregation and fusion; ranks results purely based on their position in each retrieval path, discarding original similarity scores. Rule-based and easy to implement. Suitable for performance-critical scenarios.	Average	Fast (suitable for pre-ranking)
ColBERT	Late interaction mechanism that captures rich token-level interaction between queries and documents while maintaining reasonable inference speed. Balances accuracy and efficiency.	Moderate	Moderate (suitable for pre-ranking)
Cross Encoder (BGE-Reranker)	Powerful re-ranking model that evaluates fine-grained relevance between each query-document pair via deep interaction. Best suited for scenarios requiring high accuracy.	Good	Slow (suitable for fine-grained ranking)

```
answer, partial insight, or just background.
Base your judgment only on the text
provided—no external assumptions.
### Relevance Score (choose one):
0 - Irrelevant
0.2 - Slightly Relevant
0.4 - Partially Relevant
0.6 - Fairly Relevant
0.8 - Very Relevant
1.0 - Perfectly Relevant
```

A LlamaIndex implementation can be found in LLMRerank ofllama_index.core.postprocessor.[5]   The LlamaIndex code in

---

[5]More details can be found at the following URL: https://docs.llamaindex.ai/en/stable/module_guides/querying/node_postprocessors/node_postprocessors/. Accessed at 20/06/2025.

ch8\cohere_rerank.py demonstrates this improvement by comparing two approaches:

- **Direct Retrieval:** The system directly retrieves the top N nodes from the index without any additional processing. This often results in inaccurate or suboptimal retrieval because the top nodes might not capture the full context or the most relevant information.
- **Reranked Retrieval:** Instead of directly retrieving only the top N nodes, the system first retrieves the top M (M>N) nodes. It then applies a reranking process using CohereRerank to evaluate and sort these nodes, finally returning the best N nodes. This two-step approach ensures that the most relevant and contextually accurate information is selected, leading to better performance and more precise responses from the LLM. See the code snippet below:

```
    cohere_rerank = CohereRerank(api_key=api_
key, top_n=2) # return top 2 nodes from
reranker

    query_engine = index.as_query_engine(
        similarity_top_k=10, # we can set a
high top_k here to ensure maximum relevant
retrieval
        node_postprocessors=[cohere_rerank],
# pass the reranker to node_postprocessors
    )
```

Furthermore, retriever performance can be enhanced by knowledge filtering. While the metadata may provide a filtering mechanism, and reranking models aid in prioritizing relevant content, these tools don't always prevent irrelevant or misleading information from being included in the knowledge used by LLMs, negatively affecting the final output. This limitation highlights the need for a knowledge filtering module that acts as a post-ranking checkpoint to further assess the relevance of retrieved content to the given question. One effective approach is to frame this as a Natural Language Inference (NLI) task, where the system determines whether a retrieved passage logically supports the question.

The key advantage of using a knowledge filtering module is that it allows for the training of a binary classification model based on task-specific or domain-specific data. Compared to the cost and complexity of training a full-scale ranking model, this is significantly more efficient and cost-effective. Moreover, the filtering module can be designed to be plug-and-play, making it easy to activate or disable depending on the use case or business requirements. This modularity enhances the flexibility and adaptability of the RAG pipeline, especially in dynamic enterprise environments. A simple module using prompt can be described as follows.

```
Your task is to solve an NLI problem. Given a
  premise (retrieved knowledge) and a
  hypothesis (the question), determine whether
  the knowledge contains a reliable answer
  that supports answering the question.
## [Instruction]
You should classify the relationship as one
  of the following:
Entailment (supports the question)
Contradiction (conflicts with the question)
Neutral (unrelated or insufficient)

## [Question]:
{Insert the user question here.}
## [Knowledge]:
{Insert the retrieved knowledge passage here.}
## [Format]:
Explanation: {Your reasoning here.}
## NLI Result:
{Entailment / Contradiction / Neutral}
```

8.2.2 *Ranker Fine-Tuning*

A reliable method to enhance the accuracy of your retrieved results is to finetune a cross-encoder for re-ranking, especially if you do not fully trust your base encoder. The process works as follows:

1. **Retrieval of Top k Chunks:**
   o Retrieve the top k text chunks based on the initial query using the base encoder.
2. **Cross-Encoder Processing:**
   o Pass the query and each of the top k retrieved text chunks to the cross-encoder, separated by a SEP token.

3. **Fine-Tuning:**
   o Fine-tune the cross-encoder to output 1 for relevant chunks and 0 for non-relevant chunks. This step involves training the cross-encoder on labeled data to learn the distinction between relevant and non-relevant chunks.

In practice, users can easily adopt the LlamaIndex or LangChain's rerank code with the different models. For example, the approach detailed in "Boosting RAG: Picking the Best Embedding & Reranker models"[6] provides valuable insights into selecting the most effective embeddings and rerankers to improve retrieval quality. Additionally, if higher precision is needed, you can fine-tune a custom reranker for your specific use case. The implementation guide, "Improving Retrieval Performance by Fine-tuning Cohere Reranker with LlamaIndex"[7] offers a step-by-step methodology for tailoring a reranker to your retrieval pipeline. This customization can lead to even better retrieval performance by adapting the model to the nuances of your data, ensuring that the most relevant information is accurately captured and passed on to the LLM. More information can be also found in Chapter 12.

8.3 Long Context and Compression

In RAG systems, practitioners often face two significant challenges on context:

1. **Context Length Limit of LLMs:** LLMs typically have a maximum context length they can process effectively. As the input text lengthens, the processing becomes more time-consuming and costly, posing efficiency and scalability issues.
2. **Irrelevance of Retrieved Contexts:** Often, the retrieved information may not be entirely relevant to the query at hand. It's common that only a segment of a larger chunk directly pertains to the user's question. Moreover, to construct a comprehensive answer, it may be necessary to synthesize information from multiple different chunks.

---

[6]https://blog.llamaindex.ai/boosting-rag-picking-the-best-embedding-reranker-models-42d079022e83.

[7]https://blog.llamaindex.ai/improving-retrieval-performance-by-fine-tuning-cohere-reranker-with-llamaindex-16c0c1f9b33b.

This challenge can persist even with advanced re-ranking strategies which aim to improve the relevance of retrieved documents.

To mitigate these issues, context or prompt compression has been introduced as a strategic solution. This technique focuses on distilling the key information from the extensive prompt data, thereby increasing the value of each input token. This streamlined approach not only enhances the LLM's performance by ensuring it processes only the most critical information but also reduces computational costs and improves response times. Prompt compression can also be applied directly to the retrieved contexts. This application directly condenses the extracted data before it is even input into the LLM, further optimizing the efficiency of the RAG process.

By implementing prompt compression, RAG systems can handle larger volumes of data more effectively, ensuring that the LLMs focus on the most salient information, thus making the retrieval and generation processes more precise and cost-effective. This method is particularly useful in scenarios where precision and efficiency are paramount, enhancing the overall utility of RAG in practical applications.

Overall, prompt compression methods can be divided into four main types:

- Methods based on information entropy, such as Selective Context, LLMLingua, LongLLMLingua [1]. These methods use a small language model to calculate the self-information or perplexity of each token in the original prompt. They then delete tokens with lower perplexity.
- Methods based on soft prompt tuning, such as AutoCompressor and GIST [2]. These methods require fine-tuning of the LLM parameters to make them suitable for specific domains, but cannot be directly applied to black-box LLM.
- Methods that carry out data distillation from LLM first, then train models to generate more interpretable text summaries. These can be transferred between different language models and applied to black-box LLMs that do not require gradient updates. The representative methods are LLMLingua-2 and RECOMP [3].
- Methods based on token merging or token pruning, such as ToMe and AdapLeR [4]. These methods usually require model fine-tuning or generating intermediate results during the inference process.

8.3.1 *Longllmlingua*

Prompt compression for long-context scenarios was first introduced in the LongLLMLingua paper [1]. Through its integration with LlamaIndex, this technique can now be applied as a node postprocessor. In practice, LongLLMLingua compresses retrieved context before passing it to the LLM, resulting in improved performance at a significantly reduced cost. Moreover, this optimization accelerates the overall system's processing speed.

Ch8\longllmlingua.py is a sample code demonstrating the setup of the **LongLLMLinguaPostprocessor**, which leverages the **longllmlingua** package to perform prompt compression. For more in-depth details and a complete walkthrough of the process, please refer to the full notebook on LongLLMLingua.[8] Below is the code snippet:

```
node_postprocessor =
  LongLLMLinguaPostprocessor(
    instruction_str="Given the query, please
  identify and keep the most relevant
  information.",
    target_token=300,
    rank_method="longllmlingua",
    device_map="cpu",
    additional_compress_kwargs={
        "condition_compare": True,
        "condition_in_question": "after",
        "context_budget": "+100",
        "reorder_context": "sort",
    },
  )
```

8.3.2 *LongContextReorder*

LLMs tend to process information more effectively when it is located at the beginning or end of an input context, often overlooking crucial data situated in the middle. This phenomenon, commonly referred to as the

---

[8]https://docs.llamaindex.ai/en/stable/examples/node_postprocessor/LongLLMLingua.html#longllmlingua.

"lost in the middle" problem, can lead to significant performance degradation, especially in tasks that require the model to extract or reason over information from extensive textual inputs.

To address this issue, the **LongContextReorder** technique has been developed. This method involves reordering retrieved documents or data segments such that the most relevant pieces are positioned at the beginning and end of the input context, while less pertinent information is placed in the middle. By restructuring the input in this manner, LLMs are better able to access and utilize essential information, thereby mitigating the "lost in the middle" effect.

Implementing LongContextReorder is particularly beneficial in scenarios where a large number of documents are retrieved (a high top-k value). In such cases, without reordering, critical information might be buried in the middle of the context, leading to suboptimal model performance. By strategically rearranging the input, LongContextReorder enhances the model's ability to focus on and effectively process the most relevant information, improving overall performance in tasks involving long-context inputs.

```
from llama_index.core.postprocessor
import LongContextReorder

reorder = LongContextReorder()

reorder_engine = index.as_query_engine(
    node_postprocessors=[reorder],
similarity_top_k=5
  )

reorder_response = reorder_engine.
query("how many scores did the basketball
player get in the first round?")
```

8.3.3 *AutoCompressor*

The AutoCompressor [2, 5] represents an innovative approach to managing long documents within language models, utilizing a soft prompt-based mechanism to enhance the efficiency and efficacy of information processing. This method smartly refines an existing model by introducing new techniques for condensing information, crucial for handling extensive texts in RAG applications. Here's how the AutoCompressor works, as depicted in the diagram of its architecture:

- **Expand Vocabulary:** The first step involves enriching the model's existing vocabulary with "summary tokens." These special tokens are designed to encapsulate substantial information into condensed, manageable units, essentially allowing the model to handle more data without overwhelming the system.
- **Split Document:** In this stage, the document is segmented into smaller pieces. Each segment is subsequently supplemented with summary tokens which inherit and carry forward the summary information of all previous segments. This method ensures that each part of the document is contextualized, maintaining a continuous thread of summarized data throughout the document.
- **Fine-tuning Training:** AutoCompressor applies an unsupervised training approach, specifically utilizing a "next word prediction" task. This training is aimed at fine-tuning the model to predict subsequent words based not just on the immediate tokens but also on the summary vectors derived from earlier segments of the document. This step helps the model to integrate and understand the condensed context provided by the summary tokens and vectors.
- **Backpropagation:** To efficiently manage the computational load, AutoCompressor employs backpropagation through time (BPTT) along with gradient checkpointing for each segment. This strategy reduces the size of the computational graph needed for processing. By performing backpropagation across the entire document, the model is enabled to learn and understand the associations across the full context of the document, thereby improving its overall predictive and contextual capabilities.

Through these steps, AutoCompressor effectively reduces the cognitive load on the model while ensuring that no critical information is lost. This makes it an invaluable tool for enhancing LLMs tasked with processing extensive texts, ensuring they remain efficient and effective even with increased data volume. The code is available at Github.[9]

---

[9]https://github.com/princeton-nlp/AutoCompressors.

8.3.4 *RECOMP*

RECOMP (Retrieve, Compress, Prepend) [2] is a sophisticated approach designed to enhance the efficiency and effectiveness of LLM in processing extensive text data. This method leverages two specialized types of compressors—extractive and abstractive—to streamline the input data before it is fed into an LLM. Below is an overview of the RECOMP system and its components, as shown in the accompanying figure which outlines its architecture.

Extractive Compressor: The extractive compressor within RECOMP functions by selecting the most useful sentences from a set of retrieved documents. It operates using a dual encoder model where each sentence, denoted as (s_i), from the input document set [(s_1, s_2, …, s_n)] is embedded alongside the input sequence (x). These embeddings are then evaluated through the inner product, which determines the relevance and utility of adding (s_i) to (x) for generating the desired output sequence. The top (N) sentences, ranked based on their inner product scores with the input, are compiled into a final summary (s), effectively condensing the essential content to be processed by the LLM.

Abstractive Compressor: On the other hand, the abstractive compressor utilizes an encoder-decoder model to synthesize information from the entire set of retrieved documents. This model processes the concatenation of the input sequence (x) and the retrieved documents, producing a comprehensive summary (s). The training process for this compressor involves generating a preliminary dataset using a powerful LLM, such as GPT-3, which is then refined through filtering to enhance the relevance and quality of the training data. The encoder-decoder model is subsequently trained on this curated dataset to optimize its summarization capabilities.

Implementation and Enhancement: In the context of applying these methodologies, the LongLLMLingua approach has been identified as particularly effective and has been integrated into our ongoing research project. Additionally, an alternative model, LLMLingua-2, could also be considered due to its benefits in processing speed and memory efficiency. These advanced tools within the RECOMP framework significantly enhance the LLM's ability to handle and interpret large volumes of text by pre-processing and condensing the information, thus ensuring more accurate and contextually relevant outputs. The code is available at Github.[10]

8.3.5 *Rolling Summary*

Note that in the multi-round chat system or agentic workflow, we might encounter similar issues. A naïve approach is to rely on a summarizer node that condenses all accumulated dialogue into a shorter form. However, this itself can fail: once the conversation grows too long, the summarizer is also forced to process an oversized input, leading to inefficiency, timeouts, or even system freezes. In essence, attempting to solve the long-context problem with another long-context process creates a paradox.

A more effective solution is the rolling summary technique. Instead of feeding the entire dialogue history into the model, the conversation state is split into two parts: (1) a continuously updated summary capturing long-term memory, and (2) a small buffer of the most recent exchanges. Each time summarization is triggered, only the prior summary and the latest few messages are merged into a new compact summary, and the short-term buffer is cleared. This incremental update ensures that no component ever processes the full, unbounded dialogue history.

For decision-making, the orchestrator or reasoning module no longer consumes raw, lengthy logs. It bases its actions on the rolling summary for context and the most recent user input for immediacy. This dramatically reduces token usage, prevents overload, and allows conversations to continue indefinitely. By design, every LLM call deals only with small, bounded input, making the system both efficient and robust.

The advantages extend beyond performance. Rolling summaries enable scalable memory management, reduce costs, and support production-level dialogue agents that can engage in unlimited conversations. While summarization inevitably risks some information loss, the approach can be refined with techniques such as intent recognition, retrieval augmentation, or hierarchical memory structures, depending on the business scenario. This makes rolling window techniques a practical foundation for building sustainable multi-round chat systems. Below is a

---

[10]https://github.com/carriex/recomp.

simplified version on the summary part only. A detailed pseudo-code applied to agent (tool calling) is available at /codes/ch8 /rolling_summary.md.

```
STATE:
  summary     // accumulated long-term summary
  messages[] // recent short-term messages

FUNCTION SummarizeIfNeeded():
  if messages is empty:
      return

  if messages.length > N OR
  TokenCount(messages) > MAX_TOKENS:
      // Merge old summary with new messages
      prompt = BuildSummarizerPrompt(summary,
messages)
      resp    = LLM(prompt)

      summary  = ExtractSummary(resp)    //
update long-term memory
      messages = []                              //
clear short-term buffer
```

Context enhancement techniques are applied when we need to perform fine-grain document chunking such as splitting text at the sentence level. Such techniques can enhance retrieval precision by enabling the system to match queries with highly relevant content. However, this approach may lead to a loss of contextual information as individual sentences often lack sufficient background to fully inform the LLM responses. To address this challenge, some advanced context enrichment techniques, such as Sentence Window Retrieval and Auto-Merging Retrieval, may be utilized. These techniques were previously introduced in Chapter 7.

8.3.6 *REFRAG*

A REFRAG-style RAG argues that "most computations over the RAG context during decoding are unnecessary and can be eliminated with minimal impact on performance." whose pipeline contains the following steps [6]:

- **Compress:** Turn each document chunk into a tiny vector (an embedding) so long contexts don't bloat the prompt. Prompt-compression is a known speed/cost lever.
- **Project:** Map those chunk vectors into the **decoder's token-embedding space**, so the decoder can attend to them like compact "tokens."
- **Sense & select:** Score which compact chunks look most useful; keep only the best few (policy/heuristics).
- **Expand:** For the chosen chunks, expand from compact vectors to richer, per-token representations the decoder can use directly (design choice to recover detail while staying lean).
- **Decode & measure:** Generate the answer while tracking **Time to First Token (TTFT)** for responsiveness and **throughput/tokens-per-second** for capacity; most tooling also uses **inter-token latency (ITL/TPOT)** rather than "TTIT."

Its implementation is available in github.[11] Another interesting idea is to compress long contexts by converting text into high-resolution images ("optical 2D mapping") and decoding them with a vision-language model [7].

8.4 Concluding Remarks

This chapter primarily operates after retrieval to maximize answer quality within a query-based fusion. Chunk re-ranking and filtering move from fast recall (e.g., vector + BM25) to precision with RRF, cross-encoders or entailment checkers and deduplicating near-copies. Context enhancement enriches evidence by stitching entities and timelines, normalizing tables, propagating citations, and adding light metadata (source, time, jurisdiction) so the generator can reason with structure, not just text. Content compression then distills the remaining context (e.g., via salience-aware extraction, query-focused summaries, key-sentence selection, and citation-preserving pruning) so the LLM receives compact, high-signal "evidence atoms." Together, these steps yield tighter grounding, lower

---

[11]https://github.com/simulanics/REFRAG.

latency, and more stable answers in production. Note that the multi-query rewriting (sub-questions, perspectives) and hypothesis-driven decomposition introduced in Chapter 6 can also improve coverage of long-tail or composite queries while enabling selective retrieval per sub-task.

There are many interesting existing augmentation techniques that are not discussed in deep detail here, e.g.,

- **Hybrid recall with diversity control:** Combine dense + sparse retrievers; enforce diversity via Maximal Marginal Relevance (MMR),[12] semantic clustering, or de-dup at passage and sentence levels to avoid redundancy and topic collapse. Or use tiered rerankers (bi-encoder $\rightarrow$ cross-encoder/entailment). Expect measurable lift on multi-hop QA and fewer near-duplicate passages.
- **Cross-encoder re-ranking & entailment filtering:** Use strong cross-encoders/rerankers to score relevance; add **entailment/consistency filters** to drop passages that conflict with known facts or contradict each other.
- **Query-hierarchy augmentation** (RAPTOR-style) to surface section-level evidence and reduce brittle chunking assumptions; pair with faithful compressors for stable latency.
- **Graph- and schema-aware augmentation:** Build lightweight knowledge graphs (entities, relations, time) or table schemas; inject relation-level snippets (e.g., Subject–Relation–Object tuples) that guide reasoning.
- **Salience-aware compression:** Query-conditioned extractive summaries, sentence-level scoring, section headers as anchors, and citation-preserving pruning (keep the claim + the proof) to reduce tokens without losing support.
- **Answer-sketch prompting:** Create a minimal outline (claims, constraints, required citations) and let retrieval fill the outline slots; improves faithfulness and keeps generation on-rails.

---

[12]MMR is a re-ranking strategy designed to balance relevance to the user's query with diversity among selected results, thereby reducing redundancy and improving the breadth of information. The algorithm begins with an initial relevance-based ranking and proceeds iteratively: at each step, it selects the document that maximizes a combined score, favoring items that are both highly relevant and dissimilar to those already chosen. This process yields a re-ordered list that maintains query relevance while ensuring varied and non-redundant results. See the formulation in supplementary material.

- **Cost/latency-aware routing:** Dynamic top-k, early-exit on high-confidence hits, or tiered rerankers (cheap → expensive) to meet SLOs while preserving quality.
- **Safety & provenance augmentation:** Inline PII redaction, jurisdiction tagging, and **source signatures** (URL/hash/time) so every sentence can be traced back to evidence.

For logits-based fusion RAG, augmentation will occur at decoding time by mixing token distributions from multiple contexts or sub-agents, beyond concatenating text. Promising techniques include mixture-of-contexts at the logits level (weighted by retrieval confidence, recency, or entailment scores), token-gating (suppressing tokens unsupported by any evidence stream), dynamic logit aggregation (weak-to-strong and logits arithmetic methods show that learned, time-varying weights over token distributions can outperform static blending, suggesting practical knobs for evidence-conditioned decoding) and variance-reduction ensembles or consensus decoding (agreement-weighted logits across reranked context sets). These methods keep prompts lean while letting the model "listen" to multiple evidence streams during generation.

For latent fusion RAG, we will increasingly fuse in embedding/attention space rather than in tokens, e.g., latent pooling over retrieved passages (attention-weighted centroids or learned aggregators) to form a compact task-specific latent, constraint-aware latent editing (e.g., pushing the latent toward time/jurisdiction vectors) or hierarchy-aware latent (RAPTOR's abstraction levels naturally serve as multi-scale latents; pairing them with compressors gives controllable trade-offs between faithfulness and cost), and graph-attentive fusion where message passing over an evidence graph yields a condensed latent prior to decoding. Latent fusion reduces prompt bloat and enables fine-grained control over which evidence dimensions drive the answer.

We may also expect some promising augmentation directions, e.g.,

- **Unified policy for augmentation:** Expect end-to-end, learnable policies that select *how much to retrieve, which to compress, and how to fuse*, trained on multi-objective rewards (faithfulness, latency, cost, privacy). This replaces static heuristics with adaptive decisions per query and user.
- **Uncertainty-aware fusion:** Calibrate retrieval and generation confidence; expand context only when uncertainty is high; trigger stronger

rerankers or secondary modalities (tables, images) when risk of error exceeds a threshold.

- **Personalization with edge privacy:** Maintain per-user preferences and histories to steer rewriting/reranking, but execute compression and short-term memory on device or within a private enclave; synchronize only anonymized signals.
- **Structure-first augmentation:** More table-/graph-/temporal-aware snippets will become default so the LLM reasons over entities, relations, and intervals rather than free text alone—especially in regulated domains.
- **MoA-style multi-agent fusion, with caution:** Mixture-of-Agents can lift quality via layered critics or specialists, but recent analyses warn performance is sensitive to component quality; "Self-MoA" (ensembling a *single* top model's diverse samples) is a competitive, simpler baseline.

References

[1] H. Jiang, Q. Wu, X. Luo, D. Li, C.-Y. Lin, Y. Yang and L. Qiu, *LongLLMLingua: Accelerating and Enhancing LLMs in Long Context Scenarios via Prompt Compression,* arXiv:2310.06839v2, 2024.

[2] A. Chevalier, A. Wettig, A. Ajith and D. Chen, *Adapting Language Models to Compress Contexts,* arXiv:2305.14788v2, 2023.

[3] F. Xu, W. Shi and E. Choi, *RECOMP: Improving Retrieval-Augmented LMs with Compression and Selective Augmentation,* arXiv:2310.04408v1, 2023.

[4] A. Modarressi, H. Mohebbi and M. T. Pilehvar, AdapLeR: Speeding up inference by adaptive length reduction, in *Proceedings of the 60th Annual Meeting of the Association for Computational Linguistics,* 2022.

[5] S. Verma, *Contextual Compression in Retrieval-Augmented Generation for Large Language Models: A Survey,* arXiv:2409.13385v1, 2024.

[6] X. Lin, A. Ghosh, B. K. H. Low, A. Shrivastava and V. Mohan, *REFRAG: Rethinking RAG based Decoding,* arXiv:2509.01092v2, 2025.

[7] H. Wei, Y. Sun and Y. Li, *DeepSeek-OCR: Contexts Optical Compression,* arXiv:2510.18234, 2025.

Chapter 9

Generation Techniques

9.1 Introduction

Generation in RAG systems usually involves synthesizing an answer by combining the user's query with the context retrieved in the previous step (and into a pre-defined prompt template in some cases). There are usually three ways in terms of the synthesis or integration [1–2]:

- **Input-Layer Integration:** This method combines retrieved documents or external information with the original query (input prompt), passing them jointly to the generation model. For example, models like In-Context RALM [3] concatenate the input and retrieved material into one long prompt. This simplest way directly influences the model's context and prompt, effectively guiding attention from the very first token, subject to the model's context window limit and may lead to truncation or degraded attention if inputs are too long.
- **Output-Layer Integration:** This post-hoc method involves blending or combining outputs after generation. For instance, the generation model produces its result, which is then integrated with retrieved content via reranking, posterior fusion, or subsequent filtering. A further generation might be applied. This method offers flexibility to re-weigh or adjust outputs dynamically. A typical example is Fusion-in-Decoder (FiD) [4–5].
- **Intermediate-Layer Integration:** A more advanced approach, this integrates retrieved information within the generation model's internal layers, such as inserting a memory-augmented module or side-network

that interacts with intermediate hidden states. This approach enables the model to incorporate retrieved knowledge during the reasoning process: not just at the beginning or end. It can potentially enhance coherence, factual grounding, and learning efficacy, by aligning retrieved context with latent representations. However, it also introduced additional architectural complexity and requires bespoke module training or fine-tuning.

Input-layer and output-layer generation are loosely coupled with retrieval, while intermediate-layer integration tightly coupled. Table 9.1 provides a brief comparison. In this chapter, we mainly discuss some techniques applied in the first two approaches, while leaving the third approach for Chapter 12.

For the first and second approach, as the complexity and length of retrieved contexts grow, more sophisticated generation techniques become necessary to improve answer quality and manage prompt limitations. Here are some advanced approaches:

1. **Iterative Refinement:** Instead of processing the entire context in one pass, the system divides the retrieved context into manageable

Table 9.1.   Comparison of generation methods.

Method	Pros	Cons
Input-Layer Integration	• Straightforward to implement. • Influences generation from the start. • Compatible with black-box models.	• Limited by prompt/context window. • Risk of truncation or lost context. • Information may be overwhelmed by prompt noise.
Output-Layer Integration	• No internal model changes needed. • Flexible post-processing. • Easily tunable.	• Doesn't guide internal generation steps. • Potentially misaligned reasoning. • Can be less robust in factual grounding.
Intermediate-Layer Integration	• Integrates knowledge deeply into model reasoning. • Can improve coherence, grounding, and response quality. • Customizable control over layer interactions.	• High architectural and training complexity. • Requires model internal access and additional training. • Risk of instability or overfitting in intermediate modules.

chunks. The LLM processes each chunk sequentially, generating partial answers or intermediary summaries. The system then iteratively refines these partial outputs, combining them into a final answer. This method is particularly effective when handling long documents that exceed the LLM's token limit, as it allows the model to gradually build up a coherent response. Employing chain-of-thought strategies helps the LLM reason through the information step-by-step, leading to more transparent and accurate intermediate reasoning paths that contribute to a better final answer. The similar technique has been discussed in Chapters 7 and 8.

2. **Context Summarization or Compression:** When the total retrieved context is too large for a single prompt, summarization techniques come into play. The system first compresses the retrieved information into a concise summary that retains the critical details and factual content. This summary is then combined with the original query to generate the final answer. Summarization not only helps manage prompt length but also reduces noise by focusing on the most relevant information. This technique has been discussed in Chapter 8.

3. **Multiple Answer Generation and Consolidation:** Another strategy involves generating multiple candidate answers from different segments or variations of the retrieved context. For instance, the system can process separate context chunks independently and generate an answer for each. These multiple answers are then either concatenated, re-ranked, or further summarized to form a single, coherent response. This approach leverages diversity in responses, helping to mitigate biases or errors that may arise from any single context segment. One of those approaches is named Self-Consistency Mechanisms. By generating multiple outputs with slight variations, then selecting the most consistent answer, self-consistency approaches improve reliability and mitigate hallucination issues (it will be discussed in detail in Chapter 13). The re-ranking of post-processing introduced in Chapter 8 is also helpful, e.g., after generating multiple candidate responses, advanced post-processing techniques can be used to re-rank and refine the final output based on quality metrics such as coherence, factuality, and context adherence.

Therefore, this chapter will spend more spaces on the response synthesis, generated output parsing for the downstream applications and the generator context after a glance at the simple implementation of input layer integration.

9.2 Response Synthesis

For the implementation of input-layer integration, let's use LlamaIndex as an example, which provides several ways in its Response Synthesizer component.

- **simple_summarize:** This method truncates all text chunks to fit into a single LLM prompt, offering a quick summary but potentially losing detail due to truncation.
- **no_text:** This strategy only operates the retriever to fetch nodes, without sending them to the LLM. The fetched nodes can be inspected by examining response.source_nodes.
- **context_only:** This approach simply returns a concatenated string of all text chunks, providing a straightforward aggregation of the retrieved content.
- **accumulate:** Applies the same query to each text chunk individually, accumulating the responses into an array, and then returns a concatenated string of all responses. This is suitable when distinct responses from each text chunk are needed.
- **compact_accumulate:** Similar to the accumulate method, but it compacts each LLM prompt similar to the compact method before applying the same query to each text chunk. This ensures a more concise summarization while maintaining individual responses.

For example, we can use "compact" response mode of Response Synthesizer in the code ch9\response_conpact.py:

```
response_synthesizer = get_response_
  synthesizer(response_mode=
  "compact")

response = response_synthesizer.
  synthesize("how to build a machine learning
  model efficiently?", nodes=nodes)
```

Customized prompt templates are also possible with additional variables using PromptTemplate and TreeSummarize of LlamaIndex. A sample code is available in ch9\custom_prompt_template.py:

```python
qa_prompt_tmpl = (
    "Context information is below.\n"
    "---------------------\n"
    "{context_str}\n"
    "---------------------\n"
    "Given the context information and not
 prior knowledge, answer the query in the
 tone of {tone_name}.\n"
    "Query: {query_str}\n"
    "Answer: "
)
qa_prompt = PromptTemplate(qa_prompt_tmpl)

# initialize response synthesizer
summarizer = TreeSummarize(verbose=True,
 summary_template=qa_prompt)
```

9.3 Fusion in Decoder

FiD [4] is a RAG pattern where each retrieved passage is encoded independently and the decoder fuses evidence at generation time. Instead of concatenating all passages into one long input, FiD keeps separate encoder memories per document and lets the decoder's cross-attention look across the concatenated memories when predicting every next token. This makes fusion an output-layer operation: the model decides which pieces of evidence matter exactly when it is choosing the next word.

During decoding, the model repeatedly performs self-attention over the partial answer and cross-attention over all document memories, so different tokens can draw on different passages. Document boundary markers and per-chunk positional resets help the model learn which memory came from which source, improving controllability and provenance. Many systems also bias the fusion using the retriever's scores, e.g., lightly scaling keys/values from higher-scoring passages, so that stronger evidence is more likely to influence the next token without hard-discarding complementary context. This method may be further improved with confidence [6].

FiD shines when answers must synthesize information spread across multiple passages and you want token-level evidence mixing.

Compared with early-fusion concatenation, it avoids truncation and preserves per-document structure; compared with generating per-document candidates (e.g., RAG-Sequence) it is typically more efficient because it runs a single decode step that attends to shared memories. The trade-off is computational: memory and latency grow with the number and length of retrieved chunks, so practitioners cap K, prune low-salience tokens, cache key-value states, or add a light re-ranker before FiD to stay within budget.

We will use FiD when: (1) answers synthesize facts across multiple passages, (2) you need token-level evidence mixing during generation, and (3) you can afford decoder cross-attention over multi-doc memory. Prefer re-rank-then-read or RAG-Seq when latency budgets are tight and single-document answers dominate.

Below is an implementation pseudocode.

```
# enc: encoder; dec: decoder; retrieve(x)->
  list[(d_i, score_i)]
docs, scores = retrieve(x)[:K]

memories = []
for i, (d_i, s_i) in enumerate(docs):
    tokens = pack([D_ID(i), x, SEP, d_i])
  # reset positions per pack
    H_i = enc(tokens)
  # [T_i, d_model]
    g_i = softmax(alpha * torch.
  tensor(scores))[i]   # optional gating
    memories.append(scale_kv(H_i, g_i**0.5))
  # rescale keys/values

K_cat, V_cat = concat_kv(memories)
  # across all docs

y = [BOS]
kv_cache = build_kv_cache(K_cat, V_cat)
  # per-layer cache
for t in range(MAX_LEN):
    q = dec.self_attend(y, cache=True)
```

```
    ctx = dec.cross_attend(q, kv_cache)
# fusion happens here
    logits = out_proj(ctx)
    y.append(sample_or_greedy(logits))
    if y[-1] == EOS: break
```

9.4 Output parsing and JSON Mode

There are various ways to ensure the output meets our requirements. The first approach is to enhance the output description inside the prompt, then we can utilize output parsing and function parameters.

9.4.1 *Enhance Prompt*

Clarify Instructions:

- **Be Explicit:** Clearly state the desired format in the prompt.

Example: "Provide the following information formatted as a markdown table with columns for 'Name', 'Age', and 'Occupation'."

- **Specify the Format Syntax:** Mention any specific syntax or standards to follow, such as JSON, XML, or CSV.

Example: "List the top five programming languages in JSON format with fields 'language' and 'popularity'."

Simplify the Request and Use Keywords:

- **Use Direct Language:** Employ imperative verbs like "list," "summarize," or "create a table."
- **Highlight Keywords:** Emphasize important terms by placing them in quotes or capital letters.

Example: "Please *LIST* the steps to install the software."

Provide Examples with Few-shot:

- **Sample Output:** Include an example of the expected output to guide the LLM, e.g.,

Example:
```
[
  {
    "name": "Apple",
    "average_weight_g": 182,
    "primary_nutrients": ["Dietary Fiber",
"Vitamin C", "Potassium"]
  },
  {
    "name": "Banana",
    "average_weight_g": 118,
    "primary_nutrients": ["Vitamin B6",
"Vitamin C", "Potassium"]
  }
]
```

Iterative Prompting and Follow-Up Questions:

- **Refinement:** If the initial response is incorrect, provide feedback and ask the LLM to adjust.

Example: "Can you please reformat your previous answer as a bullet-point list?"

- **Confirmation:** Ask the LLM to confirm its understanding of the instructions.

Example: "Do you understand how to format the response as a CSV file?"

9.4.2 *Utilize Output Parsing*

To transform unstructured outputs into a structured format, one can utilize post-processing scripts that parse and convert the data accordingly. Additionally, employing regular expressions allows for the extraction of specific data patterns from the language model's responses.

Output interpretation can be applied in these manners to ensure the expected outcome:

- to provide instructions for formatting for any command/request
- to provide "interpretation" for LLM responses

LlamaIndex supports connections with interpretation modules offered by other systems, such as Guardrails and LangChain.

A sample code is available in ch9\output_parsing.py using LangchainOutputParser of LangChain's interpretation modules that you can utilize within LlamaIndex.

```
# define output parser
lc_output_parser = StructuredOutputParser.
from_response_schemas(
    response_schemas
)
output_parser =
LangchainOutputParser(lc_output_parser)
```

A sample output is as follows.

```
{'application': 'Bitcoin', 'methodology':
 'Maintaining privacy through anonymous
 public keys, using new key pairs for each
 transaction, and the risk of revealing
 ownership through multi-input transactions.
 Describing the scenario of an attacker
 trying to generate an alternate chain faster
 than the honest chain, the Binomial Random
 Walk characterization, and the probability
 calculations related to an attacker catching
 up with the honest chain.'}
```

For further information, visit LlamaIndex's documentation on interpretation modules.[1]

---

[1] https://docs.llamaindex.ai/en/stable/module_guides/querying/structured_outputs/output_interpreter.html. Accessed on 07/09/2025.

9.4.3 *Apply Function Calling*

Function calling is a feature that enables LLMs to interact with external functions or APIs by generating structured outputs that can be programmatically parsed and utilized. This capability enables developers to define functions that the LLM can invoke, facilitating tasks such as data retrieval, computation, or integration with external services. Essentially, it handles two things:

1. **Model behavior:** the model learns when *and how* to emit a *structured action* (e.g., { "tool": "get_weather," "arguments": {...} }) instead of free text.
2. **Runtime contract:** the server/API enforces a schema, executes the tool safely, feeds results back as a message, and resumes generation.

Below is a general procedure to use function calling.

*   **Defining Functions:** Developers specify functions by providing the LLM with descriptions of the expected output structure. For instance, consider a function create_user_profile that takes parameters like name (string), age (integer), and occupation (string). By defining this function and its parameters, the LLM understands the required format for the output.
*   **Automatic Parsing:** Once the function is defined, the LLM can generate data in a structured format that aligns with the function's parameters. This structured output can then be programmatically parsed and used within applications, ensuring consistency and reliability in the data handling process.
*   **API Integration:** By leveraging on APIs that support function calling, developers can enforce specific output formats, enabling seamless integration between the LLM and external services. This approach ensures that the data exchanged between the LLM and other components of the application adheres to predefined structures, enhancing interoperability and reducing the likelihood of errors.

Ch9\function_calling.py is a simple example demonstrating how to implement function calling with an LLM using the OpenAI API:

```python
# Define the function schema
functions = [
    {
        "name": "create_user_profile",
        "description": "Create a user profile
 with name, age, and occupation.",
        "parameters": {
            "type": "object",
            "properties": {
                "name": {"type": "string",
"description": "The user's name"},
                "age": {"type": "integer",
"description": "The user's age"},
                "occupation": {"type":
"string", "description": "The user's
occupation"}
            },
            "required": ["name", "age",
"occupation"]
        }
    }
]

# Generate a response from the LLM
response = openai.ChatCompletion.create(
    model="gpt-3.5-turbo-0613",
    messages=[{"role": "user", "content":
 query}],
    functions=functions,
    function_call={"name": "create_user_profile"}
)
```

Note that OpenAI JSON mode permits us to configure response_
format as {"type": "json_object"}, enabling JSON mode for the reply.
When JSON mode is activated, the model is restricted to solely produce
strings that are interpretable as valid JSON objects. Although JSON mode

mandates the output's format, it does not aid in the validation against a designated schema. For further information, visit LlamaIndex's documentation on OpenAI JSON Mode vs. Function Calling for Data Extraction.[2]

When some LLMs, e.g., OpenAI and Anthropic, support function calling natively, some are not. Native support typically requires all of the following; if any are missing, the vendor may not "support" it:

- **SFT data on tool use:** The model was fine-tuned (and often preference-tuned) on dialogs where the assistant chooses tools, formats JSON correctly, and recovers from tool errors.
- **Schema-following & constrained decoding:** The endpoint can force JSON/BNF grammars (or at least strongly bias them) so outputs are syntactically valid.
- **Chat protocol hooks:** The API has roles/messages for tool_call and tool_result and handles multi-turn tool loops.
- **Guardrails & safety:** Built-in allowlists, argument validation, redaction, and rollback.
- **Telemetry & retries:** The stack knows when a tool failed, how to retry/repair arguments, and how to limit tool loops.

Models without native tool use generally lack one or more of: (a) supervised traces for tool calls, (b) decoding constraints, or (c) server-side plumbing. They can still "approximate" tool use with prompting, but we must own schema enforcement, validation, and loops in the application layer (see supplementary material).

Consider a progressive ladder using prompt engineering + thin orchestration:

1. **Prompt-only, schema-first routing:** Instruct the model to output *only* a strict JSON decision (tool or none) with arguments; validate server-side, execute, then re-prompt with results for the final answer. *Fastest to ship; relies on good prompts and validators.*
2. **Constrained decoding (grammar/JSON-Schema/regex):** Enforce output structure at decode time (EBNF/JSON-Schema), minimizing

---

[2]https://docs.llamaindex.ai/en/stable/examples/llm/openai_json_vs_function_calling.html. Accessed on 07/09/2025.

malformed JSON and hallucinated keys. *Higher reliability; needs an inference server that supports constrained decoding.*

3. **Planner–Executor loop (ReAct-style):** Two-stage loop: plan → call tool → reflect → (optionally) call again → finalize; capped by max steps with error-repair/abstention rules. *Best for multi-tool/multi-step tasks; adds latency and orchestration code.*

4. **Lightweight fine-tuning/adapters for tool use:** SFT/LoRA on tool-use traces to improve *when-to-call* judgment, argument fidelity, and chaining; optionally add a tiny classifier head for call vs. abstain. *Most robust; requires data and training but reduces prompt fragility and retries.*

A sample prompt-only snippet might be as follows.

```
System: You are a precise tool router. Output
 ONLY valid minified JSON.

User question: {{query}}
Tools: [
  {"name":"search_docs","args":{"q":"string",
 "k":"integer<=10"}},
   {"name":"get_rate","args":{"ccy":"string",
 "date":"YYYY-MM-DD"}}
]

If a tool helps, return:
{"tool":"<one of the
 names>","arguments":{...}}
Otherwise return:
{"tool":"none","final_answer":"..."}
```

In general, native function-calling is a combination of training and runtime enforcement. If your model lacks it, you can still achieve production-grade tool use with schema-first prompts, constrained decoding, and a thin planner (executor loop) plus standard safety and observability. This is often sufficient until you decide to SFT a "tool-use head" into the model.

9.4.4 *Validate LLM Output*

Ensuring the integrity and correctness of data is paramount in software development. Python offers several libraries tailored for data validation and error handling, in particular, for open source small LLMs. Below is a comparison of some notable libraries:

Library	Description	Key Features
Pydantic	Utilizes Python type hints for data validation and settings management.	Fast, integrates with IDEs, supports complex data types, and auto-generates JSON schemas.
Cerberus	Lightweight, extensible data validation library.	Simple syntax, customizable validation rules, and no external dependencies.
Marshmallow	Converts complex data types to and from Python data types.	Serialization, deserialization, and validation with schema definitions.
Pandera	Statistical data validation for Pandas data structures.	Seamless integration with Pandas, supports dataframe schemas, and checks for data integrity.
Great Expectations	Provides extensive data validation capabilities, especially for data pipelines.	Declarative expectations, data profiling, and detailed validation reports.

Here, we look at Pydantic,[3] which is a robust data validation and settings management library in Python that leverages type annotations to define and validate data structures. By specifying data models using standard Python types, Pydantic ensures that the data conforms to the expected types and constraints, facilitating both parsing and validation. Its integration with Python's type hinting system allows for clear, concise, and maintainable code. Pydantic is known for its speed and efficiency, as its core validation logic is implemented in Rust, making it one of the fastest data validation libraries available. Additionally, Pydantic can automatically generate JSON schemas from the defined models, aiding in the documentation and serialization processes. LlamaIndex provides several types of Pydantic applications that help structure and validate the outputs generated by LLMs. These include:

---

[3]https://github.com/pydantic/pydantic.

- **LLM Text Completion Pydantic Applications:** These applications process input text by sending it to a text completion API. The raw text response is then interpreted and converted into a structured entity that matches the user-defined Pydantic model. This approach ensures that the unstructured output from the LLM is validated and seamlessly transformed into a data structure suitable for downstream tasks.
- **LLM Function Calling Pydantic Applications:** In these applications, the input text is used to trigger function calls via an LLM function calling API. The response is then parsed and converted into a structured entity as specified by the user's Pydantic model. This method is particularly useful when the LLM is used to execute specific functions or commands, with the output being automatically structured in a predictable format.
- **Prepackaged Pydantic Applications:** These applications come with predefined Pydantic models that dictate the structure of the output. They are designed to directly convert input text into these predefined structured entities without requiring additional customization. This offers a plug-and-play solution for quickly integrating structured output capabilities into your workflows.

Each of these applications leverages Pydantic's powerful data validation features to ensure that the LLM output conforms to a desired schema, thereby increasing reliability and simplifying integration with other systems. For more details, you can refer to LlamaIndex's documentation and sample notebooks which demonstrate these concepts in practice.

In scenarios where the LLM's output doesn't meet the specified format, it's essential to implement error handling strategies. One approach is to use the tenacity[4] library to retry the LLM call upon encountering validation errors. This involves defining a retry mechanism that attempts to regenerate the output until it conforms to the expected structure or a maximum number of retries is reached. Another approach combines regular expression and LLM retry.[5]

Let's look at an example in ch9\output_validating,py, where the validation and error handling mechanisms are implemented within the get_validated_user_profile function. The output from the LLM is parsed and

---

[4]https://github.com/jd/tenacity.

[5]https://github.com/junxu-ai/llm_output_parser.

validated against the UserProfile model. If the output doesn't conform to the expected structure, a ValidationError is raised, triggering a retry.

For additional information, refer to LlamaIndex's documentation on the pydantic application[6] for links to the notebooks/guides of the various pydantic applications.

9.5 Context Window in Models

The evolution of LLMs with extended context windows, from initial 4K or 8K tokens to current capacities of 1M or even 10M tokens,[7] has significantly influenced the design and application of RAG systems. Table 9.2 provides a brief comparison of some typical models. This progression presents both opportunities and challenges, necessitating a nuanced understanding of their interplay. The extended context windows of LLM potentially provide the benefits for

1. **Enhanced Retrieval Capabilities:** Larger context windows enable LLMs to process more extensive retrieved content in a single pass. This capacity allows for the inclusion of multiple relevant documents, facilitating comprehensive responses and reducing the need for multiple retrieval cycles.

2. **Potential Redundancy in Certain Scenarios:** In cases where the required information fits within the model's context window, the necessity for a separate retrieval step may diminish. For instance, tasks involving summarization of a single long document might be handled directly by the LLM without external retrieval.

3. **Complementary Strengths:** Despite the capabilities of long-context LLMs, RAG remains valuable, especially when dealing with dynamic or proprietary data not present in the model's training corpus. RAG's ability to fetch up-to-date information ensures that responses remain current and relevant.

---

[6]https://docs.llamaindex.ai/en/stable/examples/output_parsing/openai_pydantic_program.html.

[7]The approximate conversion between characters and tokens in language models is as follows: 1) 1 token $\approx$ 4 English characters or 0.75 English words. 2) 1 Chinese character $\approx$ 0.6 tokens.

Table 9.2. Typical LLM context size.

Model	Input Context Length	Output Token Limit	Notable Details
Anthropic Claude 3.7/4.5 Sonnet	200 K	up to 128 K (thinking mode)	State-of-the-art long context for reasoning
DeepSeek V3 (deepseek-chat) / R1 (deepseek-reasoner)	64 K	8 K	Reasoner variant allows 32K reasoning chain
Gemini 2.5 Pro	1 M	64 K	Latest Google Gemini update
Gemini 2.0 Flash	1 M	8 K	Lightweight Gemini variant
Gemma 3 (Google)	128 K (4 B, 12 B, 27 B); 32 K (1 B)	8 K	Released on HuggingFace
Grok 3 (X.ai)	1 M	16 K	Elon Musk's Grok model
Llama 4 Scout	10 M	—	Approx. 20 million words or 20 hours video context
Llama 4 Maverick	1 M	—	Larger expert model
OpenAI GPT-4.5 Preview / GPT-4o	128 K	16 K	Latest OpenAI mainstream models
OpenAI GPT-5.4	1 M	128 K	Extreme long context experimental model
Qwen2.5 / QwQ (based on Qwen2.5)/ QVQ-Max (Qwen variant)	128 K	8 K	Alibaba's latest LLM / Max input 96K, reasoning chain 32K / Vision + reasoning, max single image input 16,384

When choosing a proper LLM for generation, we should have some strategic considerations, e.g., dynamic vs static data, resource allocation and system complexity. The decision hinges on various factors, including the nature of the task, data characteristics, and system constraints. Table 9.3 is a comprehensive table delineating criteria and corresponding scenarios to guide this selection. Chapter 12 will further discuss Context Engineering.

Table 9.3.　Model selection criteria.

Criterion	Short Context (<8 K tokens)	Medium Context (8 K–64 K tokens)	Long Context (64 K–1 M + tokens)
Task Complexity	Simple queries, FAQs, brief summaries	Moderate complexity tasks, multi-turn dialogues, standard document Q&A	Complex reasoning, extensive document analysis, long-form content generation
Data Freshness	Static or infrequently updated data	Periodically updated data requiring occasional refresh	Frequently changing data necessitating real-time retrieval
Document Length	Short documents or snippets	Medium-length documents, articles, or reports	Long documents, comprehensive reports, entire knowledge bases
Retrieval Granularity	Fine-grained retrieval with precise context injection	Balanced retrieval with moderate context size	Coarse-grained retrieval encompassing large context windows
System Latency Requirements	Low latency applications requiring swift responses	Moderate latency tolerance	High latency acceptable for in-depth processing
Computational Resources	Limited resources, cost-sensitive environments	Moderate resource availability	Abundant resources, high-performance computing environments
Use Case Examples	Chatbots, simple customer support, basic information retrieval	Technical support, educational tools, standard enterprise applications	Legal document analysis, medical research, large-scale knowledge management

9.6　Concluding Remarks

RAG systems leverage a diverse set of generation techniques to enhance factual grounding, reasoning, and output coherence. FiD processes all retrieved passages through the encoder in parallel and combines their representations during decoding, enabling comprehensive multi-document reasoning. Retrieval-conditioned generation ensures the generation process is grounded in the most relevant passages by incorporating learned scoring or re-ranking mechanisms to reduce hallucinations. Contextualized or iterative retrieval with generation schedules interleaved retrieval and generation steps, allowing the system to dynamically refine queries and evidence, particularly effective for multi-hop reasoning tasks (see Chapter 8). Prompt chaining and feedback loops use generated outputs to inform further retrievals or dynamically adjust the prompt, enhancing adaptability when initial retrievals are insufficient (see Chapter 6). Additional strategies such as iterative refinement, context compression, chain-of-thought or multiple-answer generation, and structured output parsing (e.g., with citations or reranking) collectively improve both the relevance and interpretability of RAG outputs.

This main component of RAG continues to advance through a range of promising methods that both refine the current generation process and chart the path forward, including:

- **FiD and its optimized variants:** FiD processes multiple retrieved documents in parallel and fuses their encoder outputs during decoding to improve multi-document coherence. Its enhanced versions, e.g., **FiDO** [4], which achieves up to a 7× speedup in inference, and **KG-FiD** [5], which leverages knowledge graphs and graph neural networks to rerank and reduce noisy passages, significantly elevate both performance and efficiency.
- **Dynamic and parametric retrieval and generation strategies:** Models like **FLARE** actively anticipate upcoming content and iteratively retrieve as needed, allowing adaptive and context-aware generation, while Dynamic & Parametric RAG frameworks determine *when* and *what* to retrieve during generation and explore injecting knowledge at the parameter level for enhanced effectiveness [7].
- **Self-reflective looping through feedback:** Techniques such as Prompt Chaining with Retrieval Feedback employ a structured chain of prompts and real-time feedback to adaptively refine responses based on evolving context, enhancing accuracy especially in interactive or troubleshooting scenarios. Similarly, Self-RAG incorporates reflection tokens that allow the model to critique and adjust its retrieval and generation behaviors mid-process to boost factuality [8–9].
- **Multimodal RAG (MRAG):** Extending beyond text, MRAG systems integrate images, videos, and structured data into retrieval and

generation workflows. This multimodal grounding has demonstrated improved accuracy in tasks demanding both visual and textual understanding.

- **Federated RAG for Privacy and Security:** Federated Retrieval-Augmented Generation employs federated learning to perform distributed retrieval without exposing raw data—ideal for privacy-sensitive domains like healthcare and finance.

- **Dynamic & Parametric RAG:** Dynamic RAG adapts retrieval patterns in real time, determining when and what content to fetch during generation. Parametric RAG advances knowledge injection from the input level into the model's parameters, boosting efficiency and retrieval effectiveness.

References

[1] W. Fan, Y. Ding and et al., *A Survey on RAG Meets LLMs: Towards Retrieval-Augmented Large Language Models*, arxiv:2405.06211v3, 2024.

[2] Y. Gao and et al., *Retrieval-Augmented Generation for Large Language Models: A Survey*, arxiv:2312.10997, 2023.

[3] O. Ram and et al., *In-Context Retrieval-Augmented Language Models*, arXiv:2302.00083v3, 2023.

[4] M. de Jong and et al., FiDO: Fusion-in-Decoder optimized for stronger performance and faster, in *Findings of the Association for Computational Linguistics: ACL 2023*, Toronto, Canada, 2023.

[5] D. Yu and et al., *KG-FiD: Infusing Knowledge Graph in Fusion-in-Decoder for Open-Domain Question Answering,* arXiv:2110.04330v2, 2022.

[6] Y. Fu, X. Wang, Y. Tian and J. Zhao, *Deep Think with Confidence,* arXiv:2508.15260, 2025.

[7] W. Su, Q. Ai, J. Zhan, Q. Dong and Y. Liu, *Dynamic and Parametric Retrieval-Augmented Generation,* arXiv:2506.06704, 2025.

[8] A. Asai, Z. Wu, Y. Wang, A. Sil and H. Hajishirzi, *Self-RAG: Learning to Retrieve, Generate, and Critique through Self-Reflection,* arXiv:2310.11511, 2023.

[9] D. Wu, J.-C. Gu, K.-W. Chang and N. Peng, *Self-Routing RAG: Binding Selective Retrieval with Knowledge Verbalization,* arXiv:2504.01018v1, 2025.

Chapter 10

Evaluation Methodology

10.1 Introduction

Establishing a robust and coherent evaluation framework is critical to the success of any LLM project. This framework should be comprehensive, generalizable across different use cases, and free from contradictions to ensure consistent and reliable results. In a banking system, the fusion of LLMs and RAG enables models to ingest proprietary data like client records, contracts, and regulatory documents. This allows them to provide precise answers to employee queries, whether for resolving client concerns or extracting key terms from dense agreements. Unlike generic chatbots, the tailored approach ensures responses are both contextually rich and strictly confined to verified internal sources, balancing innovation with operational rigor.

- **Precision-Testing for Real-World Impact:** To validate performance, we're stress-testing the system on two fronts: technical benchmarks and human judgment. Quantitative metrics like precision (minimizing irrelevant answers) and recall (capturing all critical details) are paired with qualitative feedback from bankers piloting the tools. Imagine a wealth manager using the system to swiftly compile a client's cross-border transaction history: does the AI surface obscure tax clauses buried in decade-old PDFs? Can it summarize a 200-page M&A contract without omitting force majeure implications? This dual-lens

247

evaluation ensures the technology doesn't just work in theory but excels in the messy reality of financial workflows.

- **Guardrails for Growth:** In banking, trust is non-negotiable. We're wrapping the system in fortress-like security: encryption mimics asset vaults, access controls mirror tiered financial clearances, and audits track every data interaction like a blockchain ledger. But safeguards alone aren't enough as the models shall evolve through iterative learning loops. Each misstep (e.g., misclassifying a GDPR clause) becomes a training moment, refining the AI's grasp of niche financial jargon and shifting regulations. This isn't just about launching a tool; it's about cultivating a self-improving resource that grows sharper with every query, empowering teams to focus less on data hunting and more on strategic client partnerships.

Recent research has underscored the importance of moving beyond traditional evaluation methods. While classical statistical metrics remain valuable, the advent of LLM-based judgment techniques now offers scalable and explainable alternatives that better capture nuanced aspects of language quality. Note that the evaluation may be applied to the individual step (e.g., retrieval or generation) or the overall pipeline or even specifically for Agent.

10.2 Evaluation Methodology

Evaluating the quality of RAG systems and LLM outputs in general can be broadly divided into two approaches: statistical evaluation and LLM-as-a-Judge.

10.2.1 *Statistical Evaluation*

The statistical evaluation method is usually based on quantitative measures comparing the generated output to a predefined "ground truth." Standard metrics include[1]:

---

[1]Please refer to supplementary material for the explanation and formulations of the metrics that mentioned in this chapter.

- **ROUGE Scores:** These measure the overlap between the generated summary and reference texts by evaluating n-gram, word sequence, and longest common subsequence matches. ROUGE variants (e.g., ROUGE-1, ROUGE-2, ROUGE-L) are widely used in summarization tasks.
- **BLEU Scores:** Originally designed for machine translation, BLEU evaluates the precision of n-gram matches between the generated text and reference translations. Higher BLEU scores suggest that the generated output closely resembles the reference text.
- **Perplexity**, a measure indicating how well the model predicts a sample. Lower perplexity values suggest better predictive performance and coherence in the generated text.
- **Precision, Recall, Accuracy and F1 Score:** Precision assesses the ratio of relevant information among the generated content, whereas recall measures the completeness with which the relevant information is captured. The F1 score, as their harmonic mean, provides a balanced overall performance metric.
- Other commonly used metrics include Discounted Cumulative Gain (DCG), Normalized Discounted Cumulative Gain (NDCG), Mean Reciprocal Rank (MRR) and Mean Average Precision (MAP). For every query, Average Precision (AP) is the mean of the precisions observed each time we encounter another relevant document as we scan down the ranked list. MAP is the arithmetic mean of AP over all queries, capturing both the completeness and the order of relevant documents. See Table 17.1 of supplementary material for the comparison of DCG, NDCG and MRR.

These statistical measures provide a robust approach for assessing how well the generated text aligns with expected outputs. They are particularly useful during early development stages and for scenarios where objective, reproducible metrics are essential.

10.2.2 *LLM-as-a-Judge*

LLM-as-a-Judge is to use LLMs themselves as evaluators. This approach leverages a separate LLM to assess the quality of the generated responses based on qualitative criteria such as coherence, relevance, fluency, and adherence to context.

Key characteristics of the LLM-as-a-Judge approach include:

- **Context-Aware Judgment:** By providing an evaluation prompt that includes the original query, retrieved context, and the generated response, the LLM judge can deliver nuanced assessments that go beyond simple word overlap, such as Faithfulness, Answer Relevancy and Correctness.
- **Scalability and Consistency:** Unlike human evaluators who are subject to variability and resource constraints, LLM judges can evaluate thousands of outputs rapidly and consistently. They are particularly valuable for continuous monitoring in production systems, such as Context Relevancy, Context Recall and Context Precision.

It often uses iterative feedback for improvement. The insights provided by an LLM judge (often along with an explanation) can be used to fine-tune both the retrieval and generation components. This feedback loop is crucial for ongoing improvements and aligns with modern evaluation-driven development practices.

Both evaluation methods have their merits. Statistical metrics offer clear, numerical benchmarks that are easy to compare and track over time, whereas LLM-as-a-Judge methods capture the subtle, qualitative dimensions of language that are often overlooked by traditional metrics.

10.3 Evaluation Framework

Evaluating LLM requires a comprehensive framework that integrates both statistical metrics and qualitative assessments to ensure their outputs are accurate, coherent, and user-friendly. This section discusses two frameworks. One is process-based, and the other is layer-based.

10.3.1 *Process-Based Evaluation*

Evaluating RAG holistically requires multiple perspectives: 1) Retrieval phase: Does the system fetch the right context? 2) Generation phase: Given a certain context, does the LLM produce correct, relevant, and faithful answers true to it? 3) End-to-End phase: Does the system satisfy user needs in real-world tasks (holistic utility)?

This structure aligns with academic papers (e.g., RAGAS, TruLens, HELM [1]) and enterprise evaluation frameworks, where evaluation moves from low-level retrieval accuracy $\rightarrow$ mid-level generation quality $\rightarrow$ high-level task usefulness. It is both logical and practical.

a) **Retrieval Phase (Document/Chunk Retrieval):** The goal is to ensure that the retriever fetches documents that are both relevant and sufficient to answer the query. Poor retrieval cascades into poor generation. As discussed in Chapter 7, this phase uses sparse search, vector search, hybrid search or graph-based retrieval. And the retrieval performance is evaluated by comparing retrieved documents with gold-standard relevant documents (if available) or via proxy metrics (similarity scores, coverage). The biases may exist in chunk size, embedding quality, re-ranking strategies, etc. Some typical example metrics are as follows:

   - **Recall@k:** % of queries where at least one ground-truth document is in top-k.[2]
   - **Precision@k:** Fraction of retrieved docs that are truly relevant.
   - **nDCG (Normalized Discounted Cumulative Gain):** Relevance weighted by position in ranked list.
   - **Coverage:** % of ground-truth answer content found in retrieved documents.
   - **Context Diversity:** Avoids redundant chunks.
   - **Latency/Efficiency:** Time to retrieve top-k documents.

b) **Generation Phase (Answer Synthesis & Faithfulness):** The Goal is to ensure the LLM produces high-quality answers using retrieved content, without hallucinations or omission. In this phase, the LLM must ground its answer on retrieved context. Evaluation often requires automatic metrics plus human-in-the-loop verification. The key risks are hallucination, verbosity, and a lack of faithfulness to the retrieved context. Some typical example metrics are as follows:

   - **Faithfulness Score:** % of answer statements grounded in retrieved context. (e.g., RAGAS factual consistency check)

---

[2]The K means the retrieved top K results. In a 10-chunk database, 3 of the retrieved K = 5 is correct, then accuracy@5 = 3/5, recall@5 = 3/10. Precision@K is similar to accuracy@K.

- **Answer Completeness:** Whether all key aspects from retrieved docs are included.
- **Relevance Score:** Semantic similarity between query and generated answer.
- **Conciseness/Fluency:** Measured by readability scores or perplexity.
- **Citation Accuracy:** % of answer sentences that correctly cite supporting context.
- **Hallucination Rate:** % of content not verifiable in retrieved docs.

c) **End-to-End Phase (Task/Utility Evaluation):** The goal is to assess overall usefulness of the RAG system in real-world settings, not just retrieval or generation individually. This phase is often measured with human judgments or task success benchmarks, which includes evaluation of user satisfaction, domain correctness, and business KPIs. One note is that even if retrieval and generation are imperfect, the system can still be useful if it solves user tasks. Some typical example metrics are as follows:

- **Answer Accuracy (Task-level):** Does the final answer solve the query? (compared with gold answers if available).
- **Exact Match (EM)/F1:** For QA tasks with structured answers.
- **Human Satisfaction Scores:** Likert ratings from human evaluators.
- **Task Success Rate:** % of queries solved correctly end-to-end.
- **Robustness:** Performance under adversarial/noisy queries.
- **Cost & Latency:** API calls, GPU usage, response time.
- **User Retention/Engagement:** Real-world feedback metrics.

Table 10.1 summarizes the process-based evaluation methods. To strengthen it, one should use automatic metrics (e.g., recall, faithfulness, F1) for scalability, including human evaluations (satisfaction, task success) for completeness and track efficiency metrics (latency, cost) since they matter in production.

10.3.2 *Layer-Based Evaluation*

The three layers (Foundational Verification → Business Logic/Compliance → User Experience/Communication) form a progressive hierarchy that mirrors how LLMs/RAGs are used in real applications: correctness →

Table 10.1. Process-based evaluation.

Layer	Goal	Example Metrics
Retrieval	Fetches correct and sufficient context	Recall@k, Precision@k, nDCG, Coverage, Context Diversity, Latency
Generation	Produces faithful & relevant answers	Faithfulness Score, Completeness, Relevance Score, Conciseness, Citation Accuracy, Hallucination Rate
End-to-End	Ensures task-level usefulness	Answer Accuracy, EM/F1, Human Satisfaction, Task Success Rate, Robustness, Cost & Latency, User Retention

safety/compliance → user satisfaction. First, check for factual correctness (baseline truth). Then, assess domain-specific logic and compliance (business/regulation alignment). Finally, optimize for human interaction quality (usability, tone, resonance).

1. **Foundational Verification Layer:** This layer validates the factual accuracy and consistency of the LLM-generated content with following typical metrics:
   - **Statistical Metrics:** Utilize the common metrics, e.g., Perplexity (fluency baseline), Precision/Recall@k, and BERTScore (semantic similarity to reference answers)
   - **LLM-as-a-Judge:** Employ advanced models like GPT-4o/ DeepSeek-v3 to cross-verify factual elements within the text, such as numerical data and dates, ensuring alignment with known truths, e.g., faithfulness, context recall, answer relevancy, correctness, etc.

Example: When summarizing a financial report, this layer checks figures like revenue and profit margins match the original document, ensuring factual consistency.

2. **Business Logic and Compliance Layer:** This layer assesses the logical coherence and regulatory compliance of the generated content.
   - **Statistical Metrics:** Implement **BLEU** and **ROUGE** scores and COMET or BARTScore (semantic adequacy for summarization/ translation) to evaluate the quality of text by comparing it to reference texts, focusing on precision, recall, and the quality of summaries.

- **LLM-as-a-Judge:** Deploy LLMs to analyze the text for adherence to industry regulations and internal policies, ensuring that the content meets specific compliance standards., e.g., *Business Rule Consistency* (Does reasoning follow domain logic e.g., finance, law, medicine?), *Regulatory Alignment* (Any GDPR/PII violations, financial reporting breaches, toxic or biased outputs?), *Toxicity Detection Metrics* (Perspective API scores or toxicity classifiers) and Fairness.[3]

Example: In generating a legal document, this layer verifies that all clauses comply with current laws and that the language used is appropriate for the legal context.

3. **User Experience and Communication Layer:** This layer evaluates the clarity, tone, cultural appropriateness, empathy and overall effectiveness of the communication in the generated content. It is also useful for personalized chatbots.
   - **Statistical Metrics:** Use METEOR (semantic match, synonym-aware), GLEU (Google's variation for fluency in responses), and Readability scores (Flesch–Kincaid, Coleman–Liau), which provides a more nuanced assessment of text quality.
   - **LLM-as-a-Judge:** Implement models to assess the appropriateness of language, tone & politeness (Professional vs casual, empathetic response), *engagement* (Is it persuasive, concise, or human-like enough?) and emotional resonance, ensuring the content aligns with the intended audience's expectations and cultural norms.

Example: For customer service responses, this layer ensures that the language is empathetic and professional, enhancing customer satisfaction and engagement.

By integrating these layers with both statistical metrics and LLM-as-a-Judge approaches, organizations can systematically evaluate LLM outputs, ensuring they are accurate, compliant, and effectively tailored to user needs. Table 10.2 gives a brief overview.

---

[3]Bias may appear in multiple stages in RAG, e.g., the retrieval model, the retrieval process, and re-ranking algorithms. Readers may refer to [8] [7] for more details.

Table 10.2.   Layer-based evaluation.

Layer	Goal	Statistical Metrics	LLM-as-a-Judge Checks
Foundational Verification	Factual correctness, consistency	Perplexity, Recall@k, Precision@k, BERTScore	Faithfulness, Numerical consistency, Correctness
Business Logic & Compliance	Logical coherence, regulatory adherence	BLEU, ROUGE-L, COMET, Toxicity classifiers	Compliance check, Domain logic validation, Bias
User Experience & Communication	Clarity, tone, empathy, resonance	METEOR, GLEU, Readability indices	Tone alignment, Politeness, Cultural fit, Engagement

10.4 Evaluation Tools

There are several tools used to evaluate the metrics, e.g., DeepEval, Truelens, Ragas, LangSmith, rag_evaluator llama pack and etc. When they share some commonalities, they also have their unique features. Table 10.3 provides a brief comparison of those tools.[4]

- **LangChain's LangSmith Framework:** LangChain offers an advanced evaluation framework, LangSmith, where custom evaluators can be implemented. It also monitors the traces running inside your RAG pipeline, making your system more transparent and easier to debug.
- **LlamaIndex's rag_evaluator:** For those using LlamaIndex, the rag_evaluator llama pack provides a quick tool to evaluate your pipeline using a public dataset, facilitating rapid assessment and optimization. A sample code can be found here.[5]
- **TruLens** offers a comprehensive suite of tools designed for developing and monitoring neural networks, including TruLens-Eval for evaluating LLM-based applications and TruLens-Explain for deep learning

---

[4]See more details at https://github.com/junxu-ai/RAG-Collection.

[5]https://github.com/openai/openai-cookbook/blob/main/examples/evaluation/Evaluate_RAG_with_LlamaIndex.ipynb.

Table 10.3.　Evaluation tool comparison.

Tools	Core Features	Core Advantages	Applicable Scenarios
RAGAS	– Automated, dual-mode (statistical + LLM-based) evaluation – Customizable metrics (e.g., context precision, recall, semantic similarity)	– Fine-grained, robust evaluation – Combines quantitative and qualitative analysis	– Complex RAG pipelines – Private/specialized domain evaluations
Trulens	– Fairness and bias evaluation tools – Ethical and compliance metrics	– Additional metrics on fairness and ethical assessment – Diagnoses model bias	– Deployments requiring regulatory compliance – Fairness assessments in sensitive contexts
Deepeval	– Detailed performance analysis – Metrics on semantic similarity, factual correctness, logical consistency	– In-depth insights into model behavior – Uncovers nuanced performance issues	– Research and development – Critical applications demanding high accuracy
RAGChecker	– Holistic evaluation for RAG systems – Diagnostic metrics for retrieval & generation – Claim-level entailment	– Comprehensive pipeline evaluation – Identifies areas for targeted improvements	– RAG systems with separate retrieval and generation components – Iterative model refinement
Phoenix	– Efficient and robust evaluation framework – Combines statistical and self-knowledge metrics	– Balances speed and thoroughness – Scalable for large-scale deployments	– Enterprise-level evaluations – Dynamic, large-scale LLM performance assessments

explainability. Both components can be used independently and are essential for systematically evaluating performance and enhancing the transparency of model decisions. This framework supports developers in identifying and improving failure modes in LLM applications, moving beyond basic assessments to ensure both efficacy and reliability.

- **Ragas**, short for RAG Assessment, is a framework specifically designed to evaluate RAG pipelines. While there are various tools and frameworks available to construct RAG pipelines, assessing their performance and quantifying outcomes can be challenging.
- **RAGChecker** library is a sophisticated automatic evaluation framework tailored for assessing and diagnosing RAG systems. It provides an extensive array of metrics and tools for thorough analysis of RAG performance. The framework necessitates ground truth data to produce evaluation metrics, including both statistical measures and LLM-based qualitative assessments.
- **DeepEval** is an evaluation framework focused on performing a deep analysis of language model performance. It provides a comprehensive suite of metrics designed to assess semantic similarity, factual accuracy, and logical consistency in generated text. By leveraging both automated statistical measures and qualitative assessments, Deepeval aims to offer detailed insights into the strengths and weaknesses of LLM outputs, particularly in scenarios where fine-grained analysis is critical.
- **Phoenix** is an open-source AI observability platform designed for more than evaluation: tracking, experimentation, playground and troubleshooting as well.

10.5 Evaluation Procedure

To further illustrate the evaluation procedure, we use a climate risk assessment RAG as an example. In this case, we use RAG to answer some predefined assessment questions based on clients' documents, such as sustainable report, annual report, ESG report, etc. The questions are generally answered with "Yes" or "No," and several different options. For example, when asking if a company has a plan for carbon emissions reduction, it may have several options like "Yes, we have a plan under implementation," "Yes, we have a plan but it's not fully implemented,"

"No, we don't have a plan, but we will have one within next 2 years,"
"No, we don't have a plan, and no intend to do so within next 2 years" etc.
In addition, the plan might be long-term or short-term, and the plan might
be based on specific industry standards or industry-specific standards.
Since we already have some historical results created by human experts,
we may follow the procedure below for evaluation.

1. **Create benchmark dataset for evaluation:** Collect historical human assessment results, e.g., document, question/query, and response (Yes/No with comments).
2. **Split benchmark dataset** into evaluation and test sets.
3. **Generate responses** using the RAG pipeline (batch mode) for evaluation.
4. **Parse responses into 3 parts:** Answer (Yes/No with options), explanation/summary, and reference (chunks from documents).
5. **Evaluate semantic/context precision/recall/accuracy** using the parsed data and tools (RAGAS, RAGChecker).
6. **Analyze failed cases** using diagnostic metrics of RAGChecker or RAGAS.
7. **Refine RAG pipeline** based on diagnostic feedback and evaluate again. Possibly, we shall revise the system prompt.
8. **Report results** based on test sets.
9. **Conduct other evaluations, e.g., fairness** using additional tools (e.g., Trulens).

As we cannot share the data and questions of CRA, we use the public dataset "explodinggradients/amnesty_qa." We then follow the structured approach that involves creating a benchmark dataset, generating responses, parsing and evaluating these responses, analyzing failures, refining the system, and reporting results. Below is a Python program that implements this evaluation procedure using tools like RAGAS, RAGChecker, and TruLens. The code in ch10\evaluation.py provides the runnable procedure. Below are a few steps to highlight.

- **Generate Responses Using the RAG Pipeline**

  Implement your RAG pipeline to generate responses for the evaluation dataset. This involves retrieving relevant documents and generating answers using the LLM.

```python
def rag_pipeline(question, retriever, llm):
    # Retrieve relevant documents
    contexts = retriever.retrieve(question)
    # Generate response using the LLM
    response = llm.generate(question,
    contexts)
    return response, contexts

# Example usage (assuming you have a
 retriever implemented)
# responses = [rag_pipeline(q, retriever, llm)
 for q in eval_dataset.questions]
```

- **Evaluate Accuracy of the Answer**

  Evaluate the accuracy of the answers using confirmative precision and recall metrics.

```python
# Define evaluation metrics
metrics = [Faithfulness(llm=llm),
 AnswerRelevancy(llm=llm)]

# Evaluate the RAG pipeline
results = evaluate(eval_dataset, metrics)

# Calculate mean scores
for metric in metrics:
    scores = results[metric.name]
    mean_score = np.mean(scores)
    print(f"{metric.name} Average Score:
 {mean_score:.4f}")
```

- **Evaluate Semantic/Context Precision and Recall**

  Assess the semantic and contextual relevance using additional metrics.

```python
# Add context-related metrics
context_metrics = [ContextPrecision(llm=llm),
 ContextRecall(llm=llm),
 NoiseSensitivity(llm=llm)]
```

```python
# Evaluate the RAG pipeline
context_results = evaluate(eval_dataset,
  context_metrics)

# Calculate mean scores
for metric in context_metrics:
    scores = context_results[metric.name]
    mean_score = np.mean(scores)
    print(f"{metric.name} Average Score:
  {mean_score:.4f}")
```

- **Analyze Failed Cases Using Diagnostic Metrics**

  Utilize RAGChecker to diagnose and analyze failed cases in the RAG
  pipeline.

  ```python
  # Initialize RAGChecker
  rag_checker = RAGChecker(llm=llm)

  # Analyze failed cases
  diagnostics = rag_checker.analyze_
    failures(eval_dataset, results)

  # Review diagnostics to identify issues
  for issue in diagnostics:
      print(issue)
  ```

- **Refine RAG Pipeline Based on Diagnostic Feedback**

  Based on the diagnostic feedback, make necessary adjustments to your
  RAG pipeline to address identified issues. This may involve improving
  document retrieval strategies, fine-tuning the LLM, or enhancing
  response parsing methods.

- **Conduct Fairness Evaluation Using Additional Tools**

  Utilize tools like TruLens to conduct fairness evaluations of your RAG
  system, ensuring that responses are unbiased and equitable.

```
# Initialize TruLens
tru = Tru()

# Conduct fairness evaluation
fairness_results = tru.
  evaluate_fairness(test_dataset)

# Review fairness results
print(fairness_results)
```

10.6 Concluding Remarks

Evaluating RAG systems usually requires a dual-pronged approach assessing both the *retrieval* and *generation* components, alongside their integration as a unified pipeline. A recent survey highlights the inherent complexity in evaluating hybrid systems [1]. Both retrieval and generation metrics in terms of statistical model and LLM are discussed. We may also consider the following holistic and explainable evaluation frameworks:

- **THELMA** introduces six interdependent metrics for reference-free, task-centric evaluation, which is useful for real-world, end-to-end RAG QA applications without needing labeled answers [2].
- **RAGBench** presents a large-scale industry-oriented benchmark (100 K examples) and introduces the **TRACe** framework: explainable, actionable metrics designed for consistent performance analysis across domains [3].
- **SePer** (Semantic Perplexity Reduction) offers a novel automatic approach that quantifies retrieval utility by measuring how much retrieved content reduces the model's semantic uncertainty closely matching human preference [4].
- **Auepora** (A Unified Evaluation Process of RAG) uses a structured framework to analyze existing benchmarks and guide future development [1].
- **IRSC** provides new benchmark tasks and metrics: **SSCI** (Similarity of Semantic Comprehension Index) and **RCCI** (Retrieval Capability Contest Index), to evaluate the semantic quality of retrievers, especially in multilingual RAG contexts [5].

These frameworks enable precise breakdowns of component-level performance and inform targeted improvements, particularly by distinguishing whether failures stem from poor retrieval or faulty generation; moreover, automated metrics such as precision, recall, and SePer facilitate scalable and reproducible evaluation pipelines necessary for rapid iteration (e.g., evaluating retrieval relevance and generation fidelity). Additionally, explainability frameworks like TRACe and THELMA offer interpretable and actionable insights, making evaluation more than just scoring, but a tool for debugging and refinement, which enables diagnosis functionalities. Finally, alignment of automated metrics with human judgments, particularly through novel approaches like SePer that reflect reductions in semantic uncertainty, helps close the gap between system performance and real-world user trust.

All these methods and tools motivate a self-evaluation framework as shown in Figure 10.1.[6] Each request flows through the RAG pipeline, where we log the query, retrieval results, packed context, model output, citations, and runtime metrics into an observability store. The UI collects explicit user feedback (e.g., solved/partial/wrong, evidence flags, pinned bad chunks) and passive signals (retries, dwell). When feedback arrives or confidence drops below a threshold, an evaluation pass runs in one of two modes: with ground truth, it computes EM/F1 and semantic similarity against the reference; without ground truth, it uses an LLM-judge with a strict rubric plus faithfulness checks that match claims to cited spans and scan for contradictions. The diagnosis module then maps metric patterns to likely root causes (such as low recall, weak reranking, poor context packing, or prompt/faithfulness issues) and proposes remediations. If policies permit, the system applies safe, reversible tweaks (for example, raising top-k, enabling query expansion and reranking, switching to evidence-first prompting, or reserving tokens for tables), and records outcomes for A/B comparison.

On a scheduled batch loop, the system recomputes comprehensive metrics across cohorts (by intent, domain, document set), surfaces the top failure modes, and identifies content actions like re-indexing, re-chunking, or adding summary chunks. A lightweight optimizer (e.g., a bandit) tunes context-engineering policies against a composite reward that blends judged correctness, attribution quality, user-solved rate, and cost/latency, while governance guardrails pin model versions, redact sensitive data in feedback, and enforce fallbacks when faithfulness degrades.

---

[6]A reference implementation is available at https://github.com/junxu-ai/RAGdx.

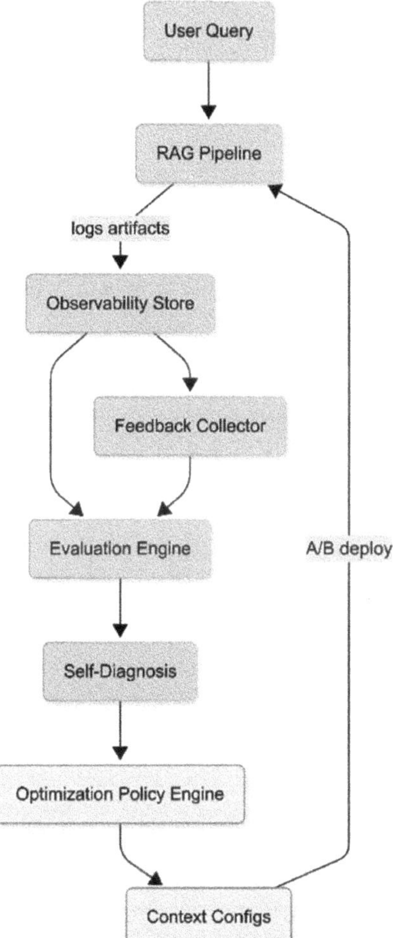

Figure 10.1. RAG self-evaluation workflow.

The resulting playbooks, dashboards, and policy updates feed the next deployment cycle, closing the loop so that retrieval, generation, and verification continually improve even when ground truth is sparse.

The future of RAG evaluation will focus on creating more robust, nuanced, and efficient methodologies that can keep pace with the rapid advancements in RAG architecture.

- **Unified Composite Metrics:** Develop integrated measures that combine retriever precision/relevance and generator faithfulness into a single interpretable score.

- **Multi-Modal Evaluation:** As RAG systems incorporate more than just text—such as images, audio, and video—evaluation metrics must evolve to assess multi-modal outputs. This includes evaluating the correctness of information retrieved from images or the coherence of a generated response that incorporates visual and textual data.
- **Human-in-the-Loop Feedback Integration:** While automated metrics are scalable, human judgment remains the gold standard. Future evaluation systems should integrate a continuous feedback loop where human annotations on a small, representative sample of data can be used to calibrate and improve the automated evaluation models, ensuring they remain aligned with real-world user preferences and expectations.
- **Domain-Adaptive Thresholds:** Tailor sufficiency and relevance metrics to the fidelity required in different domains, e.g., healthcare vs. customer support.
- **Task-Specific Metrics:** While question-answering (QA) is a common RAG task, systems are being applied to more complex problems like summarization, content creation, and dialogue. Future evaluation frameworks need to develop task-specific metrics that go beyond simple correctness. For example, for summarization, metrics could assess information density or readability. For dialogue, they could evaluate conversational flow and persona consistency.
- **Fine-Grained and Causal Analysis:** Moving beyond aggregate scores, there's a need for more fine-grained, component-level analysis that can pinpoint the exact cause of an error. This involves tracing an incorrect answer back to a flawed retrieval step or a misinterpretation by the LLM. Techniques like causal tracing could be adapted to understand the influence of each retrieved document on the final output, allowing developers to debug and optimize the pipeline more effectively.
- **Explainable Per-Sample Diagnostics:** Move beyond summary statistics to per-query diagnostic tools that flag whether issues are retrieval, generation, or integration-related [6].
- **Dynamic Benchmarking:** Develop evolving benchmarks that reflect changes in grounded knowledge (e.g., news, regulations), ensuring retrievers are evaluated for freshness as well as accuracy.

References

[1] H. Yu, A. Gan, K. Zhang, S. Tong, Q. Liu and Z. Liu, *Evaluation of Retrieval-Augmented Generation: A Survey,* arXiv:2405.07437v2, 2024.

[2] U. Patel, R. Mulkar and *et al.*, *THELMA: Task Based Holistic Evaluation of Large Language Model Applications-RAG Question Answering,* arXiv:2505.11626v2, 2025.

[3] R. Friel, M. Belyi and A. Sanyal, *RAGBench: Explainable Benchmark for Retrieval-Augmented Generation Systems,* arXiv:2407.11005v2, 2025.

[4] L. Dai, Y. Xu, J. Ye, H. Liu and H. Xiong, SePer: Measure retrieval utility through the lens of semantic perplexity reduction, in *ICLR 2025 Spotlight,* Singapore, 2025.

[5] H. Lin, S. Zhan, J. Su, H. Zheng and H. Wang, *IRSC: A Zero-shot Evaluation Benchmark for Information Retrieval through Semantic Comprehension in Retrieval-Augmented Generation Scenarios,* arXiv:2409.15763v2, 2024.

[6] H. ZHAO and *et al.*, Explainability for large language models: A survey, *ACM Transactions on Intelligent Systems and Technology,* vol. 15, no. 2, p. 20, 2024.

[7] I. O. Gallegos and *et al.*, *Bias and Fairness in Large Language Models: A Survey,* arXiv:2309.00770v3, 2024.

[8] X. Wu, S. Li, H.-T. Wu, Z. Tao and Y. Fang, *Does RAG Introduce Unfairness in LLMs? Evaluating Fairness in Retrieval-Augmented Generation Systems,* arXiv:2409.19804v2, 2025.

Chapter 11

Serving and Monitoring

11.1 Introduction

As RAG systems transit from research prototypes (PoCs) to production environments, the challenge extends far beyond model design. The need to serve these systems reliably and monitor their performance becomes paramount. Serving a RAG system involves not only handling user queries with low latency but also ensuring that retrieval and generation components integrate seamlessly under varying loads and diverse input scenarios. The serving layer must balance efficiency, scalability, and compliance, especially in enterprise contexts where sensitive or domain-specific data is at stake.

Equally important is monitoring, which ensures that deployed RAG systems remain trustworthy, efficient, and aligned with user expectations. Unlike traditional ML monitoring, RAG monitoring must address additional dimensions such as retrieval accuracy, knowledge freshness, context relevance, and user interaction quality. Without continuous oversight, RAG applications risk suffering from performance drift, degraded reliability, or misleading outputs. In production environments, these risks are magnified: downtime, inaccurate responses, or governance failures can have far-reaching implications for both users and organizations.

Together, serving and monitoring form the operational backbone of production-grade RAG systems. Serving provides the infrastructure to deliver responses reliably at scale, while monitoring safeguards quality and trust over time. This dual focus is essential for organizations seeking to harness RAG not only as a research innovation but as a dependable foundation for real-world applications.

11.2 RAG System Serving

Serving RAG systems in production is not just the LLM serving issues. It introduces several challenges that are unique to their hybrid nature. One of the primary difficulties is maintaining the delicate balance between retrieval efficiency and generation quality. The retrieval module must quickly sift through vast amounts of data to find the most relevant documents, while the generative model must then synthesize these into coherent and contextually accurate responses.

Scalability is another key challenge. As user demands grow, the system must adapt to increased loads without sacrificing performance. This requires careful architecture design and resource management, ensuring that both the retrieval and generation components scale together harmoniously. Additionally, latency is a critical factor; developers must optimize each component so that the time taken from query input to final output remains within acceptable limits for end users.

To design a proper serving framework, we must first define a reasonable goal with SLOs. For example,

- **Primary SLOs:** p95 latency, answer faithfulness, cost/$, and throughput.
- **Secondary SLOs:** data-residency/compliance and tenant isolation.

The framework is usually divided into a control plane and a data plane for clearer separation of concerns. The control plane includes a Profiler that analyzes queries for complexity, retrieval needs, and expected answer length; a Policy Engine that makes resource-aware and compliance-aware decisions; a Joint Scheduler that orchestrates multi-stage pipeline execution to meet SLOs; an Autoscaler that dynamically scales components; and an Observability & Evaluation stack to track performance and quality. In contrast, the data plane handles per-request logic: from query normalization and profiling, through hybrid retrieval and reranking, to LLM generation with batching and caching, finishing with answer delivery and feedback logging [1].

Let's look at a sample implementation, which relies on a modular serving fabric such as Ray Serve or KServe with per-component deployment and autoscaling. LLM inference is managed via open source vLLM, SGLang or Xinference with KV optimizations and speculative decoding or commercial databricks, Azure, AWS, etc. Retrieval is powered by Milvus or OpenSearch, with optional custom FAISS backends. Reranking uses BGE or cross-encoders. Policy rules control hybrid routing and configuration decisions, informed by compliance requirements. We also set up infrastructure for trace collection, dashboards, and policy dashboards. Below is the detailed explanation.

11.2.1 *LLM Serving*

LLM inference can be significantly optimized by improvements at the scheduling and runtime layers. Prefix caching enables reuse of previously computed key/value (KV) pairs across shared prompt segments, reducing redundant computation especially in multi-turn conversations. Chunked prefill scheduling further enhances GPU utilization and throughput by breaking large prompts into smaller chunks for staged processing. On the high-level runtime side, speculative decoding leverages a smaller draft model to propose multiple tokens in parallel, which are then verified by the main model via cutting token latency by up to 2–3× while preserving output quality. Additionally, Multi-LoRA adapter support allows runtime loading of multiple fine-tuned adapters in lightweight formats to efficiently support multi-user or multi-task workloads.

Scaling model execution across hardware resources relies on parallelism strategies. Tensor parallelism partitions large model weights across multiple GPUs to distribute matrix computations for ultra-large models. Pipeline parallelism executes different model layers sequentially across dedicated computing stages, improving throughput for deep. Meanwhile, Mixture-of-Experts (MoE) or expert parallelism selectively activates only relevant expert sub-networks per input token, optimizing compute usage and fostering scale-efficient inference.

At the hardware-kernel and interface level, a number of strategies help reduce memory movement and communication overhead. Fused attention kernels (e.g., FlashAttention)[1] combine computation of $Q \cdot K^T$,

---

[1] https://github.com/Dao-AILab/flash-attention.

softmax, and V projection into one operation—reducing memory usage from $O(L^2)$ to $O(L)$ and accelerating attention layers substantially. Simultaneously, quantization methods (such as INT8, FP8, or 4-bit precision) shrink model memory and bandwidth requirements while maintaining acceptable performance. Finally, optimized collectives using high-performance libraries like NCCL[2] enable efficient distributed communication for tensor and pipeline parallel workloads, further boosting throughput.

When integrated, these techniques deliver full-stack performance gains: runtime scheduling (e.g., prefix caching, speculative decoding) reduces latency and avoids redundant work; dynamic model-level strategies (quantization, LoRA adapter loading) compress and personalize execution; parallelism (tensor, pipeline, MoE) enables distributed scaling; and hardware-aware kernel fusion and optimized collective communication underlie high efficiency in real-world serving infrastructure. Collectively, these optimizations enable high-throughput, low-latency LLM inference that fulfills the operational requirements of production-grade applications.

We can deploy LLM models using optimized serving engines such as vLLM, Xinference and TensorRT-LLM. These engines support efficient techniques like continuous batching, prefix/KV caching, and PagedAttention to reduce waste and latency. For further speed, optional speculative decoding (draft + verify) or multivariate "Medusa-style" decoding heads may be enabled. Model serving is grouped into pools (e.g., small, medium, large) that autoscale independently, allowing resource-efficient routing based on request requirements, such as preferred model size or decoding behavior.

11.2.2 *Retrieval Subsystem*

The retrieval subsystem may start from utilizing a hybrid of lexical (e.g., BM25) and dense vector retrieval, combining results through methods like reciprocal rank fusion (RRF) or normalized score fusion as discussed in Chapters 7 and 8. Depending on scale and precision needs, we select suitable indexing backends (e.g., FAISS, ScaNN,[3]

---

[2]https://github.com/NVIDIA/nccl.

[3]https://github.com/google-research/google-research/tree/master/scann.

Milvus, or OpenSearch) to optimize resource use and recall. To improve relevancy of the retrieved contents to query further, a reranker (e.g., BGE-Reranker v2 or distilled cross-encoder) narrows top-k documents. For handling long-context or multi-hop reasoning, optional hierarchical indexes (e.g., RAPTOR-style recursive summarization) serve as an advanced alternative.

11.2.3 *Query Profiling for Configuration Adaptation*

A lightweight Profiler, such as a small LLM or a rule-based model, analyzes each query to estimate its complexity, need for joint reasoning, number of information chunks required, and expected answer length. This smart routing technique is discussed in Chapter 6 (and will further be discussed in Chapter 15). Using this profile, the framework prunes the enormous configuration space such as retrieval depth, reranking choice, synthesis method (e.g., map-reduce vs. stuff), or model size, to a compact set of promising options. This enables per-query optimization and avoids one-size-fits-all configurations.

11.2.4 *Resource-Aware Policy & Joint Scheduling*

Decisions are taken based on both profiling results and dynamic system state (e.g., GPU memory, queue depths, cache heat). Starting from high-quality configurations, the policy engine selects the most efficient one that is expected to meet SLOs within resource constraints; if none match, the framework falls back to the lightest viable option. The Joint Scheduler then coordinates pipeline stages (batching retrievals and reranks, micro-batching LLM decoding, prioritizing near-deadline requests, and leveraging cache locality) to optimize throughput and reduce tail latency [2].

11.2.5 *Graceful Fallback Hierarchy*

When resource constraints or high load threaten to breach SLOs, the framework implements progressive fallback strategies as discussed in Chapter 12. It may turn off reranking or switch to simpler retriever models, reduce retrieval depth or adjust fusion weights, downgrade synthesis method (e.g., map-reduce → stuff), select a smaller or speculative-enabled LLM pool, or finally provide a partial "answer skeleton" that can be enriched later. Each step is guided by the original query profile to minimize quality degradation.

11.2.6 *Hybrid On-Premises and Cloud Routing*

To address data sensitivity and burst scaling, the system routes queries based on policy: sending sensitive or regulatory-bound queries to on-prem infrastructure, while cloud services handle elastic capacity or larger models. A security filter checks for PII or regulatory constraints before cloud routing. This hybrid architecture enables flexibility, cost savings, and compliance, especially in regulated industries.

11.2.7 *Smart Caching Strategy*

We can implement caching at multiple levels to improve latency and reduce cost: a semantic response cache keyed on prompt embeddings to serve repeated queries; retrieval caches that store top-k results or warm up ANN indexes for frequent queries; and KV-cache reuse in LLM inference, with instance routing that aligns with hot chunks (like KVLink[4] or Cache-Craft ideas [3]). See more in Chapter 12.

11.2.8 *Security and Risk*

The most common and highest-impact threats that organizations are actively grappling with are prompt injection (leading to data leakage), hallucinations, and over-reliance by users.

- **Prompt-level risks, such as** jailbreaks/prompt-injection (including RAG-borne injection); policy violations (e.g., suitability, market abuse advice, discrimination); data leakage (PII, PCI/PHI, confidential deal or client data).
- **Modeling/answer-quality risks, such as hallucinations** in research, advice, or regulatory interpretation; incomplete/opaque citations; mis-calculated figures; mis-summaries.
- **Operational risks and Human factors**, such as tool misuse (unapproved trades, unsanctioned data pulls); logging gaps (violating 17a-4),

---

[4]https://github.com/UCSB-NLP-Chang/KVLink.

weak kill-switches, and unreviewed incidents; over-reliance on the model, "authority bias," and work-around behavior (off-channel communications).

If we focus solely on the data poisoning paths of RAG, we will obtain a long list of poison categories:

- **Source-level poisoning** occurs when attackers tamper with data before it even enters the RAG pipeline. This can happen through poisoned web pages, malicious documents in third-party repositories, misinformation in partner-shared data, or user-uploaded files that contain hidden instructions, zero-width characters, or steganographic payloads. Even APIs can be targeted by returning adversarial fields designed to mislead the system.
- **Ingestion and preprocessing poisoning** exploits vulnerabilities in how raw data is transformed into usable text. Adversaries may craft HTML, Unicode tricks, or malformed tables to confuse parsers. OCR and ASR systems can be manipulated with adversarial fonts, image perturbations, or audio backdoors. Attackers can also fabricate metadata, such as authorship or timestamps, and design text to manipulate chunking or segmentation rules.
- **Embedding and index poisoning** targets the vector space and indexing process that underpins retrieval. Malicious documents can be engineered so their embeddings collide with popular queries, thereby hijacking retrieval. Attackers may flood the index with near-duplicate content, manipulate metadata like boosting scores, or poison graph structures by inserting false relations and centrality manipulations in knowledge graphs.
- **Retrieval-time poisoning** happens when malicious documents are surfaced at the point of query execution. Attackers can bias query reformulation to over-retrieve harmful content or manipulate reranker training data so that adversarial content is consistently ranked higher. They may also exploit freshness signals by publishing poisoned updates just in time to match user queries, or use multilingual tricks to sneak in seemingly relevant but misleading documents.
- **Augmentation and orchestration layer poisoning** leverages the fact that retrieved documents directly influence the LLM's generation. Poisoned text may contain indirect prompt injections that instruct the model to ignore safety rules or exfiltrate sensitive data. In Text-to-SQL

contexts, adversarial schema comments can steer queries toward attacker-preferred outcomes. Similarly, poisoned planning data can misdirect routers or supervisors in multi-agent RAG workflows.

- **Generation-time and post-generation poisoning** exploits the fact that models often cache or learn from their own outputs. Attackers can seed cache entries with poisoned Q&A pairs that later get reused as "trusted" results. User feedback mechanisms can also be brigaded to promote adversarial content, while self-training or continual learning processes risk absorbing poisoned passages directly into model weights.

- **Evaluation and guardrail poisoning** undermines the systems meant to keep RAG safe and reliable. Attackers may corrupt evaluation sets or seed benchmarks with misleading facts so that models are rewarded for repeating falsehoods. They might also tamper with policy libraries or safety filter rules, effectively whitelisting malicious triggers. When synthetic data is generated by smaller models, poisoning them can propagate bad patterns into evaluations.

- **Supply-chain and model-artifact poisoning** takes place when adversaries target the third-party tools and models integrated into the RAG stack. Poisoned embedding models may contain backdoors, rerankers may be trained on compromised datasets, and even parsers or OCR libraries can normalize text in adversarial ways. Content delivery networks and storage systems are also vulnerable, where attackers can tamper with cached content or object tags.

- **Access-control and tenancy poisoning** exploits weaknesses in multi-tenant setups and ACL configurations. Attackers can manipulate entitlement metadata such that their malicious documents inherit privileged tags or gain unintended visibility. Cross-namespace joins or shadow indices may allow poisoned content from one tenant to leak into another tenant's retrieval results, creating a serious cross-contamination risk.

- **UI and human-factor poisoning** focuses on manipulating the human curators and operators of RAG systems. Attackers might social-engineer operators into pinning poisoned content as "top answers" or compromise annotation pipelines so labelers are primed with misleading context. In crowd-sourced annotation platforms, consensus mechanisms can be gamed to slip poisoned examples into training or evaluation datasets.

- **Multimodal and file-format poisoning** introduces adversarial payloads hidden in complex document formats. PDFs and Office files may contain invisible layers, malicious annotations, or embedded external links. Vision-based RAG can be attacked with adversarial images, QR-code-like triggers, or manipulated EXIF metadata. Similarly, audio or video sources can hide ultrasonic triggers or poisoned subtitles that distort transcription outputs.

- **Infrastructure and observability poisoning** targets the operational data and metrics that shape system decisions. Attackers can inject adversarial strings into logs, manipulate popularity counters to bias retrieval, or coordinate content delivery to exploit freshness heuristics. By poisoning telemetry, they can steer monitoring and automated safeguards in ways that amplify rather than mitigate malicious influence.

Many of them can be relieved by some proper guardrails as stated in the next subsection.

11.2.9 *Guardrails*

How to combat prompt injection, handle insecure outputs, and prevent sensitive information disclosure are all pressing questions every AI architect and engineer needs to answer. For LLM guardrails, we might mainly consider the following two types:

- **Input rails:** These are applied to user input; an input rail can either reject the input, halting further processing, or modify the input (e.g., to obscure potentially sensitive data or to rephrase).
- **Output rails:** Applied to the LLM-generated output; an output rail can reject the output to prevent it from being shown to the user or modify it (e.g., to remove sensitive data).

Some variants exist for RAG in chatbot setting:

- **Retrieval rails:** In a RAG scenario, these are applied to the retrieved data segments; a retrieval rail can discard a segment to prevent its use in prompting the LLM or modify the segments (e.g., to obscure sensitive information).
- **Execution rails:** These are applied to the input and output of custom actions (also known as tools) that the LLM needs to call.
- **Dialog rails:** These influence the prompting of the LLM; dialog rails operate on canonical form messages (for more details, see the Colang

Guide) and decide whether an action should be executed, whether the LLM should generate the next step or a response, or whether a pre-defined response should be used. The other four types of rails may interact with dialog rails, and it may also include historical chat content.

Table 11.1 lists some commonly used guardrails for LLMs. Some general PII tools, e.g., Presidio,[5] can be used as guardrails too. In practice, some over-regulated guardrails often lead to performance drop, e.g., some non-sensitive names or series numbers might be mis-filtered, and some normal output might be marked as harmful contents by mistake. A layered guardrail architecture (prompt + guardrails by design) is proposed here for the balanced performance and security issues, as shown in Figure 11.1. It represents an enterprise-grade, defense-in-depth approach that is well-suited for a high-stakes environment like financial institutions.

- **Policy-to-Code Layer:** At the foundation, financial institutions must translate their regulatory, compliance, and internal governance policies into machine-readable rules. This is typically achieved through a "policy-as-code" approach, where conduct risk policies, KYC/AML boundaries, and data confidentiality rules are encoded into executable logic. By doing so, the organization ensures that guardrails are both auditable and enforceable, allowing automated decisions to be traced back to well-defined regulatory or internal policies. Tools like Open Policy Agent (OPA)[6] are built for this exact purpose. The main challenge is not technical but organizational: it requires close collaboration between compliance, legal, and engineering teams to accurately translate nuanced regulatory text into precise code.
- **Input Control Layer:** Before any data reaches the language model, an input control layer validates and sanitizes the request. This involves using classifiers to detect sensitive or inappropriate requests, applying prompt-hardening techniques to enforce hierarchical roles and permissible intents, and screening against prompt injection or malicious instructions. For RAG systems, governance is added to ensure that only approved and sanitized sources are allowed into the prompt. This can be built using a combination of techniques: standard input sanitization

---

[5]https://github.com/microsoft/presidio.
[6]https://github.com/open-policy-agent/opa.

Table 11.1.   Comparison of some guardrails.

Feature	NVIDIA NeMo-Guardrails	Guardrails (Guardrails.ai)	LLM Guard (Protect AI)	Llama Guard (Meta, Purple Llama)
Integration	Works with many LLMs/providers and LangChain (RunnableRails).	Python framework + managed service; integrates in-app; supports I/O guards.	Library/API; **model-agnostic** across providers and frameworks.	Safety classifier model usable with any pipeline; part of Purple Llama tools.
Customization/ Function	Define rails/policies and flows in Colang; steer inputs/outputs and tools.	User-defined guards; structured output validation; guard packs via Guardrails Hub.	Prebuilt scanners (PII, injection, toxicity, etc.) + configurable policies.	**Classifies** prompts/ responses against a safety spec (Llama Guard 2 categories).
Monitoring/ Observability	OTel tracing, not a built-in UI; plug into Datadog/ Elastic/Fiddler, etc.	Offers observability in managed service; OSS supports OTel for metrics/traces.	Library focuses on scanning; no native monitoring/ alerting UI.	Classifier model only; no monitoring layer.
Documentation	Extensive docs (configuration, LLM support, LangChain integration).	Docs + hub with examples and risks catalogue.	Docs and examples on site/GitHub.	Model card and research write-up.
Community	Active OSS repo and issues/discussions.	OSS + Discord/ GitHub; active updates.	Active OSS toolkit from Protect AI.	Backed by Meta's Purple Llama initiative.

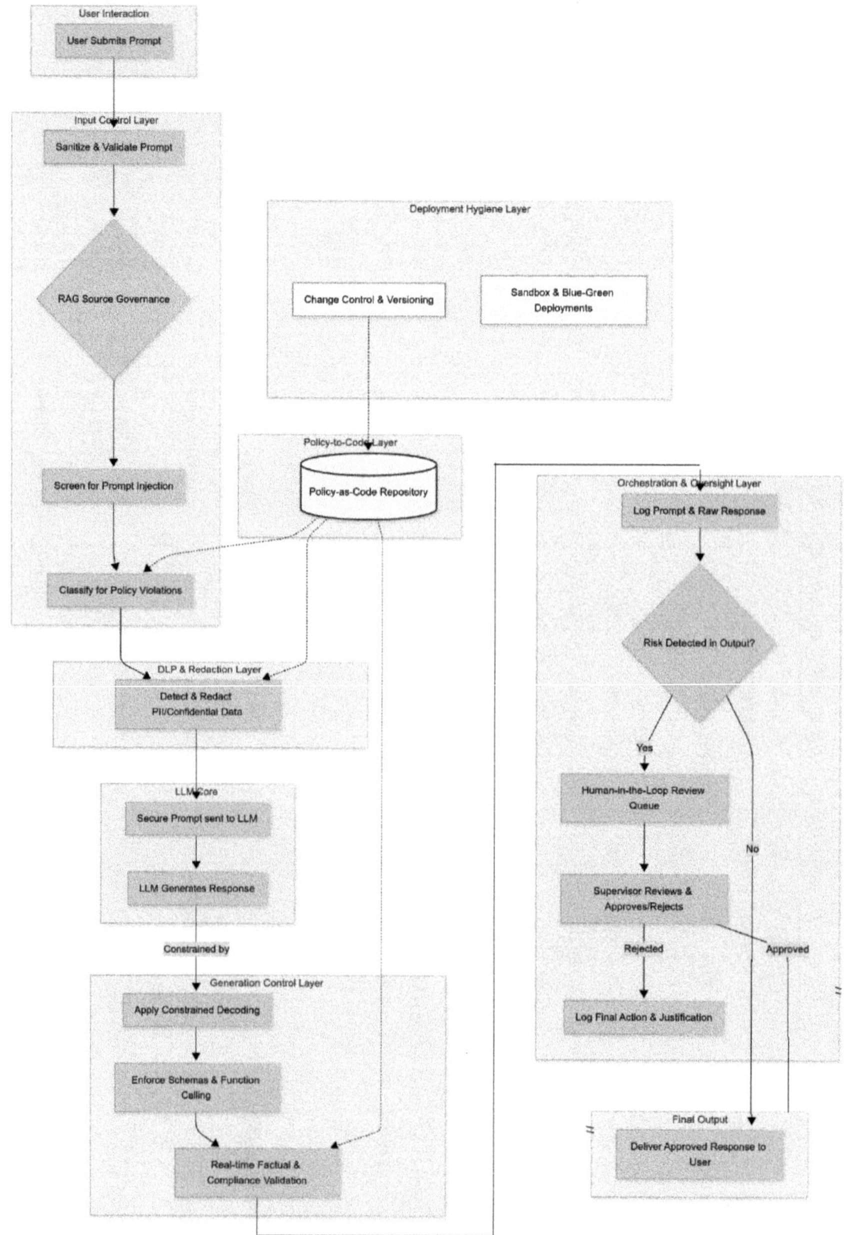

Figure 11.1.   Guardrail framework.

libraries, allow-lists for RAG documents, and custom classifiers (which can be smaller, fine-tuned models) to detect malicious intent or policy violations (e.g., a user asking for guaranteed investment returns).

- **Generation (Output) Control Layer:** During the response generation process, the model is constrained by predefined rails and structured safeguards. This layer applies constrained decoding methods, enforces JSON schemas or function-calling restrictions, and integrates programmable guardrails such as NeMo Guardrails to guide the flow of conversation. Validators are also applied in real time to confirm that generated outputs adhere to regulatory requirements, maintain factual accuracy, and comply with defined structural or numeric constraints.

- **Data-Loss Prevention and Redaction Layer:** To prevent sensitive information from leaving controlled environments, a dedicated redaction and data-loss prevention layer detects personally identifiable information (PII), client identifiers, or confidential terms. When such data is found, it is either masked or replaced with placeholders before being sent outside the organization's boundary. This ensures that cloud APIs or external models do not receive any internal or sensitive inputs.

- **Orchestration and Oversight Layer:** The orchestration layer manages the overall workflow and ensures that every decision point is supervised. When risks are detected, the system can escalate responses for human review, require explicit approvals, or trigger overrides with justification. At the same time, comprehensive logging of prompts, model decisions, and guardrail actions is maintained in an audit-compliant storage format. This satisfies requirements for supervisory review and audit readiness in heavily regulated environments. The key is to carefully tune the "Risk Detected?" logic to avoid overwhelming supervisors with too many false positives.

- **Deployment Hygiene Layer:** This layer ensures that guardrails and prompts are consistently managed across development, testing, and production. Separate sandbox and production environments are enforced, and all changes to prompts, rules, or guardrails are subject to formal change control. Blue-green deployment strategies allow updates to be tested safely before full rollout, and version histories are maintained to support model risk management and regulatory validation. This is the standard, best-practice MLOps/DevOps. Using tools like Git, CI/CD pipelines, and separate testing/production environments is the industry standard for managing software reliably.

11.3 Methodology and Metrics for Monitoring

Monitoring the metrics discussed in Chapter 10 is equally complex. A systematic approach to serving and monitoring RAG systems starts with establishing a robust methodology. This involves setting clear performance baselines and continuously measuring the system against these benchmarks. Developers need to implement monitoring at multiple levels: infrastructure, retrieval performance, and generative output quality.

At the infrastructure level, traditional metrics (such as request throughput, system latency, and resource utilization) remain relevant. For the retrieval component, key performance indicators include query response time, relevance of retrieved documents, and recall/precision metrics that quantify how effectively the system is filtering and ranking data.

On the generation side, output quality can be monitored using both automated and manual evaluations. Automated metrics might include perplexity, BLEU scores, or more advanced quality indicators based on user feedback and continuous evaluation frameworks. In production systems, it is also common to incorporate A/B testing to compare different model configurations or parameter settings. Developers can always build some metrics based on their own application requirements. For example, besides those introduced in Chapter 10, below are some other examples in each level:

Retrieval Layer Metrics

1. **Chunk Relevance Score:** Measures the percentage of retrieved chunks containing query-relevant information using techniques like sentence-level relevance classification. Implemented through LLM-powered classifiers or semantic similarity thresholds.
2. **Positional Impact Metric:** Quantifies the performance degradation when relevant context appears in different positions within the context window.
3. **Index Freshness:** Tracks the time delta between document updates and their availability in search indices.

Generation Layer Metrics

1. **Source Attribution Rate:** Calculates the percentage of generated claims traceable to retrieved contexts.

2. **Hallucination Density:** Measures hallucinations per generated token using contradiction detection models.
3. **Context Utilization Efficiency:** Analyzes what percentage of provided context informs the final response.

System-Level Metrics

1. **End-to-End Latency Percentiles:** Tracks latency distributions across the entire RAG pipeline.
2. **Query Success Rate:** Measures the percentage of queries successfully processed without fallbacks.
3. **Cost Per Quality-Adjusted Response:** Combines operational costs with quality metrics for ROI analysis.

11.3.2 *Monitoring Tools and Operational Practices*

Modern development ecosystems offer a rich set of tools that facilitate both the serving and monitoring of RAG systems. For deployment, container orchestration platforms such as Kubernetes enable scalable and resilient service management. These platforms, combined with robust API gateways, ensure that both the retrieval and generation endpoints can be managed efficiently. On the monitoring side, integrating logging frameworks like ELK (Elasticsearch, Logstash, and Kibana) or Grafana allows developers to visualize performance metrics and quickly identify anomalies. There are also some specialized tools for model monitoring of LLM and RAG systems, that can track model drift, performance degradation, or unexpected changes in data distribution.

Developers may also consider using automated testing frameworks to simulate user interactions under various load conditions. This not only helps in stress-testing the system but also in validating that the integration between the retrieval and generation components remains intact over time.

Below, we divide the specialized RAG monitoring ecosystem into three layers with the implementation of typically used tools addressing different operational needs:

1. **Pipeline Observability**
   - *LlamaIndex*: Provides instrumentation hooks for tracing retrieval and generation operations.

- *Langfuse/portkey/helicone*: Implements cross-component tracing with context propagation.
- *Arize Phoenix*: Offers visualization dashboards for embedding drift and retrieval patterns.

2. **Quality Assurance**
   - *RAGAS Framework*: Automated testing suite for benchmarking retrieval and generation quality.
   - *Weights & Biases*: Experiment tracking with custom RAG-specific metrics.
   - *MLflow*: Model registry with version-controlled evaluation benchmarks.

3. **Infrastructure Monitoring**
   - *Prometheus/Grafana Stack*: Time-series monitoring of resource utilization and latency metrics.
   - *Elasticsearch Logging*: Centralized log analysis with RAG-specific parsing rules.
   - *Kubernetes Operators*: Custom controllers for autoscaling vector databases and embedding pipelines.

Effective operational practices combine these tools with process safeguards:

- **Canary deployments** for document corpus updates.
- **Shadow mode testing** of new retrieval strategies.
- **Circuit breakers** for fallback to cached responses during system overload.
- **Multi-armed bandit** approaches to balance exploration of new indexing strategies with exploitation of known-good configurations.

Comprehensive observability is baked in using OpenTelemetry and Prometheus to capture metrics like stage latencies, queue times, cache hit rates, and SLO misses. For quality evaluation, we integrate frameworks like Phoenix and Ragas to measure faithfulness, citation accuracy, and helpfulness. Logged profiling decisions are periodically retrained via offline tuning loops (e.g., bandit-style or cost models) to update pruning, fusion weights, and scheduling heuristics.

Table 17.3 in supplementary material provides an overview of various LLM logging, monitoring, and evaluation tools, detailing their descriptions, unique features, and relevant links. Each tool has its own niche focus ranging from full-stack product platforms (like Braintrust Data) to specialized observability or evaluation frameworks (like PromptFu). In many cases, only a few are open source (e.g., MLflow and PromptFu), while most follow a proprietary or source-available model.

For example, we can use portkey-ai in a rather simple way:

```python
# pip install -qU portkey-ai
from portkey_ai import Portkey

# OpenAI compatible client
client = Portkey(
    provider="openai", # or 'anthropic',
  'bedrock', 'groq', etc
    Authorization="sk-***" # the provider API
  key
)

# Make a request through your AI Gateway
client.chat.completions.create(
    messages=[{"role": "user", "content":
"How to build a machine learning model?"}],
    model="gpt-4o-mini"
)
```

You can easily find the logs from http://localhost:8787/public/logs, once you run the program.

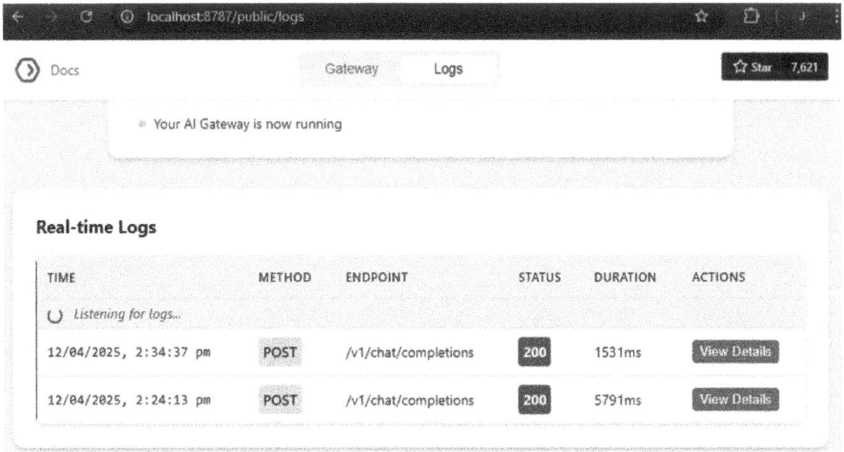

You may also use helicone by calling either the SaaS URL or a self-hosted URL:

```
import OpenAI from "openai";
import os
from dotenv import load_dotenv
load_dotenv()

const openai = new OpenAI({
  apiKey: os.getenv("OPENAI_API_KEY"),
  baseURL: 'https://oai.helicone.ai/v1',
# To create a user go to http://
 localhost:54323/project/default/auth/users
 and add your account. You can use this
 account to sign into Helicone at
 localhost:3000 via your browser.
  # baseURL: 'https://localhost:8787/v1/
 gateway/oai/v1',
  defaultHeaders: {
    "Helicone-Auth": 'Bearer ${process.env.
 HELICONE_API_KEY}',
  },
});

response = openai_client.completions.create(
    model="gpt-3.5-turbo-instruct",
    prompt="Count to 5",
    stream=False,
)
assert response.choices[0].text is not None
```

11.4 Concluding Remarks

Effective serving and monitoring of RAG systems are central to ensuring that RAG pipelines deliver accurate, efficient, and trustworthy responses in production.

On the aspect of serving, recent advancements have focused on resource-aware scheduling, dynamic query profiling, and hybrid routing strategies that balance on-premise and cloud APIs. Techniques such as joint scheduling of retrieval and generation, fallback mechanisms, and multi-index fusion (combining keyword, semantic, and graph-based retrieval) represent promising directions for improving both efficiency and output quality. Emerging work on voice-enabled conversational interfaces (integrating speech-to-text and text-to-speech) also broadens the accessibility of RAG systems, paving the way for multimodal deployments.

On the aspect of monitoring, the field is moving beyond traditional evaluation metrics to embrace observability frameworks that track performance at multiple layers: retrieval recall, generation coherence, latency, cost efficiency, and user satisfaction. Open-source monitoring tools, coupled with LLM-as-a-Judge evaluation, enable more robust quality assurance. At the same time, integration with enterprise observability stacks ensures that RAG systems are not "black boxes" but measurable components within larger IT operations. Techniques like continuous feedback loops, canary testing, and dynamic retraining signals from monitoring data are becoming essential to sustain system reliability.

The multi-step nature of RAG systems means that anomalies can arise from either the retrieval or the generation side, or the interplay between the two. Detecting and diagnosing issues demands comprehensive logging, real-time analytics, and sometimes even automated alerting mechanisms to address problems before they impact users. Below are some additional points to resolve:

1. **Embedding Space Telemetry:** Continuous monitoring of vector space density and separation characteristics to detect concept drift.
2. **Automated Relevance Tuning:** Reinforcement learning approaches that adjust retrieval thresholds based on observed generation quality.
3. **Adaptive and Self-optimizing Systems:** Serving layers that automatically adjust retrieval depth, model selection, or response generation strategies based on context, workload, and observed quality.
4. **Cross-Layer Optimization:** Joint optimization of chunking strategies and generation prompts based on end-to-end metrics.
5. **Causal Analysis Tools:** Root cause analysis systems that trace quality regressions to specific document updates or code changes.

As RAG systems continue to evolve, several future directions emerge in both serving and monitoring that will shape their deployment at scale. On the serving side, new advances in dynamic retrieval orchestration are likely to expand beyond static query refinement or rule-based scheduling.

We can expect the rise of adaptive pipelines where retrieval, caching, and LLM serving are jointly optimized in real time, guided by user context, latency budgets, and evolving knowledge graphs. Integration with agentic orchestration frameworks will further enable RAG services to perform complex, multi-step reasoning and tool usage, while maintaining efficiency through cost-aware scheduling and hardware-aware serving strategies.

For monitoring, the emphasis will shift from static metrics to continuous, adaptive observability. Future systems will not only track performance, latency, and accuracy, but also proactively detect hallucinations, bias, and data drift across retrieval and generation stages. Guardrails will increasingly need to integrate multi-layered security checks that blend prompt-level validation, retrieval filtering, and policy-aware response shaping. Beyond technical monitoring, there is also a growing need for governance dashboards that bring together compliance, explainability, and business KPIs, ensuring RAG applications meet both regulatory and operational standards in critical sectors such as finance and healthcare.

Finally, a key direction is the co-evolution of serving and monitoring. Closed-loop self-optimization, where monitoring insights directly inform serving strategies, will become essential. For example, if monitoring detects recurring retrieval failures or unsafe responses, the serving pipeline could automatically adjust retrieval sources, update caching rules, or trigger fine-grained context engineering. Such self-healing capabilities will push RAG systems closer to autonomous reliability, making them robust enough for enterprise-critical adoption.

References

[1] B. Hu, L. Pabon, S. Agarwal and A. Akella, *Patchwork: A Unified Framework for RAG Serving,* arXiv:2505.07833v1, 2025.

[2] S. Ray, R. Pan, Z. Gu, K. Du, G. Ananthanarayanan, R. Netravali and J. Jiang, *RAGServe: Fast Quality-Aware RAG Systems with Configuration Adaptation,* arXiv:2412.10543v1, 2024.

[3] S. Agarwal and *et al.*, *Cache-Craft: Managing Chunk-Caches for Efficient Retrieval-Augmented Generation,* arXiv:2502.15734, 2025.

Chapter 12

Pipeline and Orchestration

12.1 Introduction

In RAG systems, the pipeline, also known as orchestration integration technique, plays a pivotal role in streamlining the process from retrieval to generation, ensuring efficient and effective responses. The pipeline typically starts with a retrieval phase where relevant data is extracted from a comprehensive datastore using advanced search algorithms. This is followed by an augmentation or fusion phase where the retrieved information is seamlessly integrated into the generation process. Different techniques, such as query-based, latent, or logits-based fusions, are employed to enhance the inputs or internal states of the generation model, leveraging the context to produce more accurate and relevant outputs. Subsequently, the refined input or enriched latent vectors are fed into a generative model, often an LLM, which synthesizes the final response based on the enhanced information. Throughout these stages, the system must maintain a balance between retrieval efficiency and generation quality, optimizing for speed and accuracy to meet user expectations in real-time applications. This orchestration of components within the RAG pipeline exemplifies the complex interplay of retrieval and generation needed to harness the full potential of AI in understanding and responding to user queries.

To further improve the overall performance, efficiency and user experience, multiple techniques can be applied. For example, caching can store the previous requests and responses so that the next same inquiry is not necessary to go through the whole pipeline again. Streaming can get the

Figure 12.1. Actions in pipeline and orchestration.

"real-time" response instead of waiting until the final token is generated. When API calling rate is limited or errors appear, we might think about fallback models. Model optimization (hyperparameter tuning without changing model weights and the finetuning with weights changed) can significantly impact the performance. Finally, the context engineering covers the context across the whole pipeline. In this chapter, we will discuss those techniques as shown in Figure 12.1 in detail.

12.2 Iterative RAG

Iterative RAG refers to RAG systems that employ a feedback loop or multiple steps of retrieval and generation to refine the output of one or several LLMs. It is usually designed for complex, multi-step, or ambiguous queries that cannot be effectively answered through a single retrieval step.

The key idea behind iterative RAG is to break down complex queries into manageable sub-tasks via multiple cycles of retrieving external information and reasoning over it. After each retrieval phase, the accumulated knowledge is used to refine the next set of retrieval queries, progressively narrowing the context or filling information gaps until the system is confident that the answer is comprehensive and correct. This process continues until a stopping criterion is met, such as having sufficient evidence or reaching a maximum number of iterations. The basic concept is related to Query Rewriting in Chapter 6 and extended from Iterative and Recursive Retrieval in Chapter 7.

A generic iterative RAG workflow comprises the following steps, as illustrated in Figure 12.2:

1. **Initial Query Submission:** The user submits a complex or ambiguous question.
2. **First Retrieval & Reasoning:** The system retrieves top-k relevant documents and synthesizes insights, possibly identifying sub-questions.
3. **Context Update:** Intermediate findings are stored in memory modules, which may record both global and step-wise (local) information relevant to the solution.
4. **Assessment & Next Query Generation:** The model determines if sufficient information has been gathered. If not, it formulates a new, more targeted sub-question to address gaps or ambiguities and triggers another retrieval step.
5. **Iteration:** Steps 2–4 repeat, with accumulating evidence and increasingly focused queries each cycle, until the stopping criterion is met (information sufficiency, confidence threshold, or iteration limit).
6. **Final Answer Generation:** The system compiles all critical findings from memory modules and generates a comprehensive answer to the original question.

A LangChain implementation can be found in github.[1]

---

[1] https://github.com/NirDiamant/RAG_TECHNIQUES/blob/main/all_rag_techniques/retrieval_with_feedback_loop.ipynb.

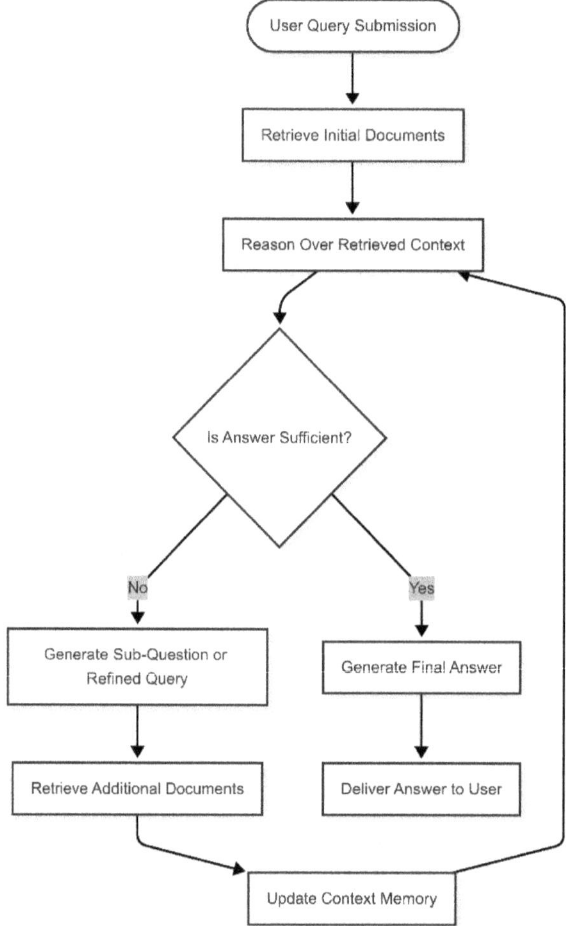

Figure 12.2.   Iterative RAG workflow.

12.3 Adaptive RAG

Adaptive RAG dynamically tailors retrieval and generation strategies to each user query's complexity and information requirements, resulting in a flexible, resource-efficient, and highly accurate system for applications ranging from customer support to conversational AI and defense analytics. It generally contains the following components:

- **Dynamic Strategy Selection:** Instead of always performing a fixed retrieval routine, the system determines on a per-query basis whether to respond using an LLM's internal knowledge, perform a single-pass (one-shot) retrieval, engage in iterative (multi-step) retrieval, or skip retrieval entirely.
- **Query Complexity Awareness:** At its core, adaptive RAG employs a query complexity classifier. This model analyzes incoming questions and predicts their complexity or informational demand, determining the optimal retrieval strategy.
- **Self-Correcting Feedback:** Adaptive RAG systems often include mechanisms to evaluate the quality and completeness of an LLM's response. If the output is incomplete, incorrect, or generic, the system adapts: rewriting the query, rerunning the retrieval, or switching to a more involved retrieval strategy

Its main idea is similar to Query Routing discussed in Chapter 6 without considering the full lifecycle. We can create a 3-stage workflow for adaptive RAG:

- **Query complexity classification:** When a user submits a query, the system employs a dedicated classifier—typically a lightweight, efficient language model—to analyze the complexity and informational demand of the question. This classifier assigns labels such as "simple," "moderate," or "complex" based on pre-trained data or heuristic rules. The goal of this stage is to ensure that each query is accurately assessed and routed to a retrieval strategy that best matches its difficulty and requirements.
- **Dynamic Retrieval Strategy Routing:** Once the complexity of the query has been determined, the system dynamically selects an appropriate retrieval strategy. For straightforward, fact-based questions, the system may bypass external retrieval entirely and rely solely on the language model's internal knowledge for an immediate response. More challenging queries trigger a single-step retrieval from an external knowledge base before generating the answer, while highly complex or ambiguous queries activate multi-step or iterative retrieval. In these cases, the system may perform several rounds of information retrieval, chain together pieces of evidence, or refine the query, progressively building up to a comprehensive and accurate response.

- **Quality Grading and Feedback:** Dedicated graders evaluate the relevance and accuracy of the retrieved documents, check for factual correctness (to prevent hallucinations), and ensure that the response fully addresses the user's intent. If the response is found to be lacking, incomplete, or incorrect, the system initiates a self-correcting loop: it may adjust the retrieval strategy, rewrite the query, or conduct further retrieval rounds. This iterative evaluation and adaptation process ensures that the final output delivered to the user is as accurate, relevant, and complete as possible.

Figure 12.3 illustrates the general workflow. The implementation can be found in github,[2] which supports both LlamaIndex and LangChain. A more complicated tool can be obtained from FlashRAG, which also comes with iterative RAG techniques.[3]

12.4 Caching

When working with large volumes of data and multiple transformation steps, caching is an essential optimization tool to speed up processing. It provides the following benefits:

1. **Efficiency:** Caching avoids re-processing data when configurations and inputs have not changed.
2. **Scalability:** Large datasets often require significant time and computation, and caching helps scale by reducing repeated workloads.
3. **Reproducibility:** Cached outputs ensure that previous runs can be reloaded and reused without discrepancies.

Below, we use LlamaIndex's IngestionPipeline as an example to explain the procedure. IngestionPipeline can automatically cache results by hashing each combination of input data and transformation steps. This caching mechanism ensures that repeated runs using the same input data and pipeline configuration can be executed efficiently without redundant computations.

---

[2]https://github.com/starsuzi/Adaptive-RAG; https://github.com/mistralai/cookbook/blob/main/third_party/langchain/adaptive_rag_mistral.ipynb; https://github.com/mistralai/cookbook/blob/main/third_party/LlamaIndex/Adaptive_ RAG.ipynb.
[3]https://github.com/RUC-NLPIR/FlashRAG.

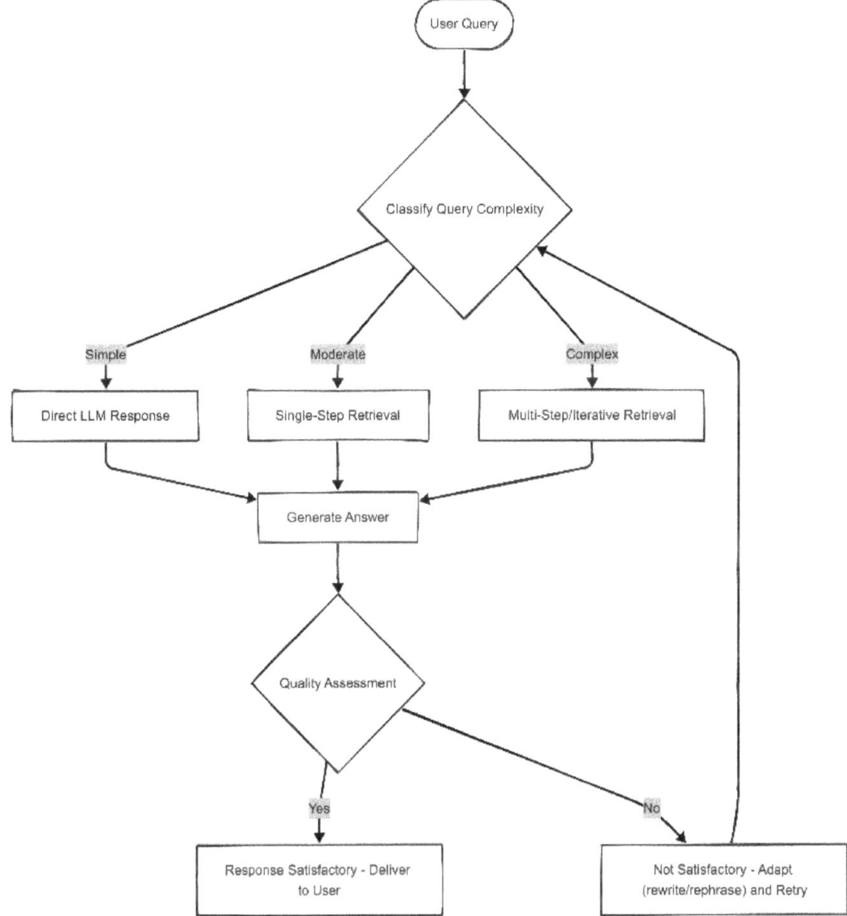

Figure 12.3.   Adaptive RAG workflow.

1. **Setting Up the Pipeline:** The IngestionPipeline consists of **transformations**, such as splitting text into chunks, extracting key metadata, or applying other preprocessing steps. Once the pipeline is configured, caching occurs automatically by hashing the nodes (data chunks) and transformations.

2. **Saving and Persisting the Cache:** Once a pipeline has processed data, you can save the cache to disk for future use. This makes it easy to restart from a known state without re-executing all transformations.

3. **Loading and Reusing the Cache:** You can load the previously saved cache to ensure that subsequent runs benefit from cached outputs without recalculating them.

Let's look at the sample code in ch12\caching.py, where we store the processed pipeline state on disk using the persist() method and restore the saved pipeline using the load() method to quickly access cached data. If the cache grows too large or needs to be refreshed, use the clear() method to remove stored data.

An enhanced version is called Cache-Augmented Generation (CAG) [1] has emerged, changing how knowledge-intensive tasks are addressed. CAG harnesses the immense information retention capabilities of long-context LLMs by seamlessly integrating preloading and caching techniques. The operational flow of CAG[4] can be divided into three key stages, each with its own distinct technical logic:

- **External Knowledge Preloading:** Before the CAG system begins responding to user queries, it first preprocesses a curated collection of documents relevant to the target application. This involves steps such as text cleaning, formatting standardization, and conversion into a format suitable for LLM consumption. The documents are then processed by the LLM, which encodes them into internal semantic vector representations through complex neural network computations. These representations are stored in a **key-value (KV) cache**,[5] which acts as a "knowledge repository" containing the LLM's precomputed understanding of the document set. A key advantage of this approach is that the document encoding process occurs only once, regardless of the number of subsequent queries. This dramatically reduces the system's overall computational load.
- **Inference:** When a user submits a query, the system loads both the precomputed KV cache and the query into the LLM. The model's internal attention mechanism dynamically links the user query to the relevant knowledge stored in the cache. By rapidly retrieving and matching semantic vectors from the preloaded knowledge, the LLM extracts the information most relevant to the query. It then generates a response

---

[4]https://github.com/hhhuang/CAG.

[5]Similar idea is also applied to LLM inferences. See Chapter 11.

by combining this extracted knowledge with its native language generation abilities. Because all the necessary external knowledge is already embedded in the cache, the inference process avoids real-time document retrieval, thereby eliminating retrieval delays and minimizing the risk of retrieval errors. This allows for fast, accurate, and consistent response generation.

- **Cache Reset:** To ensure the CAG system remains efficient across multiple inference sessions, an intelligent cache reset mechanism is employed. The KV cache grows incrementally as new tokens are processed. Over time, this can lead to excessive memory usage. When the cache exceeds a predefined threshold or when a new document set needs to be loaded, the system efficiently resets the cache. This is accomplished by truncating the most recent tokens rather than fully reloading the cache from disk, significantly saving both time and system resources. As a result, CAG maintains a high level of responsiveness and operational stability needed for sustained use.

12.5 Streaming

Streaming enhances user experience by delivering generated text incrementally as it's produced token by token, rather than waiting for the entire response to be ready. This approach, similar to the ChatGPT web interface, provides immediate feedback, making interactions feel more responsive and natural.

In traditional response generation, users experience a delay while the model processes the input and generates the complete output. Streaming mitigates this latency by transmitting each token as it's being generated, allowing users to receive and process information in real-time. This method is particularly beneficial in applications like chatbots, live translations, and interactive storytelling, where prompt feedback is crucial.

Several LLM providers offer APIs with streaming capabilities:

- **OpenAI:** Provides streaming support in their API, enabling developers to receive partial responses as the model generates them.
- **Hugging Face:** Offers streaming features for various models, allowing real-time token generation.

- **LangChain:** Supports streaming through its integration with multiple LLMs, facilitating seamless real-time interactions.
- **LlamaIndex:** Implements streaming in its query engine, allowing responses to be processed and displayed as they are generated.

To illustrate the streaming mechanism, we provide the implementation code using OpenAI, LangChain, and LlamaIndex: An example can be found in ch12\streaming.py.

- **OpenAI:** To utilize streaming with OpenAI's API, set the stream parameter to True in your API request. This code streams the response from the model, printing each token as it's received.
- **LangChain:** It integrates with various LLMs and supports streaming. To enable streaming, set the streaming parameter to True when initializing the LLM. This setup streams the generated tokens, allowing for real-time processing.
- **LlamaIndex:** It supports streaming responses through its query engine. To enable streaming, configure the query engine with streaming=True:

This configuration streams the response tokens, providing immediate feedback to the user.

Note that UI tools such as Gradio and Streamlit provide streaming support so that we can easily build a chat interface. Implementing streaming requires careful design to handle real-time data flow through backend services and frontend interfaces. Ensure that your application architecture supports asynchronous processing and that the user interface can render partial responses smoothly. Selecting appropriate orchestration and UI frameworks that facilitate streaming is crucial for delivering a seamless user experience.

12.6 Fallback Model

When you are working with LLMs, you might encounter issues such as exceeding the rate limits of OpenAI's models. It is essential to have error handling mechanism and backup models as a contingency plan in case your main model encounters problems.

One of common handing mechanisms is Retry and Exponential Backoff, which catch a rate limit error (e.g., HTTP 429) and then retry the request after waiting an increasing amount of time. This method is often implemented using libraries such as backoff.[6]

To reduce the chance of hitting rate limits on a single endpoint, we may also consider

- **Multi-Region Deployments:** If you're using Azure OpenAI, for instance, you can deploy your models across multiple regions. This not only provides load balance but also increases the available quota.
- **Multiple API Keys or Accounts:** Some developers set up multiple API keys or even multiple OpenAI accounts (ensuring compliance with OpenAI's terms of service) so that the load can be distributed among several keys. This approach should be carefully managed to avoid abuse and maintain proper tracking.
- **Integrating Alternative LLMs as Backups:** Switch to an Open-Source or Commercial Alternative: If you rely heavily on OpenAI's models, consider integrating support for other LLMs such as Anthropic's Claude or Meta's LLaMA-3 or DeepSeek. Some companies even build systems that automatically choose the best model for a given task based on current load or pricing considerations.

To effectively switch among different APIs of models, it is suggested to adopt a proper API gateway, either commercial or open-source implementations. Table 17.4 in supplementary material lists some commonly used tools.

Each of these solutions offers mechanisms to detect errors (such as "429 Too Many Requests"), automatically retry requests, and route to alternative endpoints. Depending on your environment (e.g., clouds, containerized microservices, or self-hosted LLM deployments), you can choose one that best matches your scalability, security, and integration needs.

Most of them are general-purpose gateways without specific optimization for LLMs. Neutrino and OpenRouter are dedicated to the LLM and provide similar fallback and routing capabilities, making them good options to consider in mixed API setups. A sample code to call LLM API from the gateways is available in api_gateway.py.

---

[6]https://github.com/litl/backoff.

12.7 Chat Engine and Memory

Chat engines can integrate multiple functionalities into a single component, thus playing a pivotal role in facilitating dynamic and context-aware interactions. They enable the system to maintain dialogue continuity, handle follow-up questions, and manage references to previous interactions, thereby enhancing the overall user experience. With the engine, it provides

- **Contextual Understanding:** By maintaining a history of the conversation, chat engines can interpret user queries within the context of prior interactions, leading to more accurate and relevant responses. In many cases, the intent identification is critical important.[7]

---

[7]One of the core strategies we often adopt is the introduction of an intent recognition model at the front end of the LLM pipeline. This model plays a crucial role in analyzing incoming user queries, classifying their intent, and determining whether to invoke the vectorized knowledge base or redirect the request to other systems. For example, when a generic question is asked, such as *"What are the company's ESG strategies?,"* the intent recognition model classifies it as a standard query and directly routes it to the general-purpose knowledge base. In contrast, for a query like *"Who is responsible for net zero modeling?,"* the model identifies this as a personnel-specific request. Instead of relying on generic knowledge, it dynamically retrieves the relevant information from the internal HR database, ensuring that the response is up-to-date and contextually accurate.

Furthermore, when the user asks an operational or workflow-related question, such as *"How do I submit a travel request?,"* the model classifies it as an internal procedural inquiry. Rather than generating an answer from knowledge bases, the system redirects the user to a custom-built enterprise application (e.g., an internal travel management portal), where the actual process can be initiated or guided.

This intent-driven routing mechanism not only streamlines the question handling process, but also enforces a clear separation between general knowledge and proprietary enterprise data. As a result, it enhances both efficiency and response precision, while respecting system boundaries and ensuring that each type of query is handled by the most appropriate backend resource.

To further extend this strategy:

- For legal or compliance-related questions, such as *"What is the latest version of the anti-bribery policy?,"* the model can route the query to the compliance document database.
- For IT support questions, like *"How can I reset my VPN password?,"* the intent model can classify it as a technical issue and forward the query to a support bot or helpdesk system.

Through this structured and intelligent orchestration, the system ensures context-aware, secure, and targeted responses across a wide range of user intents.

Table 12.1.   Comparison of two chat engines.

Aspect	ContextChatEngine	CondensePlusContextMode
Mechanism	Retrieves relevant context and appends the full chat history from a memory buffer to the current query.	Condenses the chat history and last message into a compressed query, which is then used to retrieve context.
Complexity	Relatively simple to implement.	More sophisticated due to the need for effective query compression techniques.
Handling Long Histories	Directly includes all past dialogue, which can lead to longer prompts if not managed.	Reduces the prompt length by summarizing past dialogue, enhancing efficiency with lengthy histories.
Best for	Applications with moderate dialogue depth where full context is required.	Scenarios with extensive dialogue histories where redundant information needs to be minimized.

- **Efficient Query Handling:** They manage follow-up questions and anaphora (references to previous statements), ensuring that the system comprehends and responds appropriately to user inputs that depend on earlier dialogue.
- **Enhanced User Engagement:** By providing coherent and contextually relevant responses, chat engines improve user satisfaction and engagement, making interactions feel more natural and intuitive.

There are mainly two types of chat engines based on how the dialogue information is passed in Llamaindex. Table 12.1 compares them in several attributes.

Both LlamaIndex (e.g., index.as_chat_engine()) and LangChain (e.g., ConversationChain) provide the supports to chat engine. See the sample code in ch12\chat_engine.py with the memory support.

12.8 Model Hyperparameter Tuning

LLM are used in multiple steps of RAG pipeline. Hence, we shall know at least their hyperparameters during inference phase. Some typical variables are summarized with their descriptions and general implications in Table 12.2.

Table 12.2. LLM inference hyperparameters.

Hyperparameter	Description	General Implication
Temperature	Controls the randomness of predictions by scaling logits before applying softmax.	A higher temperature (e.g., >1) increases randomness, leading to more diverse outputs; a lower temperature (<1) makes the output more deterministic and focused.[8]
Top-p (Nucleus Sampling)	Limits the sampling pool to a subset of probable next tokens based on cumulative probability.	Balances randomness and coherence by considering only the most likely tokens up to a cumulative probability threshold, ensuring generated text remains contextually relevant.
Top-k Sampling	Considers only the top "k" probable next tokens during generation.	Restricts the model to choose from the "k" most probable tokens, reducing the chance of less likely tokens being selected, which can enhance coherence but may limit diversity.
Max Tokens	Sets the maximum number of tokens to generate in the response.	Limits the length of the generated output, preventing excessively long responses and controlling computational resources.
Stop Sequences	Specifies sequences that, if generated, will halt further generation.	Ensures the model stops generating text upon encountering specified sequences, which is useful for controlling output format and preventing undesired continuations.
Frequency Penalty	Penalizes tokens that have already appeared frequently in the generated text.	Discourages repetition by reducing the likelihood of selecting tokens that have been used frequently, promoting more diverse outputs.
Presence Penalty	Penalizes tokens that have already appeared in the context.	Reduces the probability of generating tokens that are already present in the input context, encouraging the introduction of new information.

[8]Setting temperature as 0 may not make the result deterministic. The blog post (by thinking machine at https://thinkingmachines.ai/blog/defeating-nondeterminism-in-llm-inference/) challenges the common belief that LLM nondeterminism is primarily caused by floating-point arithmetic and GPU concurrency. It argues that while these can contribute, the main culprit is a lack of "batch invariance." This means that the output of a single query can change depending on what other queries are processed alongside it in the same batch.

Let's look at how to adjust parameters for chunk_size and similarity_top_k only in ch12\hyper_parameter.py, as modifying these parameters affects the balance between computational speed and the accuracy of the information retrieved. We can define a function objective_function_semantic_similarity with param_dict containing the parameters, chunk_size and top_k, and their corresponding proposed values: For more details, refer to LlamaIndex's full notebook on Hyperparameter Optimization for RAG.[9]

Another important top k parameter happens in retrieval process. Selecting a fixed number of top-k results can inadvertently include outliers: results that are semantically similar but contextually irrelevant. To enhance retrieval precision, consider the following strategies:

1. **Distance Threshold Filtering:** After retrieving the initial top-k candidates, apply a similarity score threshold to exclude items that fall below a certain relevance level. This method ensures that only results with a high degree of similarity to the query are retained.
2. **Adaptive Cutoff (Autocut):** Instead of relying on a fixed threshold, analyze the distribution of similarity scores among the retrieved results. If a significant drop in similarity is observed between consecutive items, it may indicate the transition from relevant to irrelevant results. By identifying this "elbow point," you can dynamically determine the optimal number of results to include, tailoring the cutoff to each specific query.
3. **Dynamic Top-k Prediction via Cross-Encoders:** Employ a trained cross-encoder model to predict the optimal number of top-k results for each query dynamically. This method assesses the intricacy of each query and adjusts the retrieval breadth in real-time, ensuring that each question is met with a tailored, precise, and resource-efficient response.

12.9 Model Finetuning

RAG systems combine the generative power of LLMs with dynamic external knowledge retrieval. While RAG excels at grounding responses in up-to-date or domain-specific data, its effectiveness hinges on the LLM's ability to synthesize retrieved information coherently and align

---

[9]https://docs.llamaindex.ai/en/stable/examples/param_optimizer/param_optimizer.html.

with task-specific requirements. General-purpose LLMs like GPT-4/5 or Llama-3/4 are trained on broad datasets, making them suboptimal for specialized tasks. As discussed earlier, the common issues like Hallucinations, Style Mismatch and Distractor Handling, can be further relieved by a better LLM. Fine-tuning bridges the gap between retrieval and generation and then enhances retrieval-response synergy, e.g.,

- **Context Integration:** Models learn to prioritize and cite retrieved passages, improving answer accuracy.
- **Adaptation to Imperfect Retrieval:** Even with noisy retrievers, fine-tuned models can compensate by leveraging internalized domain knowledge.
- **Task-Specific Optimization:** For applications like medical diagnosis or legal analysis, fine-tuning aligns the LLM's reasoning with domain workflows.

12.9.1 *Methodology*

One of the most common approaches is Parameter-Efficient Fine-Tuning (PEFT) shown in Figure 12.4, which allows you to fine-tune LLMs like Llama3/4 by updating only a small subset of parameters. This approach significantly reduces computational and storage costs and mitigates the problem of catastrophic forgetting observed during full fine-tuning of LLMs. PEFT methods have also been shown to outperform traditional

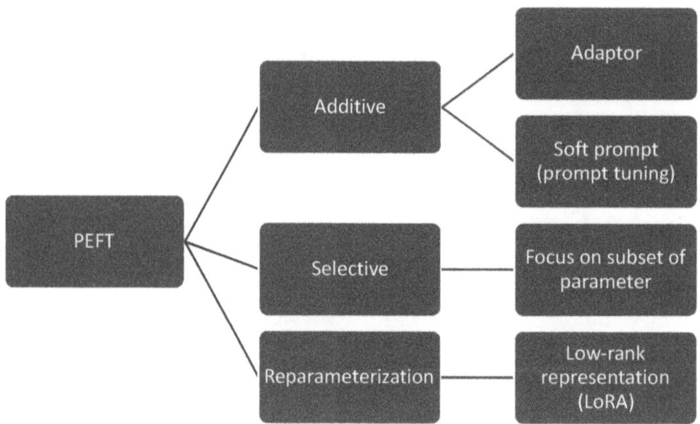

Figure 12.4. PEFT methods.

fine-tuning in low-data regimes and generalize better to out-of-domain scenarios.

1. Selective Fine-Tuning

This method involves selecting only a subset of the model's parameters to fine-tune while keeping the majority of the model's parameters frozen (unchanged). By focusing only on a specific part of the model, it reduces the number of parameters that need to be optimized, making the process faster and less computationally expensive.

- **Use Case:** Tasks where most of the model's pre-trained knowledge is useful, and only a small portion needs adjustment for the new task.

2. Reparameterization (LoRA – Low-Rank Adaptation)

In this method, the model's parameters are "reparameterized" using a low-rank approximation. LoRA applies low-rank matrices to approximate changes in the weights of the model. Rather than fine-tuning the entire weight matrix of the model, a low-rank update is made, which saves memory and computation costs. The key benefit of LoRA is that it keeps majority of the model unchanged and only makes efficient adjustments where necessary.

- **Use Case:** Resource-constrained environments where the goal is to fine-tune large models without increasing memory overhead.

3. Additive Fine-Tuning

In additive methods, new trainable parameters or layers are added to the model without altering the original model's weights. These new layers or parameters are trained while the original model remains fixed. Two common approaches in this category are:

- **Adapters:** Small trainable modules are added between layers of the pre-trained model. This design allows for task-specific adaptation and improves performance while preserving the underlying model's pre-trained knowledge. The adapters modify the information flow within the model, improving task performance without retraining the entire model.
- **Prompt Tuning/Soft Prompts:** In prompt tuning, additional prompts or embeddings are learned to guide the model's responses. Instead of modifying the model's weights, the model is given learned soft prompts that help the model better handle specific tasks.

Below are key steps in the fine-tuning procedures:

1. **Task Definition:** Identify the target application (e.g., QA, summarization) and performance metrics (e.g., accuracy, BLEU score).
2. **Dataset Preparation:** Curate labeled data (e.g., question-answer pairs) or unlabeled data for self-supervised learning.
3. **Model Selection:** Choose a base LLM (e.g., Llama-2, Mistral) based on size, licensing, and compatibility with frameworks like Hugging Face.
4. **Parameter-Efficient Fine-Tuning (PEFT):** Use techniques like LoRA (Low-Rank Adaptation) to update a subset of weights, reducing computational costs.
5. **Training & Evaluation:** Optimize using loss functions (e.g., cross-entropy) and validate with hold-out datasets.

12.9.2 *Tools*

Commercial LLM providers, such as OpenAI, Google and Anthropic, have direct APIs to fine tune some models in a relatively easy way. For example, OpenAI introduced a fine-tuning API for LLMs, and LlamaIndex has provided a tutorial on fine-tuning GPT-3.5-turbo within a RAG setting to "distill" some of the knowledge from GPT-4.[10] The process involves the following steps:

1. **Question Generation:** Use GPT-3.5-turbo to generate a series of questions based on a given document.
2. **Answer Generation:** Utilize GPT-4 to generate answers to these questions using the document's contents, effectively creating a GPT-4-powered RAG pipeline.
3. **Fine-Tuning:** Fine-tune GPT-3.5-turbo on the resulting dataset of question-answer pairs.
4. **RAG Evaluation:** Use a framework for RAG pipeline evaluation, e.g., ragas, RAGchecker.

Some popular open-source general-purpose tools, can achieve the similar goals, e.g., Hugging Face, PyTorch Lighting, LoRA, and etc.

---

[10] https://developers.llamaindex.ai/python/examples/finetuning/openai_fine_tuning/.

There are tools dedicated to RAG pipeline optimization. For example, Retrieval Augmented Fine Tuning (RAFT) by UC Berkeley optimizes LLMs for RAG on domain-specific knowledge [2]. A more sophisticated approach is detailed in **RA-DIT (Retrieval Augmented Dual Instruction Tuning)** by Meta AI Research [3]. This technique involves tuning both the LLM and the retriever (a dual encoder) into triplets of query, context, and answer.

- **Triplet Tuning:** Fine-tune the LLM and the retriever on query-context-answer triplets.
- **Implementation:** The technique was used to fine-tune both OpenAI LLMs through the fine-tuning API and the Llama2/3 open-source model.

This approach resulted in approximately a 5% increase in knowledge-intensive task metrics compared to the Llama2 65 B model with RAG. Additionally, there was a slight improvement in common sense reasoning tasks. This advanced fine-tuning method demonstrates significant improvements in the model's ability to handle complex and knowledge-intensive queries, thereby enhancing the overall performance of RAG systems.

12.9.3 *Illustration: Fine-Tuning a Medical QA Model for RAG*

This section demonstrates fine-tuning a Llama-2 7 B model to improve RAG performance in medical question answering via the sample code in ch12\fine_tune.py. The fine-tuning pipeline begins with data preparation, where question-context-answer triplets are created to train the model on retrieval-augmented generation tasks. Next, a parameter-efficient LoRA adaptation is applied to a pre-trained language model (like TinyLlama or GPT-2) to enable efficient fine-tuning with minimal trainable parameters. The pipeline then uses supervised fine-tuning with formatted prompts to teach the model how to generate accurate answers based on retrieved documents. During inference, the trained model can generate context-aware responses by taking questions and supporting documents as input. Finally, the implementation includes a comprehensive evaluation framework to assess model performance using both quantitative metrics and qualitative assessments. We may also apply Retrieval Accuracy (Recall@K) and BLEU Score for evaluation (the details can be found in Chapter 10 on Evaluation).

12.10 Context Engineering

The recent concept *context engineering* also covers the whole pipeline with an expanded definition of "context." It is not merely the immediate prompt sent to an LLM. Rather, it encompasses *everything* the model can perceive prior to generating a response. Broadly, context can be categorized into seven distinct components:

1. **System Instruction (Prompt):** The initial directives that define the behavior of the model throughout a session. These may include rules, tone guidelines, formatting preferences, or exemplar prompts.
2. **User Prompt:** The specific task or query issued by the user in the current interaction.
3. **Conversation History (Short-Term Memory):** The accumulated dialogue between the user and model during the session, providing continuity and coherence.
4. **Persistent Memory (Long-Term Memory):** The system stores knowledge from previous sessions, such as user preferences, summaries of past projects, or facts the model was instructed to retain for future use.
5. **External Knowledge Base (RAG):** Information retrieved from external sources—such as document databases, APIs, or real-time search engines—designed to supplement the model's native knowledge with up-to-date and query-specific content.
6. **Available Tools:** Definitions of callable functions or services available to the model (e.g., check_inventory, send_email, or query_weather), enabling it to interact with external systems and perform actions beyond text generation.
7. **Output Format Specifications:** Definitions of the required structure for model outputs, such as response schemas in JSON or tabular formats, ensuring compatibility with downstream systems. This can be part of system or user prompts.

While *prompt engineering* traditionally focuses on crafting a single optimal text string to guide the model's behavior, *context engineering* is a broader and more systemic discipline, i.e., context engineering is the practice of designing and building dynamic systems that delivers the right information and tools, in the right format, at the right time. In other words, it provides LLMs with everything they need to complete a task effectively.

This discipline treats context not as a static prompt, but as the *output of a structured system* that executes prior to any LLM invocation. Key characteristics include:

- **Systemic:** Context is not just a string—it results from orchestrated operations, potentially involving APIs, databases, calendars, user profiles, and more.
- **Dynamic and Task-Specific:** Context is constructed dynamically, tailored to each request. For instance, the context for a scheduling query may include calendar data, while a travel query may include recent flight prices or location history.
- **Precision Delivery:** One of the primary goals is to ensure that the model receives all essential information without overloading it. This involves curating both *knowledge* (informational context) and *capabilities* (tool access), providing them only when they are both necessary and useful.
- **Format Matters:** How information is presented can greatly influence model performance. Concise summaries are generally more effective than raw data dumps, and clearly defined tool specifications are more reliable than vague or ambiguous instructions.

In essence, *context engineering* is the foundation for building intelligent, adaptive LLM-powered systems. It shifts the focus from prompt crafting to *orchestrating an environment* in which the model can operate with precision, relevance, and reliability.

In practice, context engineering may have typical failure modes, e.g., 1) context poisoning (bad or hallucinated facts linger), 2) distraction (repetition or filler derails plans), 3) confusion (too many similar tools), 4) clash (contradicting calls). All become more likely as contexts grow.

Also, multiple independent studies now show performance degrades non-uniformly as input length increases ("context rot"), so bigger windows alone don't save you.

To alleviate those issues, some core strategies were suggested by LangChain[11] and Manus[12]:

1. **Offload context:** Push long-term state and heavy artifacts out of the prompt into durable storage (files, notes, docs). Read/write them on demand; use them as the agent's working memory (e.g., todo.md, research briefs).

2. **Reduce context:** Summarize/prune history and tool outputs. Summarize agent-agent handoffs. Be wary of aggressive compression—information loss can hurt downstream steps.

3. **Retrieve context:** Mix retrieval methods + re-ranking; assemble retrieved snippets into prompts; also retrieve the right tools from tool descriptions (not just documents).

4. **Isolate context:** Split work across sub-agents or threads to avoid mixing **unrelated histories**; keep each thread's context focused. Use with care—multi-agent coordination can fail; prefer isolation for parallelizable tasks (e.g., sub-topic research) and consolidate later.

5. **Cache context:** Treat KV-cache hit rate as a first-class metric. Keep prefix prompts stable; make histories append-only and deterministically serialized; avoid touching tool lists mid-run (it busts cache). Add mutable, recent observations at the suffix. Cached tokens can be ~10× cheaper.

Manus-style context engineering emphasizes a few concrete habits: keep the tool inventory fixed and steer behavior by masking logits (or via a context-aware state machine) instead of dynamically adding or removing tools, which preserves cache coherence and reduces model confusion; persist bulky observations outside the prompt (e.g., in a sandbox or filesystem) and reference them with compact handles to keep tokens low while remaining fully reversible; periodically "recite" objectives by rewriting a brief or to-do at the tail of the context to counter

---

[11]Context engineering in agents—Docs by LangChain, https://docs.langchain.com/oss/python/langchain/context-engineering, 2025. Accessed on 30/09/2025.

[12]Context Engineering for AI Agents: Lessons from Building Manus, https://manus.im/blog/Context-Engineering-for-AI-Agents-Lessons-from-Building-Manus, Jul 2025. Accessed on 30/09/2025.

lost-in-the-middle effects; retain failures, stack traces, and other telemetry in the run so the model and operators can self-correct; and, above all, design around the KV-cache by stabilizing prefixes and serialization because cache hits drive latency and cost in production. Finally, avoid brittle imitation from overly uniform few-shot traces by injecting small, structured variation so agents don't pattern-lock on a single style. Figure 12.5 shows a agent with the context engineering workflow (See more in Chapter 15 on Agent).[13]

12.11 Deep Search

Deep Search focuses on enhancing search itself, aiming to provide more precise, thorough, and context-aware results than traditional search engines. As it usually comes with search summary, it is generally a RAG pipeline. Rather than simply listing links, it interprets user intent at a semantic level and digs deeper across broader data sources. For example, Bing's Deep Search enhances complex queries with AI, while IBM's Deep Search applies semantic search across scientific papers and patents. An open-source implementation DeepSearcher integrates latest LLMs and vector databases for deep search, evaluation and reasoning for private data.[14] Its key techniques include:

- **Semantic query expansion:** Transforming short queries into more descriptive ones to capture intent.
- **Wide-ranging, multi-hop & layered search:** Traversing more sources than standard search, sometimes at tenfold scale. Iteratively expands, narrows, and verifies evidence.
- **Result aggregation and ranking:** Sorting results by relevance, authority, recency, and completeness.
- **AI-driven summaries:** Presenting concise lists, answers, or comparative summaries rather than raw links.

Deep search is used where queries are complex, ambiguous, or highly specialized, e.g., complex Q&A of multi-dimensional queries such as global loyalty programs, specialized research such as Mining patents,

---

[13] When the context engineering is applied to AI Agent, three new concepts, specification engineering, skills and harness engineering, are emerging. They have similar foundation from prompt engineering, but different in core irresponsibles.

[14] https://github.com/zilliztech/deep-searcher.

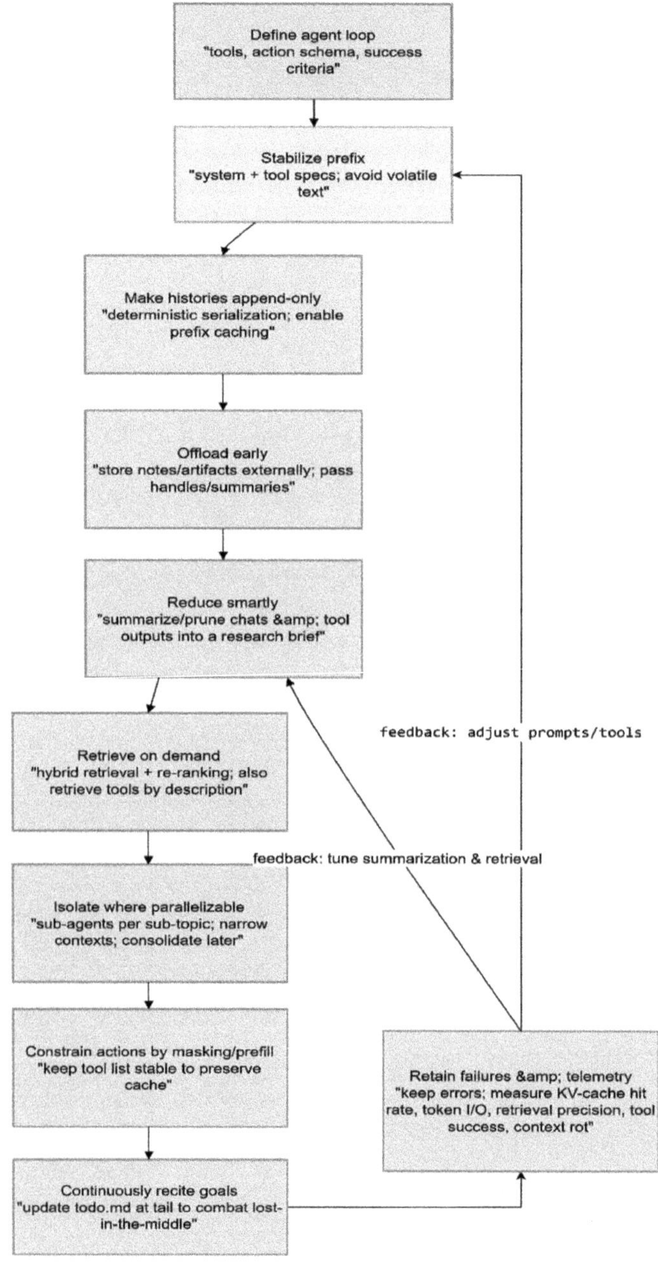

Figure 12.5.   Content engineering workflow.

scientific articles, or niche databases, personalized or long-tail queries such as travel planning, niche events, or obscure investigative tasks.

We can see this technique improves retrieval quality and breadth, making information discovery more efficient, while it consumes more resources, has slower response times, and may misinterpret ambiguous queries. It complements rather than replaces conventional search. Figure 12.6

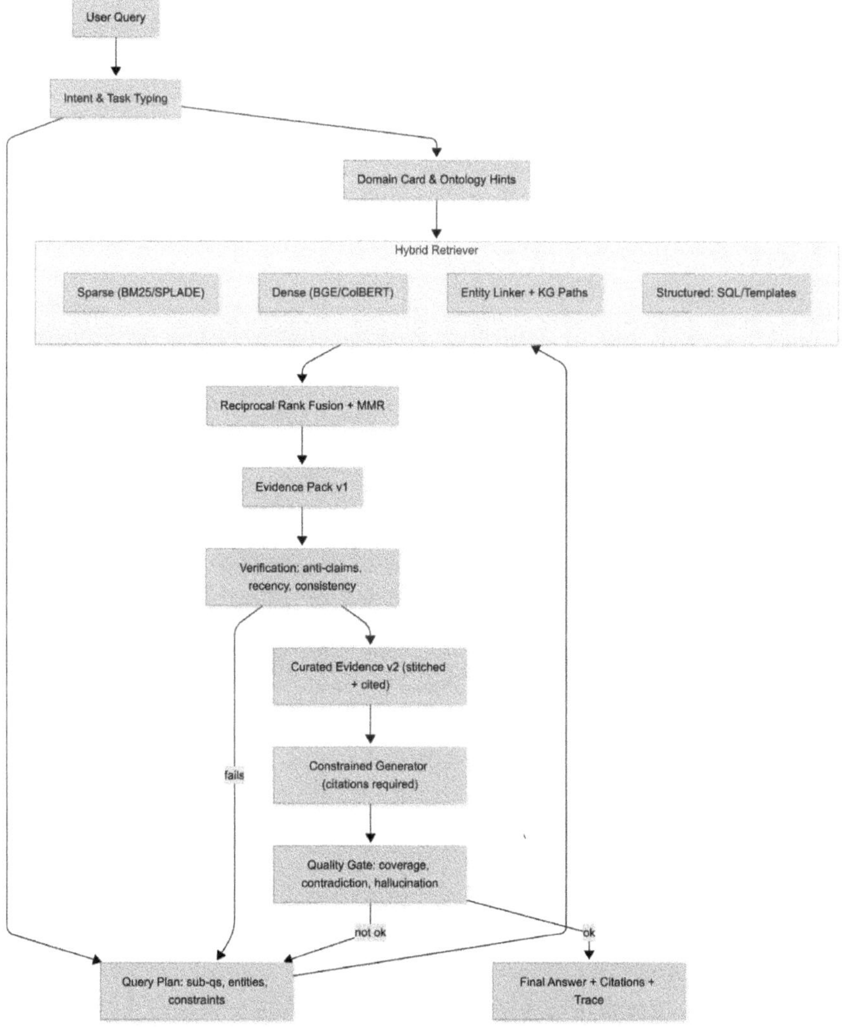

Figure 12.6. Deep search workflow.

illustrates a reference workflow. The deep search pipeline starts by classifying the user request to determine the task type and evidence requirements, then builds a structured query plan that decomposes the problem into subquestions with entities, constraints, and time windows. A hybrid retriever executes this plan across lexical (BM25/SPLADE), dense (e.g., BGE/ColBERT), knowledge-graph hops, and structured routes (SQL/templates), after which re-ranking is applied. High-scoring snippets are stitched into coherent passages with provenance, then a verification pass probes for counter-evidence, checks recency and cross-source agreement, and flags gaps. If checks fail, the plan is refined and another retrieval hop runs; if they pass, a constrained generator produces an answer that cites sources inline. A final quality gate enforces coverage and contradiction checks before returning the answer with a trace.

A releted concept "Deep Research" will be discussed in Chapter 15.

12.12 Concluding Remarks

A robust RAG system is ultimately an orchestration problem: coordinating retrieval, generation, and control logic so the right *data* reaches the right *model* with the right *constraints*. In this chapter we examined four levers. First, pipeline enhancement patterns such as *adaptive RAG*, i.e., retrieving only when and as much as needed, and *iterative RAG*, i.e., reasoning, acting, and re-retrieving in loops, improve factuality and relevance while containing cost. Second, engineering components like semantic cache, tiered memory, streaming transport, stateful chat engines, and resilient fallback/routing make applications responsive and reliable at scale. Third, model optimization via hyperparameter tuning[15] and selective finetuning aligns models to domain style and tasks without overfitting. Finally, context engineering from chunking and windowing to prompt schemas and guardrails keeps the LLM grounded and auditable. In short, great RAG is less a single model choice and more a disciplined pipeline that blends adaptive retrieval, careful context control, and production-grade orchestration. Below are few future directions

---

[15]We might view the whole pipeline as a procedure full of "hyperparameters." Therefore, we can tune chunking, top-k, similarity thresholds, rerankers, and etc.

- **Policy-driven adaptive RAG as a default:** Expect broader adoption of adaptive retrieval policies that *decide when to retrieve, how much to retrieve, and when to skip retrieval entirely*, as in Self-RAG; this reduces unnecessary calls and improves citation quality. Tooling will expose these policies as first-class, configurable components.
- **Iterative reasoning–retrieval loops with stronger planners:** ReAct-style agents [4] that interleave reasoning, tool use, and re-querying will become standard, with small "planner" models steering depth vs. latency per query. Recent work on stepwise reasoning + retrieval suggests steady gains here [5].
- **Tiered memory beyond the context window:** "Virtual context" managers (e.g., Letta/MemGPT[16]) will formalize short-term, long-term, and episodic memory, letting chat engines preserve state across long sessions and large corpora without ballooning token costs.
- **First-class semantic caching in the serving stack:** Caching will move from ad-hoc to managed (Redis semantic caches, cloud tutorials), with cache-hit telemetry shaping retrieval depth and reranking choices dynamically.
- **Resilience by design: guardrails and multi-LLM fallback:** Production stacks will standardize guardrails plus dynamic routing/ fallback across providers (latency-, cost-, or capability-aware), minimizing downtime and policy violations.
- **Platform-level orchestration patterns:** Cloud reference architectures (Azure, AWS Step Functions) will codify multi-step RAG flows (e.g., evaluation, safety checks, and human-in-the-loop) into reusable blueprints.
- **Evaluation-driven tuning loops:** Hyperparameters (chunk size/overlap, top-k, thresholds, rerankers) and fine-tuning will be optimized via continuous, task-specific evaluations wired into the pipeline, closing the gap between offline metrics and live user outcomes.

The bottom line of the next generation of RAG will look less like a static chain and more like a *controlled system* or even an operation system, e.g., policy-guided, cache- and memory-aware, resilient to failures, and continuously tuned by measurements. The winners will treat pipeline & orchestration as product surfaces, not glue code.

---

[16]https://github.com/letta-ai/letta.

References

[1] B. J. Chan, C.-T. Chen, J.-H. Cheng and H.-H. Huang, *Don't Do RAG: When Cache-Augmented Generation is All You Need for Knowledge Tasks,* arXiv:2412.15605v2, 2025.

[2] T. Zhang and *et al.*, *RAFT: Adapting Language Model to Domain Specific RAG,* arXiv:2403.10131, 2024.

[3] X. V. Lin and *et al.*, *RA-DIT: Retrieval-Augmented Dual Instruction Tuning,* arXiv:2310.01352, 2024.

[4] S. Yao, J. Zhao, D. Yu, N. Du, I. Shafran, K. Narasimhan and Y. Cao, *ReAct: Synergizing Reasoning and Acting in Language Models,* arXiv: 2210.03629v3, 2023 .

[5] Y. Li, Q. Luo, X. Li, B. Li, B. Wang, Y. Zheng, Y. Wang, Z. Yin and X. Qiu, *R3-RAG: Learning Step-by-Step Reasoning and Retrieval for LLMs via Reinforcement Learning,* arXiv:2505.23794v1, 2025.

Chapter 13
RAG with Database and Text2SQL

13.1 Introduction

Text2SQL (or NL2SQL) [1–3] systems are crucial for enabling users to interact with databases by translating natural language (NL) queries into structured query language (SQL). This capability is instrumental in democratizing data access and analysis, allowing non-technical users to extract valuable insights from structured data efficiently. In the context of RAG over structured data, Text2SQL acts as a vital bridge, connecting the natural language reasoning abilities of LLMs with the scalable computational power of database management systems. This integration is critical as traditional RAG often relies on limited point lookups, while table-augmented generation (TAG) [4] or RAG2SQL provides a unified paradigm to handle broader, more complex user questions over databases.

Text2SQL systems have evolved through several distinct stages, starting with early rule-based systems and semantic parsing techniques that, while foundational, struggled to handle the diversity of natural language queries. A significant shift occurred with the advent of deep learning and neural networks, introducing sequence-to-sequence (Seq2Seq) architectures that improved translation. Subsequently, the integration of pre-trained language models (PLMs) further advanced the field, leveraging extensive linguistic knowledge to enhance natural language understanding and SQL generation. Most recently, LLMs have revolutionized Text2SQL with their emergent capabilities, such as instruction-following and few-shot learning, and can also help mitigate hallucination issues by grounding LLM knowledge with factual database content.

Despite significant advancements, Text2SQL systems face several persistent challenges in handling the linguistic complexity and inherent ambiguity of natural language queries, which often include nested clauses, coreferences, and vague or underspecified terms. Further hurdles arise from database complexity and diversity, such as hundreds of interconnected tables with diverse naming conventions, and the inherent complexity of SQL generation due to its one-to-many mapping for complex operations. Models also struggle with cross-domain generalization, often requiring extensive retraining for new domains, compounded by the scarcity of large-scale, high-quality human-labeled datasets. Computational inefficiency and high inference costs are persistent technical obstacles for LLM-based systems, stemming from significant resource demands and high token consumption. Finally, concerns about data privacy when handling sensitive information and the interpretability of generated SQL queries remain significant challenges that require further attention.

The general methodology for Text2SQL systems, particularly when integrated with RAG, typically involves a series of interconnected steps. Foundational components include 1) Natural Language Understanding (NLU) to interpret user query intent and semantics, 2) Schema Linking to connect query elements to corresponding database tables and columns, 3) Semantic Parsing to convert queries into an intermediate logical form, and 4) SQL Generation to translate this form into a syntactically correct and executable SQL statement. Schema linking is particularly crucial in the LLM era due to input length limitations and the need to filter relevant schema elements.

The TAG/RAG2SQL framework provides a unified approach for answering natural language questions over databases, encompassing and extending traditional Text2SQL and RAG methods. It defines three key steps: Query Synthesis, which translates the user's natural language request into an executable database query (e.g., SQL); Query Execution, which runs the synthesized query on the database to efficiently compute and retrieve relevant structured data; and Answer Generation, which utilizes the original request and the computed data to generate the final natural language answer, leveraging the LM's semantic reasoning capabilities. By integrating these methodologies, Text2SQL systems for RAG aim to provide comprehensive, accurate, and efficient responses to complex natural language queries over diverse structured data sources. Due to its unique features, the evaluation of Text2SQL and RAG2SQL may have different metrics.

13.2 Text2SQL

In the digital era, data has become a core asset for enterprises, with SQL being the key to unlocking these assets. However, the complexity of SQL often poses a significant barrier for non-technical users, hindering their ability to directly extract value from data. To address this issue, Text2SQL technology has emerged, allowing users to interact with databases in natural language, significantly lowering the barriers to data analysis.

The Text2SQL task aims to transform natural language questions into equivalent SQL query statements. Traditional methods based on pattern matching and machine learning face limitations when dealing with complex and diverse queries, primarily reflected in the following aspects:

1. **Natural Language Ambiguity:** The same question can be posed in many different ways, and different questions may have similar expressions. This ambiguity presents a major challenge for the model to accurately understand the user's intent.
2. **Query Complexity:** Real-world data queries often involve multiple tables, complex condition filters, and aggregation operations. Traditional methods lack the precision needed to generate complex SQL queries.
3. **Domain Adaptability:** Databases across different domains have unique schemas. Traditional methods face challenges adapting to new domains, hindering their ability to generalize effectively.
4. **Human-Computer Interaction Limitations:** Traditional Text2SQL systems lack a deep understanding of user intent and an effective error feedback mechanism, which restricts the implementation of multi-turn dialogues and interactive queries.
5. **SQL Optimization Issues:** Although LLMs can generate correct SQL queries, there is significant room for improvement in optimizing query efficiency. Integrating knowledge of database principles and optimization rules into prompt engineering is essential to guide LLMs in learning cost estimation and query rewriting strategies.
6. **Use of Foreign Key Information:** Effectively utilizing foreign key information to enhance Schema Linking performance is a crucial challenge. This requires explicitly marking foreign key relationships in prompt templates or encoding foreign key information in the model's embeddings.
7. **Expansion of Few-shot Learning Capabilities:** The small-sample learning capabilities of LLMs in Text2SQL tasks need further

exploration. By designing more efficient few-shot Learning paradigms, the dependence of LLMs on annotated data can be reduced, enhancing their adaptability to new domains.

8. **Explainability, Fairness, and Data Security:** Developing LLM-driven Text2SQL systems requires significant attention to explainability, fairness, and data security to ensure responsible use of technology and avoid unnecessary risks. A proper entitlement mechanism is essential.

There are generally two types of approaches to generate SQL. The traditional approach includes rule-based systems using SQL templates for various scenarios with statistical language models (e.g., semantic parsers) and neural network-based models (deep learning-based approaches) mainly using encoder-decoder architecture.[1] Meanwhile, the LLM-based approach applies prompt engineering (in-context learning) and fine-tuning. Prompt engineering involves designing well-structured prompts that include task descriptions, table schemas, questions, and sometimes additional domain-specific knowledge or sample data. Techniques such as chain-of-thought (CoT) and Decomposition Strategy are used to guide the LLM through step-by-step reasoning processes, supporting zero-shot and few-shot learning with minimal or no task-specific training data.

---

[1]Typically, an encoder captures the semantics of the NL question and table schema, while a decoder generates the SQL query token by token.

- Encoding Strategies include LSTM-based Methods (Utilize bidirectional LSTMs (Bi-LSTM) to learn contextualized representations of NL questions and table schemas), Transformer-based Methods (Employ self-attention mechanisms, allowing them to handle long-range dependencies more effectively. These models often concatenate the NL question and database schema as an integrated input sequence), Graph-based Encoding (Represents NL and database schema as interconnected graphs, leveraging relational structures and interdependencies to offer richer context for each element. Examples include RAT-SQL, S2SQL, and LGESQL).

- Decoding Strategies include Sketch-based Methods (Decompose SQL generation into sub-modules (e.g., SELECT column, AGG function, WHERE value). These methods often predict the SQL query structure before filling in specific details, improving accuracy and efficiency. SQLNet and TypeSQL are early examples. Generation-based Methods (Decode SQL directly using Seq2Seq models, often producing an abstract syntax tree (AST) in depth-first traversal order. These approaches are more suitable for complex SQL scenarios.), Constrained Decoding (Applies SQL grammar rules at each step to ensure syntactic correctness, reducing invalid or incomplete SQL queries). Figure 13.1 compares the seq2seq vs structured approach.

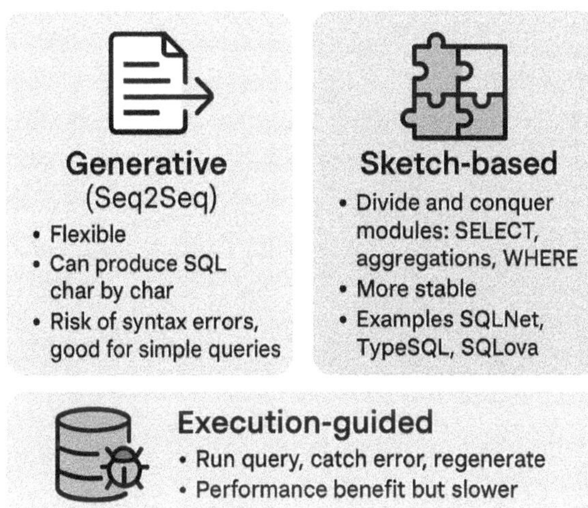

Figure 13.1.   Seq2seq vs sketch-based approach.

Fine-tuning, on the other hand, involves training pre-trained LLMs on Text2SQL datasets to adapt them to the specific task, with parameter-efficient fine-tuning (PEFT) being a preferred method due to its efficiency. We will discuss them in detail later.

There might be some post-processing strategies, which refine generated SQL queries for better accuracy and stability.

- **SQL Correction Strategy:** Identifies and corrects syntax errors or unnecessary keywords in generated SQL. DIN-SQL uses a self-correction module [5].
- **Output Consistency (Self-Consistency/Cross-Consistency):** Ensures uniformity of SQL by sampling multiple reasoning results and selecting the most consistent one, often through majority voting.
- **N-best Rankers Strategy:** Reranks top-k results generated by the Text2SQL model to enhance query accuracy.
- **Execution-Guided (EG) Strategy:** Uses the execution results of SQL as feedback to guide subsequent refinements. Errors or NULL values signal potential issues.[2]

---

[2]In EG decoding, every candidate SQL is run on the database during generation; those that raise an error are pruned and the model retries. EG delivers large gains because only syntactically and semantically valid queries survive. Yet each extra round-trip to the DB adds

13.3 RAG2SQL (TAG)

Despite the simplification of the query process by Text2SQL, it still falls short in handling complex queries and integrating external knowledge. Therefore, TAG or RAG2SQL technology was developed, combining retrieval enhancement and generative models to significantly improve the precision and capability to handle complexity in SQL queries. RAG2SQL is one of the most important technical foundations for ChatBI or data agent.[3]

Traditional Text2SQL models rely solely on pre-trained language models to convert natural language queries into SQL statements. While effective for straightforward queries, they often struggle with complex questions requiring contextual understanding or external information. RAG2SQL overcomes these challenges by implementing a two-step process using the principle of RAG, e.g., Retrieval and Generation. By combining retrieval with generation, RAG2SQL dynamically incorporates external knowledge into the query generation process, enhancing accuracy and reducing the likelihood of errors or hallucinations. RAG2SQL is designed to hide the technical complexity of querying relational data, enabling analysts, domain specialists, and other non-technical stakeholders to probe datasets conversationally and see structured results immediately.

A typical RAG2SQL system begins with a lightweight "knowledge ingestion" phase. Here, sample SQL statements, data-dictionary text, entity-relationship diagrams, and the database's Data Definition Language (DDL) are embedded and indexed. These artifacts provide the model with domain context (e.g., table names, column semantics, common joins, data types, and business vocabulary) so that later prompts can be grounded in the actual schema rather than in generic SQL patterns.

The remaining workflow is divided into two simple steps–training the RAG "model" on your data and then asking questions that return SQL

---

latency; for interactive BI dashboards the overhead may be prohibitive, so production systems often combine lightweight sketches with selective EG checks on a short list of high-probability queries. Note that some solutions suggest a simulated execution engine to reduce the overall cost.

[3]ChatBI understands user questions through AI tools and answers in natural language, processes natural language into SQL through Text2SQL, and often executes SQL in agents to display results visually. It redefines data analysis methods.

queries, which can be set to run automatically on your database, as shown in Figure 13.2.

1. **Model training/fine-tuning phase:** First, you need to train a custom model on your dataset. The higher the quality and quantity of data used, the better the model's performance. This step might be optional with the general-purpose LLMs or pre-trained SQL models.
2. **Query phase:** Once the model is trained, you can start asking questions. The system will use the model to generate SQL queries, retrieving the required data from your database.

At the query phase, the workflow unfolds in three main stages. **(1) Retrieval.** Natural-language input is matched against the indexed artifacts; the most pertinent snippets—such as similar historic queries or table descriptions—are selected and bundled with the user's question. Some implementations add a lightweight ranking or classification layer to refine that bundle, ensuring only highly relevant context is forwarded. **(2) Generation.** A language model receives the augmented prompt and produces a syntactically correct, executable SQL statement. Optionally, the platform can run the query automatically, return the result set as a Pandas DataFrame, visualize key metrics, and suggest follow-up questions that encourage deeper exploration. **(3) Execution.** The SQL is executed, possibly with human's permission. The SQL

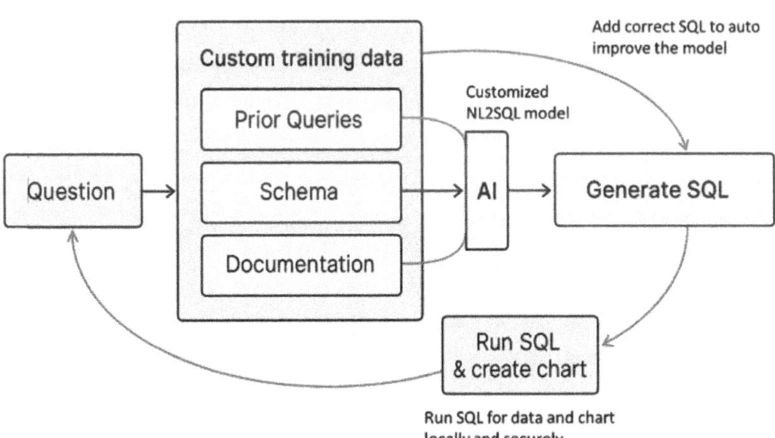

Figure 13.2.   RAG2SQL: model training + query.

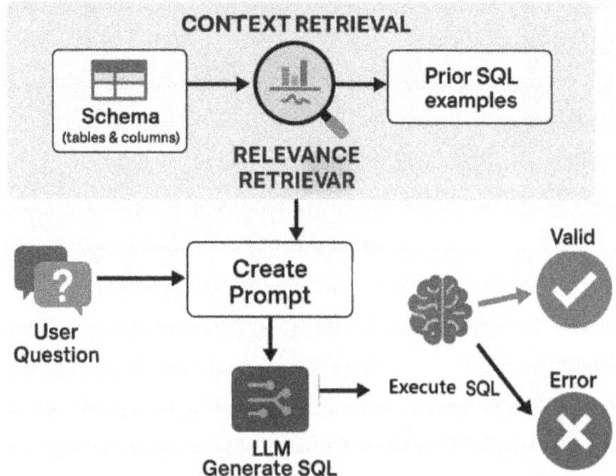

Figure 13.3.   RAG2SQL: retrieval + generation.

results might be further processed by LLM for the final answer. Figure 13.3 shows the steps. Similar to the techniques introduced in the previous chapters, the query optimization and augmentation techniques can be applied here too.

Because every generated query and its outcome can be stored, RAG2SQL systems support continual learning. Successful statements become new retrieval candidates; errors flagged by users feed fine-tuning or prompt-engineering loops; and evolving business terminology can be incorporated without retraining the core model from scratch. Over time this feedback mechanism materially improves both precision (fewer invalid queries) and recall (broader coverage of business intents).

In short, RAG2SQL offers a pragmatic bridge between conversational interfaces and structured data warehouses. By weaving schema-aware retrieval into the generation process, it delivers the twin benefits of accuracy and usability, turning everyday language into reliable SQL while steadily adapting to an organization's changing data landscape.

13.4 Evaluation

Evaluating Text2SQL systems is crucial for assessing their ability to accurately translate natural language queries into executable SQL statements and ensure their practical usability. Various metrics are employed, capturing different aspects of query generation quality. Exact Set Match Accuracy measures the proportion of generated SQL queries that precisely match the ground truth, ensuring structural correctness. Execution Accuracy (EX) evaluates whether the generated query returns the correct results when executed on the database, serving as a key performance indicator. For multi-turn conversational interactions, Question Match Accuracy assesses how well a generated query corresponds to the natural language question, while Interaction Match Accuracy specifically measures the system's ability to maintain context and generate coherent queries across dialogue turns. More advanced metrics include the valid efficiency score (VES), which considers both the correctness and execution efficiency of the generated SQL queries, and test-suite accuracy (TS), which evaluates performance on diverse and concentrated sets of test databases to measure the strict upper limit of semantic accuracy.

To provide more comprehensive evaluations, especially for real-world scenarios with linguistic and schema variations, specialized evaluation resources and frameworks have been developed. For example, Spider,[4] a large-scale, complex, and cross-domain dataset, is widely used to evaluate model generalization across diverse databases and query structures. Similarly, BIRD[5] is a comprehensive dataset designed to emphasize challenges such as handling extensive database contents and integrating external knowledge. Error analysis is also a critical part of evaluation, helping to identify limitations and guide corrective actions. Common error taxonomies categorize issues by specific SQL parts (e.g., SELECT, WHERE, JOIN), semantic misinterpretations (e.g., schema linking, database content, external knowledge), and syntax-related errors. Notably, error analysis on datasets like Spider and BIRD indicates that schema linking errors show the highest average error rates (ranging from 29–49%), highlighting a significant area needing improvement in Text2SQL tasks.

The evaluation of TAG is more difficult than that of RAG because existing Text2SQL and RAG methods are often insufficient for the broad range of user questions that require both semantic reasoning from LLMs

---

[4]https://github.com/taoyds/spider. Accessed on 17/08/2025.

[5]BIRD (BIg Bench for LaRge-scale Database Grounded Text-to-SQL Evaluation) is one of the latest cross-domain datasets for text2SQL parsing benchmark. https://bird-bench. github.io/. Accessed on 17/08/2025.

and scalable computation from databases. TAG systems are evaluated on their ability to handle queries requiring LLM knowledge (e.g., world knowledge not explicitly in the database) and LLM reasoning (e.g., logical transformations or aggregations over data).

Benchmarks for TAG often build upon existing Text2SQL datasets like BIRD, modifying queries to specifically require LM knowledge or reasoning beyond direct database lookups. Performance is typically measured by exact match accuracy and execution time. Initial evaluations highlight significant challenges: standard Text2 SQL and RAG baselines may consistently fail to achieve high accuracy on queries requiring semantic reasoning or world knowledge, demonstrating their limitations. In contrast, hand-written TAG pipelines, which leverage expert knowledge and integrate LLM-based operators within the query execution step, have shown significantly higher accuracy and improved efficiency. This performance gap underscores the substantial research opportunities for developing more effective TAG systems that seamlessly combine the strengths of LLMs and database management systems.

13.5 Tools

We can always use some general-purpose tools, such as LangChain, LlamaIndex, RAGflow and LangGraph, to build the Text2SQL and RAG2SQL. A sample prompt for a specific financial database with multiple tables is illustrated in codes\ch13\llm_sql_prompt.md. Alternatively, we can look at some specific open-source Text2SQL tools as compared in Table 13.1.

To evaluate the performance of those tools, the general steps are as follows.

1. **Prepare the Data into a Database:** First, structure your data and import it into a relational database such as MySQL or PostgreSQL. Ensure that the data is organized into properly defined tables with clear relationships between them (e.g., primary keys, foreign keys). Index the data where necessary to optimize query performance. For example, we can also utilize the Text2SQL dataset Spider, which consists of 10,181 Question-to-SQL query pairs across databases from different domains.

Table 13.1.   Comparison of Text2SQL tools.

Tool	Description	License	Implementation	Comments
DB-GPT[6]	Offers a module for fine-tuning LLM for Text2SQL tasks that can be used without its own front-end	MIT	Offers Text2SQL pipelines with LLM fine-tuning	To test if fine-tuning works
pandasai	Offers a library/ platform to easily handle Pandas Dataframes	mixed	Offers Text2SQL pipelines with visualization	Need enterprise license for production
Vanna[7]	Offers both front-end and back-end APIs. Ability to use custom LLM and embedding models	MIT	Offers Text2SQL pipelines with RAG. Able to convert output into executable visualization code using Plotly	Highly customizable, easy to integrate with a larger framework without much modification
WrenAI[8]	Designed to be used with its front-end interface. Extra efforts are needed to understand and utilize its closed back-end APIs. Ability to use custom LLM and embedding models	AGPL	Offers Text2SQL pipelines with RAG integrated	Suitable for use with minimal changes required
Superduper[9]	An all-rounded tool for building AI apps with DB integration (e.g., RAG), without out-of-the-box support for Text2SQL	Apache	Offers a data-AI integration layer for databases	Not recom- mended for Text2SQL- focused tasks

---

[6]https://github.com/eosphoros-ai/DB-GPT/.

[7]https://github.com/vanna-ai/vanna.

[8]https://github.com/Canner/WrenAI.

[9]https://github.com/superduper-io/superduper.

2. **Create Separate Agents for the LLM:** To ensure security and role-specific functionality, create distinct agents within your system, each representing a separate role (e.g., Data Analyst, Data Viewer). For each role, create a corresponding database user that only has the specific permissions necessary for that role. These users will interact with the database through the agents, executing SQL queries within their limited permission scope. This ensures that the agents can only access data relevant to their assigned responsibilities.

3. **Develop LLM Prompts for SQL Generation as a Baseline:** Create a detailed prompt for the LLM that explains the structure of the database. This prompt should describe the relevant tables, the purpose of each column, relationships between tables, and any additional context the model needs to understand the data. When the user makes a request, append this explanation to the user's query to help the LLM generate appropriate SQL.

4. **Execute and Process SQL Statements:** When the Tools or LLMs generate a SQL query based on the user's request, strip out the SQL query from the model's output. Execute this query against the database to retrieve the relevant data. After the SQL query is run, return both the dataset produced by the SQL query and a human-readable explanation generated by the LLM. This output allows the user to understand the results without being exposed to the underlying SQL code.

13.6 Advanced Techniques

Similar to the conventional RAG, we may apply multiple techniques to improve the performance of Text2SQL or TAG, e.g.,

Prompt Engineering Strategies (In-context Learning):

- **CoT Prompting:** This technique guides LLMs to generate intermediate reasoning steps before predicting SQL, which enhances both accuracy and interpretability. It can be combined with in-context learning, logical synthesis, hints, and multi-agent systems. This can be further extended to the Chain of Table strategy, which will be discussed later. The data schema (data model & data profiling) information shall be included in the prompt.

- **Decomposition or Multi-round Strategies:** These methods break down complex Text2SQL tasks into more manageable sub-tasks (e.g., schema linking, SQL generation, refinement) or divide natural language questions into intermediate sub-questions. The multi-round methods generally require multiple generations and then compare for consensus. This idea triggers the Mixed Self Consistency method [6], which we will further elaborate on in the next section. Iterative and recursive LM generation patterns for answering complex queries are still helpful.

- **Advanced Few-shot Sampling:** This involves optimizing the selection of demonstration examples through semantic similarity, diversity, or domain alignment, which significantly improves the model's performance.

Decoding Strategies:

- **Constraint-aware Incremental Decoding:** This strategy ensures the generation of syntactically correct SQL by enforcing SQL grammar rules incrementally during the decoding process.
- **Beam Search:** Instead of selecting only the single best token at each step, beam search explores a larger search space by retaining multiple candidate sequences, increasing the likelihood of generating more accurate and complex SQL queries.
- **N-best Rerankers:** These methods reorder the top-N generated SQL outputs, often by leveraging a larger model or external knowledge sources to further refine the candidates.

Specialized Fine-tuning Paradigms:

- **Parameter-Efficient Fine-Tuning (PEFT):** This approach tunes only a small subset of the model's parameters (e.g., using **LoRA** and **QLoRA**), which substantially improves training efficiency and reduces computational costs while maintaining high performance.
- **SQL-Specific Pre-training:** This involves developing LLMs specifically tailored for Text-to-SQL tasks by pre-training them on large corpora related to SQL. We may also combine Data Augmentation techniques, which employ high-quality synthetic or augmented data to boost model performance and efficiency during fine-tuning, especially in scenarios with limited data.
- **Multi-task Tuning:** Models are trained for multiple related tasks (e.g., schema linking, SQL generation, and critiquing) simultaneously. This enhances overall capabilities and multi-step reasoning.

The application of these advanced techniques for improving Text2SQL or TAG performance can be exemplified by two typical approaches: Mixed Self Consistency and Chain of Table.

13.6.1 *Mixed Self Consistency*

LLMs are equipped to analyze tabular data through two primary methodologies:

- Textual analysis via direct prompts.
- Symbolic analysis via program creation (e.g., Python, SQL, etc.)

Drawing from the study by Liu and team [6], LlamaIndex crafted the MixSelfConsistencyQueryEngine, which combines findings from both textual and symbolic analyses using a self-consistency technique (i.e., consensus voting) to attain SoTA results. Below is an example code snippet from ch13\mixed_self_consistency.py. For comprehensive information, explore LlamaIndex's complete notebook.[10]

```python
query_engine = MixSelfConsistencyQueryEngine(
df=table,
llm=llm,
text_paths=5, # sampling 5 textual reasoning
 paths
symbolic_paths=5, # sampling 5 symbolic
 reasoning paths
aggregation_mode="self-consistency", #
 aggregates results across both text and
 symbolic paths via self-consistency
 (i.e. majority voting)
verbose=True,

)

response = await query_engine.aquery("what is
 the average age of all persons")
```

---

[10]https://github.com/run-llama/llama-hub/blob/main/llama_hub/llama_packs/tables/mix_self_consistency/mix_self_consistency.ipynb, Accessed on 17/07/2025.

13.6.2 *Chain of Table*

The Chain-of-Table method works by integrating the chain-of-thought approach into tabular data processing [7]. Instead of having the LLM reason over an entire table in one go, the method breaks down a complex query into a sequence of structured operations. These operations—such as adding columns, filtering rows, selecting specific columns, grouping, and sorting—allow the system to iteratively refine the data. At each step, the updated table is passed to the LLM, which then uses this intermediate result as context to guide its reasoning toward the final answer.

At the heart of this method is the idea of sequential transformation. When a query is issued, the system begins by planning a series of table manipulations that will progressively isolate and refine the relevant data. For instance, if the query involves identifying a particular subset of rows, the pack might first filter the table to narrow down the candidate rows and then sort or group the results to further clarify the data structure. This step-by-step process leverages both the structured nature of tables and the language model's reasoning abilities, ensuring that complex tabular questions are addressed with greater accuracy.

One of the major advantages of the Chain-of-Table approach is its robustness in handling questions about tables that contain dense or multifaceted information. By exposing the LLM to intermediate transformations rather than a single, unwieldy table, the method not only improves answer accuracy but also makes the reasoning process more transparent. This has led to common opinions that such a structured refinement process is particularly effective for tabular question answering. Moreover, innovative ideas such as combining symbolic operations with chain-of-thought reasoning, open avenues for further integrating advanced retrieval strategies or dynamic query transformations, thus enhancing the overall flexibility and performance of RAG pipelines in structured data scenarios.

The ChainOfTablePack as a LlamaPack in Llamaindex implemented the "chain-of-table" study [7]. Below is the main part of the function call from ch13\chainoftable.py.

```python
from llama_index.packs.tables import
  ChainOfTablePack
# Create the ChainOfTablePack instance with
  the table and LLM
chain_pack = ChainOfTablePack(table=df,
  llm=llm, verbose=True)
```

For instructions on how to employ ChainOfTablePack for querying your structured data, visit LlamaIndex's comprehensive notebook.[11]

13.7 Concluding Remarks

By combining the conventional RAG and TAG/Text2SQL, the systematic analytics on both unstructured and structured data (i.e., ChatBI) is possible. In real-world applications, the general performance of Text2SQL has not yet reached human-level proficiency, as shown in some leaderboards,[12] e.g., humans can achieve up to 92.96% in BIRD evaluation, while the latest model score is only around 77.53%. The continuous technological advances shall fill the gaps. Recent advances illustrate a powerful shift toward dynamic, reasoning-aware query generation. One key innovation allows language models to actively interact with the database during inference by issuing exploratory SQL probes. This process enables iterative refinement of contextual understanding and significantly enhances execution accuracy, even in scenarios with no prior example-based training. In parallel, another class of methods structures SQL generation as a reasoning-driven task, guiding the model to articulate intermediate logical steps before producing the final query, and incorporating a verifier to ensure correctness. This self-improving reasoning paradigm leads to substantial gains in performance on complex benchmarks, demonstrating how transformational it can be to move from static mapping systems to autonomous, insight-driven frameworks.

Meanwhile, further innovations promote robust generalization and resilience when faced with diverse and unseen database environments. One such strategy integrates completion feedback directly into a reasoning optimization loop, which allows improvement of execution accuracy via purely self-guided iterative refinement, without reliance on external reward models. Another enables models to generate their own high-quality in-context examples, filtering them to form self-curated prompts that better guide the model on difficult or novel queries. Together, these developments underscore a trend: combining schema-conscious context, execution feedback, and self-synthesized learning signals creates the

---

[11]https://github.com/run-llama/llama-hub/blob/main/llama_hub/llama_packs/tables/chain_of_table/chain_of_table.ipynb, Accessed at 04/05/2025.
[12]https://openlm.ai/text2sql-leaderboard/, Accessed on 17/08/2025.

foundation for robust, adaptive models that thrive in real-world, dynamic database querying scenarios.

Besides technical considerations, to develop and deliver a production-grade ChatBI, we must consider the following key points.

Misaligned or Superficial Demand: A significant proportion of ChatBI projects fail because they are conceived as "leadership showcase projects" rather than solutions to genuine business needs. In many cases, the push to deploy ChatBI comes from following market hype rather than addressing high-frequency, mission-critical queries. As a result, usage is low and the system becomes a "demo-only" platform. True demand validation requires identifying specific, frequent, and high-impact use cases where Text2SQL can offer clear advantages over existing tools such as fixed BI reports or self-service dashboards. A sound approach is to conduct early-stage business value mapping, focusing on non-substitutable scenarios such as operational troubleshooting or urgent ad-hoc analytics, where ChatBI provides significant time-to-insight benefits.

Data Governance and Semantic Gaps: Text2SQL systems depend on precise mapping between natural language and database schema, yet corporate data ecosystems are often riddled with semantic mismatches, inconsistent metric definitions, and incomplete metadata. Common issues include business terms that do not match field names, variations in calculation rules across departments, and the absence of mechanisms to update metadata when new business processes are introduced. These problems lead to inaccurate SQL generation and erode user trust. Sustainable success requires instituting continuous metadata management, aligning metric definitions across the enterprise, and building semantic layers or ontology mappings that bridge the gap between human terminology and database structure.

Technical Adaptation Beyond Model-Centric Thinking: Many implementations over-invest in model selection (e.g., choosing between large, medium, or domain-specific LLMs) while underestimating the need for business-context integration. A purely model-driven strategy often fails because accuracy in SQL generation hinges as much on schema understanding and context injection as on raw language capability. Mature ChatBI design should therefore incorporate schema-aware prompting, few-shot query examples, and hybrid parsing methods that combine LLM output with rule-based SQL validation. Additionally, incorporating SQL parsers, execution planners, and result validators can help prevent execution errors and ensure that generated queries remain performant in production.

Organizational Readiness and Cultural Factors: Deploying ChatBI is not only a technical exercise but also a business transformation initiative. Without cultivating an AI-literate workforce, even the best-engineered Text2SQL system will face adoption barriers. Users must be trained not just in "how to ask questions" but also in understanding the system's limits, interpreting results critically, and feeding back inaccuracies for retraining. This requires embedding ChatBI into workflows, defining clear ownership for model and data maintenance, and ensuring that success metrics are tied to business outcomes rather than just system availability or usage counts.

Integrated Solutions for Long-Term Viability: A sustainable ChatBI solution should be built as a living system, combining robust demand validation, disciplined data governance, adaptive technical architectures, and continuous organizational engagement. Best practices include introducing a semantic middleware layer to decouple business queries from raw database schema, implementing query auditing pipelines to refine model behavior, and integrating feedback loops for iterative improvement. Ultimately, the most successful Text2SQL-based ChatBI deployments treat the technology as part of a broader decision-intelligence framework where the emphasis is on improving decision quality and operational agility, rather than simply automating SQL writing.

References

[1] X. Zhu, Q. Li, L. Cui and Y. Liu, *Large Language Model Enhanced Text-to-SQL Generation: A Survey,* arXiv:2410.06011v1, 2024.

[2] L. Shi, Z. Tang, N. Zhang, X. Zhang and Z. Yang, *A Survey on Employing Large Language Models for Text-to-SQL Tasks,* arXiv: 2407.15186v4, 2024.

[3] X. Liu, S. Shen and *et al., Survey of NL2SQL with Large Language Models: Where Are We, and Where Are We Going?,* arXiv:2408.05109v4, 2025.

[4] A. Biswal, L. Patel and *et al., Text2SQL is Not Enough: Unifying AI and Databases with TAG,* arXiv:2408.14717v1, 2024.

[5] M. Pourreza and D. Rafiei, *DIN-SQL: Decomposed In-Context Learning of Text-to-SQL with Self-Correction,* arXiv:2304.11015v3, 2023.

[6] T. Liu, F. Wang and M. Chen, *Rethinking Tabular Data Understanding with Large Language Models,* arXiv:2312.16702v1, 2023.

[7] Z. Wang, H. Z. Zhang and *et al., Chain-of-Table: Evolving Tables in the Reasoning Chain for Table Understanding,* arXiv:2401.04398v2, 2025.

Chapter 14

RAG with Knowledge Graph

14.1 Introduction

Graphs, in particular, knowledge graphs (KGs), have been regarded as a key component in enhancing retrieval-augmented generation (RAG) applications, e.g., reducing hallucinations for complex queries. These graphs not only provide a structured framework for data but also ensure the relevance and currency of the information used by RAG models within context. Therefore, a new term, GraphRAG,[1] was created [1–4]. By examining the symbiotic relationship between KG and RAG, we can better understand how they work together to address inherent challenges such as data relevance, complex query processing, and the integration of various data types. This chapter delves into the key aspects of KGs in RAG applications, the challenges they help overcome, the synergy with vectors, and the future of this technology.

14.2 Understanding KG in RAG

KGs are an advanced framework that encapsulates the interconnected nature of data. They represent a dynamic, structured representation of knowledge where entities (also known as nodes) are closely connected through relationships (or edges). This architecture not only catalogs

---

[1]We refer GraphRAG to a type of RAG which embeds Graph structure, in particular, KG, into its process; instead of a special tool.

information but also elucidates the context and complex interrelationships between data points.

KGs offer a rich, interconnected dataset that can be used for more accurate and context-aware data retrieval. For instance, in a medical knowledge graph, nodes may represent symptoms, diseases, and treatments, with edges defining relationships such as "symptom of" or "treatment for." KGs can be seen as a network of entities and relationships that model the facts and rules of a domain, enabling powerful data relationship discovery and exploration.

The utility of KGs extends to various application areas, from context-aware content recommendation to advanced drug safety analysis. They serve not only as queryable databases but also as analytical networks and knowledge bases for inference and rule-based reasoning.

14.2.1 *The Role of KGs in RAG*

KGs play a significant role in enhancing RAG applications, addressing pressing issues. Contextual relevance is a key advantage, as the structure of the graph ensures that the retrieved information is not only relevant but also contextual, providing a richer background for response generation. Table 14.1 summarizes some key features enabling RAG. We will discuss the details in the next subsection.

Table 14.1. KG Capabilities.

KG Capability	Description
Data Integration	KGs unify heterogeneous data sources (structured, semi-structured, and unstructured) into a cohesive graph model. This consolidated view enables RAG systems to draw on a richer, more complete knowledge base when generating responses.
Contextual Relevance	By encoding explicit relationships among entities, KGs ensure retrieved information is contextually appropriate. This relational context helps RAG components select and rank passages that align closely with the user's query intent.
Support for Complex Queries	The interconnected nature of a KG facilitates multi-hop reasoning. When a query spans several hops (e.g., "Which policy revisions led to reduced emissions in 2023?"), the KG's structured links allow the system to traverse paths and perform query augmentation, enforcing consistent retrieval of key concepts.

Table 14.1.   (*Continued*)

KG Capability	Description
Balance of Global and Local Knowledge	Traditional RAG can struggle with synthesizing a "global view" across multiple documents (e.g., comparing several iterations of an ESG policy). A KG can represent high-level, cross-document relationships, enabling query-focused summarization (QFS) by linking local details to broader context.
Facilitating Domain Adaptation	A domain-specific KG (e.g., one built for healthcare or finance) contains specialized entities and relations. Integrating such a KG into a RAG pipeline allows the system to leverage tailored ontologies and vocabularies, improving accuracy and relevance in niche areas.
Enhancing Interpretability	Because KGs make relationships explicit, they expose clear reasoning chains (e.g., Entity A $\rightarrow$ Relation X $\rightarrow$ Entity B). When combined with RAG, this transparency helps users understand why certain documents or facts were selected, making the system's outputs more interpretable.

14.2.2 *Integration of KG and RAG*

By integrating KGs with RAG models and enhancing them with optional node/graph embeddings,[2] we can create AI systems that understand and generate human language with unprecedented accuracy and depth.

---

[2] A graph embedding (GE) is the technique of mapping nodes, edges, or entire graphs into a continuous vector space such that semantic and structural relationships are preserved as distances or angles in that space. GEs condense complex, high-dimensional graph structures into low-dimensional, dense vectors, enabling traditional machine learning models to process network data efficiently and uncover latent patterns. By enabling tasks such as community detection, link prediction, and node classification, embeddings accelerate analysis in domains such as social networks, recommendation systems, bioinformatics, and knowledge graphs.

GE techniques fall into two main categories: node embedding methods, which learn vector representations for individual nodes based on their local context, and graph-level embedding algorithms, which model the structure of entire networks. Prominent node embedding methods include DeepWalk (https://github.com/phanein/deepwalk), which treats truncated random walks as "sentences" to learn context-sensitive representations; Node2Vec (https://github.com/aditya-grover/node2vec), which introduces a tunable bias between breadth-first and depth-first neighbor exploration; and GraphSAGE (https://github.com/williamleif/GraphSAGE), which aggregates features from sampled neighborhoods to generate embeddings inductively. On the other hand, graph-level techniques leverage neural network architectures such as graph convolutional networks (GCNs),

This approach not only addresses the inherent challenges of RAG models but also unleashes new potential for natural language processing. In other words, KGs are adept at merging structured and unstructured data, thereby overcoming a significant barrier to developing complex RAG systems [5]. Table 14.2 compares the difference between the traditional RAG and the graph-embedded RAG. Below, we discuss some details of KG's capability into RAG.

Enhancing Data Relevance

KGs ensure that responses are not only relevant but also context-rich, providing nuanced backgrounds for content generation. This is particularly beneficial for applications that require high precision, such as healthcare or financial advisory services.

Challenges faced by RAG systems, such as ensuring appropriate context and up-to-date data, are addressed by the dynamism of KGs. They flexibly handle complex queries by leveraging the interrelationships of data points, offering an in-depth understanding of context and relationships. For instance, in the case of early cancer detection, an RAG system enhanced with a knowledge graph can retrieve and integrate the latest medical research to provide informed responses.

By integrating real-time information retrieval, RAG systems adapt to dynamic environments, ensuring responses remain relevant to current

---

graph attention networks (GATs), and general graph neural network (GNN) frameworks to learn directly from both connectivity and node attributes.

Evaluating GEs generally involves measuring how well the learned vectors preserve the original network's relational and structural properties and support downstream learning tasks. Common quantitative metrics include node classification accuracy, which tests the ability to predict node labels; link prediction accuracy, which assesses how well models can reconstruct missing or future edges; and graph reconstruction error, which quantifies how accurately the adjacency matrix can be recovered from embeddings. Beyond these, evaluation frameworks often consider downstream task performance, such as clustering quality or recommendation accuracy, as well as intrinsic embedding measures like similarity preservation, dimensionality reduction efficiency, and computational overhead to ensure embeddings are both informative and practical Task-specific metrics, such as precision at K for recommendation and area under the ROC curve for link detection, further guide the selection and tuning of embedding methods to meet the needs of particular applications.

Table 14.2.   Comparison of traditional RAG and KG-enhanced RAG.

Conventional RAG	Integration of RAG and KG
• Traditional RAG chunking methods often retrieve noisy fragments. Chunks are isolated from each other, lacking interconnection, which leads to poor performance in cross-document question answering tasks. • When tasks involve aggregation, filtering, or statistical operations, vector retrieval accuracy is low, and LLMs have limited mathematical reasoning capabilities. • For questions requiring answers across multiple text blocks or documents, standard vector retrieval or even hybrid retrieval performs poorly. • Task planning with LLMs is highly uncertain and performs poorly in specific, controllable generation tasks. • LLMs tend to produce highly divergent outputs in tasks such as rewriting or recommendation.	• Introducing a KG can enhance relevance through hierarchical entity features. • Using a KG can strengthen relationships between chunks and improve retrieval accuracy. • KG can serve as an additional retrieval source to enrich context. Connecting various types of knowledge into a KG can also provide graph-based embeddings to supplement retrieval features. A knowledge graph can act as a structured rule-based knowledge base to guide large models for more controlled generation. • By leveraging structured knowledge, scene-specific KGs can be built and queried using Cypher graph search. • In domains with limited corpus availability, large models must be trained to refuse to answer when outside their scope.

events and trends. This adaptability is crucial for applications that require the latest information, such as news bots or conversational systems.

When building a RAG system, selecting an external data source that is both relevant and reliable is essential. This can range from structured knowledge bases to real-time data sources from LLMs. Seamless integration with existing workflows and AI projects is crucial for maintaining efficiency and minimizing disruptions.

Enhancing Complex Query Processing Capability

KGs significantly enhance the ability to process complex queries within RAG applications. Boldly facing the limitations of vector indexing, KGs can aggregate information in ways that vector systems cannot. For example, while vector search can identify relevant documents, it may struggle to answer queries that require counting and aggregation, such as "Which vendors supply parts to projects that are currently delayed, and who are their subcontractors?" In a graph database, this reasoning-intensive query can be expressed in Cypher:

```
MATCH (p:Project
  {status:'Delayed'})<-[:SUPPLIES_TO]-
  (v:Vendor)-[:WORKS_WITH]->(s:Subcontractor)
RETURN p.name AS delayedProject, v.name AS
  vendor, collect(s.name) AS subcontractors
```

This query requires multi-hop traversal across the graph: from projects → to vendors → to subcontractors. It doesn't just fetch documents mentioning vendors or projects, but instead leverages structured relationships to answer a question that is inherently relational.

This demonstrates the power of KGs, which can not only retrieve data but also synthesize and present it in meaningful ways. The integration of structured query languages with RAG systems makes data interaction more nuanced and complex.

By fully leveraging the strengths of vector search and KGs, RAG applications can provide more accurate and relevant responses to complex queries.

Integration of Structured and Unstructured Data
The integration of structured and unstructured data is a key step in enhancing the capabilities of RAG systems. KGs excel in this area, providing a unified repository that simplifies the complexity of handling various data types. By doing so, they enable seamless flow of information that is both contextually relevant and accessible to the underlying language models.

- **Contextual Relevance:** Ensures the relevance and context-awareness of information.
- **Complex Queries:** Facilitates the handling of complex issues.
- **Data Integration:** Provides a coordinated view by merging different data types.

KGs reduce the need for multi-language architectures, enabling a smoother data management process. This not only reduces operational overhead but also enhances the overall performance of RAG applications.

For example, unstructured text can be directly imported into the RAG workflow, while structured data may need to be converted into a format understandable by language models. KGs act as a bridge, storing structured and unstructured data in one system, thus reducing the need for extensive data preparation.

Vector Search Improvement via KG Complement
The synergy between retrieval and generation in RAG systems enables AI to go beyond simply answering queries, achieving a depth of understanding and contextual relevance that was previously unattainable. KGs play a crucial role in this process by serving as dynamic, up-to-date repositories of information. They complement vector search by adding semantic structure and relational context, addressing the gaps that vectors alone cannot fill. Graph databases, in particular, offer clear advantages over vector similarity search when handling complex, multifaceted queries that demand precision and contextual awareness (see Figure 14.1 for a sample workflow).

Vectors are highly effective for representing complex documents in a semantic space, enabling searches based on content similarity. However, their capability is limited to nearest-neighbor relationships. By integrating vectors with KGs, retrieval transcends pure similarity, incorporating the rich, interconnected relationships within the graph to deliver a more comprehensive understanding of the query.

This integration creates a powerful combination for RAG applications. KGs ground the retrieval process in structured facts, ensuring that results are not only semantically relevant but also consistent with real-world relationships. Vectors, meanwhile, capture nuanced meaning across unstructured text.

Together, they achieve a fusion of semantic and structural search. The outcome is more accurate, context-aware, and informative results, leveraging the complementary strengths of both approaches to provide a stronger foundation for generation in RAG systems.

14.2.3 *Applications of KG in RAG*

The versatility of KGs is evident in their widespread application across various industries. From enhancing semantic search to driving drug discovery, these dynamic structures have revolutionized the way we handle complex datasets. KGs automate the generation of new knowledge through data relationship discovery and exploration, enabling us to uncover previously unknown connections.

Here are some application examples of GraphRAG:

- **Context-Aware Content Recommendation Systems:** Tailoring user experiences by understanding and predicting user preferences based on interconnected data. This is particularly useful in E-commerce platforms, which construct domain-specific KGs to capture user issues, points of interest, product information, and their interrelations.
- **Investment Market Intelligence Platforms:** Analyzing trends and predicting market movements through the integration of diverse financial data sources.
- **Regulatory Document Analysis Tools:** Simplifying information discovery and compliance by mapping complex regulatory requirements and their interrelations.
- **Biomedical Research, e.g., Advanced Drug Safety Analysis:** KGs integrate data on genes, proteins, diseases, and drugs, enabling researchers to uncover new insights into disease mechanisms and potential therapeutic targets. Enhancing pharmaceutical research by identifying potential drug interactions and side effects through the exploration of biomedical data relationships.
- **Enterprise Knowledge Management:** Organizations leverage KGs to organize internal data, including documents, employee expertise, and project information. This organization enhances knowledge discovery, ensuring employees can efficiently access and utilize pertinent information across the enterprise.
- **Fraud Detection:** Financial institutions employ KGs to map and scrutinize relationships between entities like bank accounts, transactions, and individuals. This mapping facilitates the detection of fraudulent activities by revealing suspicious patterns and connections. The similar method can be applied in the social network analysis.
- **Cybersecurity:** KGs assist in identifying and analyzing relationships between entities such as IP addresses, domains, and threat actors. This helps in detecting patterns indicative of cyber threats and supports proactive defense strategies.

Each application demonstrates the transformative potential of KGs to convert vast amounts of data into actionable insights.

14.2.4 *Methodology*

There are three main types of orchestration methods in GraphRAG:

a) Rerank with Other RAG Retrieval Results

This method combines the retrieval results from GraphRAG with those from other RAG systems (such as traditional dense or sparse retrieval methods). The different sets of retrieved information are then reranked together using an LLM-based reranker or scoring mechanism to select the most relevant pieces for downstream generation.

Purpose: Leverage complementary strengths of different retrieval mechanisms.

Advantage: Balances graph-based reasoning with broader recall from other sources.

b) Agentic Graph Query Template

In this approach, an LLM agent is guided by a predefined graph query template to generate structured queries over the knowledge graph. These templates provide a scaffold or blueprint that constrains and informs how the LLM should explore the graph to answer a given question.

Purpose: Improve the reliability and structure of graph traversal via LLM.

Advantage: Balances flexibility of LLM with control of structured templates, reducing hallucination.

c) All in Graph

This method handles the entire retrieval and reasoning process within the graph itself, without relying on external RAG components. The LLM interacts solely with the graph, querying it directly and extracting all relevant information through structured graph traversals. In some cases, the vector can be embedded into graph directly, e.g., the GE approach.

Purpose: Fully capitalize on graph semantics and structure for reasoning.

Advantage: Enables high interpretability and traceability; particularly suitable for domains where relationships and entities are explicitly encoded.

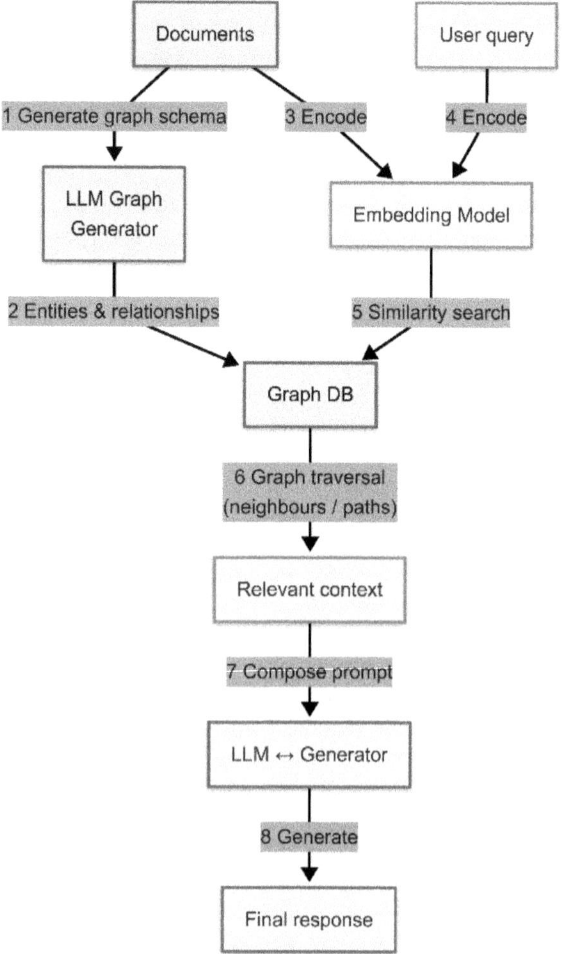

Figure 14.1.   Merge vectors and graphs in GraphRAG.

These three orchestration strategies can be individually employed or combined based on the nature of the data, the task complexity, and the desired trade-off between flexibility, precision, and control.

The retrieval methods inside GraphRAG can be roughly divided into 5 types:

1. **Subgraph Retrieval**

   This method retrieves a relevant subgraph based on the user's query. It typically involves identifying key entities or nodes mentioned in the query and extracting their local neighborhood within the graph.

   - **Use case:** Efficient when a localized context around entities is sufficient for answering the query.

   - **Advantage:** Fast and scalable, especially for large graphs.

2. **Path Retrieval**

   This method retrieves semantic paths connecting entities that are relevant to the query. It focuses on discovering explicit relationships or reasoning chains between nodes.

   - **Use case:** Useful for tasks that require explaining relationships (e.g., causal chains, reasoning over multi-hop knowledge).

   - **Advantage:** Offers more structured and interpretable evidence for reasoning.

3. **Template-based Complex Query**

   In this approach, predefined query templates are used to construct complex queries over the graph. These templates can model intricate logic, such as conditional paths, role-based relationships, or constraints.

   - **Use case:** Effective for enterprise knowledge graphs with domain-specific query structures.

   - **Advantage:** Ensures consistency, correctness, and efficiency in retrieval.

4. **Text-to-Graph Query**

   This method enables direct conversion of natural language queries into graph query languages (e.g., SPARQL, Cypher). It leverages LLMs to map text to formal query representations, enabling precise retrieval from the graph.

   - **Use case:** General-purpose interface for non-expert users to access graph data.

   - **Advantage:** Increases accessibility and flexibility while maintaining structure.

5. **Planning-based Graph Retrieval**
   This advanced method involves using LLM or symbolic agents to plan multi-step graph traversal strategies. The planner decomposes the question into a sequence of intermediate goals or subqueries, which are executed step by step.

   - **Use case:** Complex questions requiring reasoning across multiple subgraphs or logical stages.

   - **Advantage:** Enables dynamic, context-aware exploration of the graph beyond static queries.

These retrieval strategies range from localized extraction to global, agent-driven exploration, and can be combined or selected based on task complexity, user intent, and graph size.

14.3 Implementing KG-Enhanced RAG

This section presents one of the typical implementation procedures of GraphRAG. The workflow begins with document preparation and ingestion, where documents are parsed and structured into entities and relationships by an LLM-driven graph generator. Both documents and user queries are encoded via embedding models to enable similarity searches. A graph database stores and organizes entities, relationships, and document chunks, enabling efficient similarity search and graph traversal. Retrieved contextual information is then compiled into a prompt for the LLM, which generates a grounded, contextually rich final response (a simplified workflow is illustrated in Figure 14.1). The combination of dense vector retrieval and structured graph traversal ensures improved recall, reasoning over multi-hop relations, and better explainability of results. When mapping to the RAG processing, [1] shows the major phrases and techniques in Figure 14.2. Table 14.3 compares the main steps between conversional RAG and GraphRAG. Table 14.4 provides a step-by-step explanation to construct a GraphRAG.

14.3.1 *Data Collection and Preparation*

The foundation of any knowledge graph lies in a meticulous data collection and preparation phase. This stage is crucial as it determines the quality and scope of the information that will be structured within the graph.

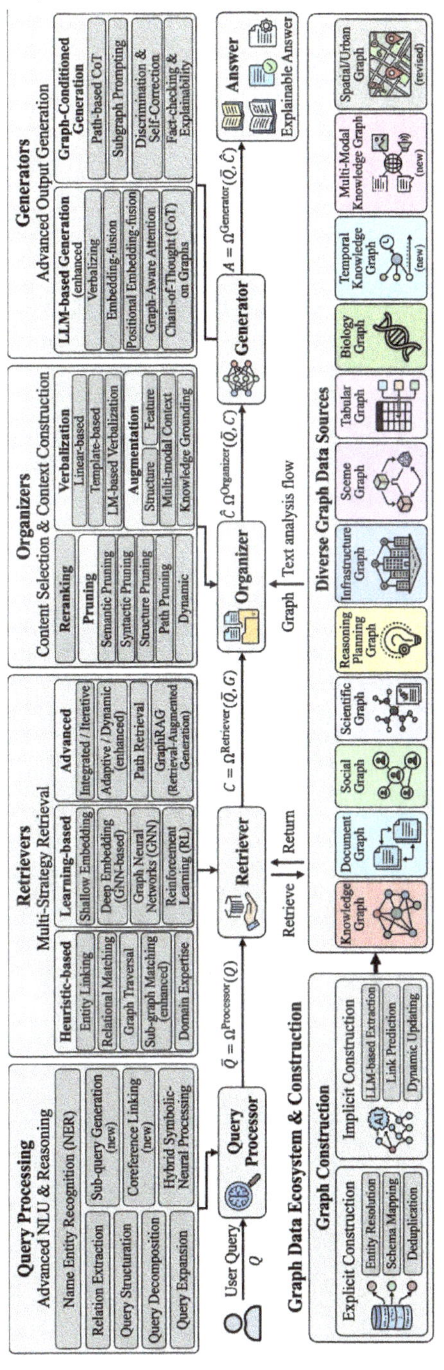

Figure 14.2.   Workflow of GraphRAG [1].

Table 14.3.   Comparison of steps in conversional and graph RAG.

Phase	Conventional RAG	GraphRAG
1. Knowledge Organization	Splits documents into isolated text chunks and stores them as embeddings. No understanding of entity or relational structure.	Extracts entities and relationships to build a structured knowledge graph, enabling interconnected knowledge representation.
2. Knowledge Retrieval	Retrieves top-k chunks based on semantic similarity to the query; lacks cross-chunk linkage and reasoning.	Locates relevant nodes and expands through graph traversal and multi-hop reasoning, capturing deeper knowledge connections. Usually, vector search is also included.
3. Knowledge Integration & Generation	Concatenates retrieved chunks as LLM input; prone to context fragmentation and limited interpretability.	Integrates retrieved knowledge subgraphs, preserves logical relationships, removes redundancy, and generates coherent, explainable answers.

Table 14.4.   Explanation of the main steps of KG-RAG.

#	Stage	Description	Key Considerations
1	**Graph construction**	After the data is prepared, a document loader and an LLM-based parser detect entities/relations and write them into a property graph.	• Ontology alignment • Incremental updates
2	**Dual encoding**	Documents and queries are converted to dense vectors using a shared (or dual) embedding model.	• Domain-specific tuning • Multi-modal embeddings (text+tables)
3	**Similarity search**	The query vector retrieves candidate nodes/edges (documents, paragraphs, triples).	• k-NN/ANN index performance • Hybrid lexical filters
4	**Graph traversal**	Starting from similar nodes, algorithms (e.g., metapath, random-walk, BFS) harvest neighborhood context, injecting structured evidence.	• Limit depth to bound latency • Edge-type weighting
5	**Prompt assembly**	The collected nodes/relationships and raw passages are formatted into a system / user prompt template.	• Token-budget control • Few-shot reasoning examples

Table 14.4.   (*Continued*)

#	Stage	Description	Key Considerations
6	**LLM reasoning**	A generative LLM answers, grounded by graph context; can call back to graph for self-refinement (agentic loop).	• Response-side citation • Policy/compliance filters
7	**Output**	User receives a fact-checked, provenance-rich answer.	• Confidence scoring • Conversation memory

For instance, in a project aimed at documenting the sources of deep learning (DL) results in biodiversity research, data was collected from a dataset previously generated through systematic literature review. The dataset was then curated by domain experts, focusing on variables related to reproducibility, which is essential for ensuring the credibility and verification of the results.

The preparation process typically involves the following steps:

1. Identifying and collecting relevant datasets or sources.
2. Collaborating with domain experts for data curation and validation.
3. Extracting and cleaning data to eliminate inconsistencies or errors.
4. Annotating or tagging data to enhance its utility within the knowledge graph.

Capturing and storing information in a structured representation is essential for effectively building a KG. Once the data is ready, it can be used to generate "capability questions" (CQs), which guide the ontology construction and subsequent knowledge graph development. The entire process from data collection to knowledge graph integration should consider ethical factors, ensuring that the final AI application adheres to ethical AI principles.

14.3.2 *Knowledge Graph Construction*

The construction of a KG is a meticulous process that requires multiple key steps. Defining entities and their relationships is the cornerstone of building a knowledge graph, as it lays the foundation for how the graph represents and interconnects data. This process often necessitates the expertise of domain specialists to ensure accuracy and relevance to the field.

After the initial setup, populating the instances of the KG is essential. This step involves integrating data from various sources, which can be a mix of structured and unstructured formats. Ensuring data quality is crucial, as it affects the reliability of the knowledge graph in applications such as semantic search or content recommendation.

The extraction workflow begins with segmenting input documents into text chunks. These chunks are combined with system and user instructions to create a structured graph-aware prompt, which may optionally include contextual information from an existing knowledge graph. The prompt is sent to a LLM API for processing. The LLM generates a response, which is received and parsed into structured data formats such as JSON or graph outputs. The extracted entities and relationships are then validated; if they meet quality and consistency criteria, they are stored in

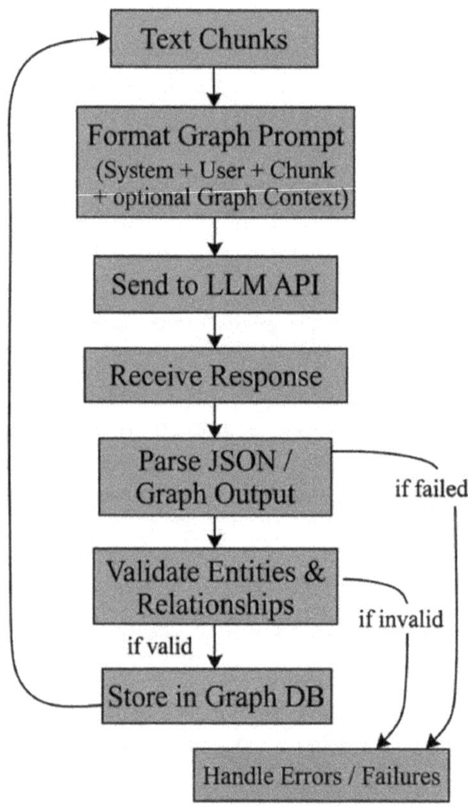

Figure 14.3. Simplified graph generation workflow.

Table 14.5.   Triplet in 3 levels.

	Document Metadata-Level	Document Chunk-Level	Document Entity-Level
Nodes	Represent document names, topics, or other metadata.	Represent individual document chunks, including titles, paragraphs, documents, tables, figures, etc.	Represent specific entity types extracted from documents (schema-based) or build free-form keyword networks (schema-free).
Edges/ Relations	Capture relationships such as document similarity, parent-child hierarchy, or referencing links between documents.	Describe parent-child relationships, co-occurrence, semantic similarity, or structural dependencies between chunks.	Define relationships between entities according to either a pre-defined schema or based on contextual co-occurrence patterns.

the graph database. If any step fails, including response parsing or validation, the system routes the issue to a dedicated error handling process for diagnosis and correction. The general procedure is illustrated in Figure 14.3.

Finally, keeping the KG updated and verified to ensure its currency and comprehensiveness is vital. This often involves contributions from the community and open-source projects, like those seen on GitHub. This workflow enables robust, automated extraction of structured knowledge from unstructured text with built-in mechanisms for quality control and fault tolerance. Note that the triplet (e.g., Subject-Predicate-Object, SPO) in the graph can be extracted in three levels as listed in Table 14.5.

In Langchain, we can simply call the toolkit "LLMGraphTransformer" to generate the KG and then store it into Neo4j database. In the example of ch14\graphrag_neo4j.py, we use the content from the famous book "Harry Potter" and then apply LLMGraphTransformer to convert the input text into structured graph documents (nodes and relationships). The extracted graph structure is stored into a Neo4j database as a node or edge in the graph. Below is code snippet.

```
llm_transformer_filtered =
  LLMGraphTransformer(llm=llm)
graph_documents = llm_transformer_filtered.
  convert_to_graph_documents(documents)
```

```
graph.add_graph_documents(
      graph_documents,
      baseEntityLabel=True,
      include_source=True
  )
```

We can also create embeddings inside the graph for complex semantic search, e.g., creates a hybrid vector index on top of Neo4j. This enables semantic search over both structured graph data and unstructured text content.

```
embed = OpenAIEmbeddings(model="text-
  embedding-3-large",api_key=os.
  getenv("OPENAI_API_KEY"))
vector_index = Neo4jVector.
  from_existing_graph(
      embedding=embed,
      search_type="hybrid",
      node_label="Document",
      text_node_properties=["text"],
      embedding_node_property="embedding",
      url="neo4j+s://d2ebd34e.databases.neo4j.io",
      username="neo4j",  #default
      password="xFqnLkaXUkmI8jnKZ3qPppFT-
  CWnhcqONGf6Pqktwbg"  #change accordingly
  )
vector_retriever = vector_index.as_retriever()
```

14.3.3 *Query in GraphRAG*

Now we can make queries to the graph DB with the designed prompt. In this case, we intend to find the organization and person information, as well as their relations, from "Harry Potter."

```
## Define the prompt to extract entities
prompt = ChatPromptTemplate.from_messages([
        ("system", "Extract organization and
  people entities from the text."),
```

```
        ("human", "Extract entities from:
  {question}")
    ])
# Define the LLM chain to extract entities
entity_chain = prompt | llm.with_structured_
  output(Entities, include_raw=True)
response = entity_chain.invoke({"question":
  "What's the relation between Harry Potter
  and Voldemort?"})
entities =  response['raw'].content
```

The output of the relation extracted as follows.

```
Harry Potter - RAISED_BY -> Muggles
Harry Potter - LINKED_BY -> Lord Voldemort
Harry Potter - PARENT_OF -> James Potter
Harry Potter - PARENT_OF -> Lily Potter
Harry Potter - ATTENDS -> Hogwarts School Of
  Witchcraft And Wizardry
Harry Potter - PARTICIPATES_IN -> Battle Of
  Hogwarts
Harry Potter - DEFEATS -> Lord Voldemort
Harry Potter - REPRESENTS -> Hogwarts School
  Of Witchcraft And Wizardry
```

Once the relationship database is built, it's easy to explore the relationship in Neo4j.

14.3.4 *GraphRAG Tools*

Multiple tools have been introduced to simplify the GraphRAG creation [6]. Besides Langchain, llamaindex has its corresponding component too. Here, we mainly introduce two tools, GraphRAG[3] from Microsoft [4] and LightRAG[4] from University of Hong Kong [7] in the content below.

---

[3]https://github.com/microsoft/graphrag.
[4]https://github.com/HKUDS/LightRAG.

14.3.4.1 *GraphRAG*

GraphRAG enables global sensemaking over large text corpora. Its main idea is to use an LLM to construct a knowledge graph from the corpus, partition the graph into communities, generate summaries for each community, and then use these summaries to answer queries in a map-reduce fashion.

Its high-level overview of methodology also includes Text Chunking, Entity & Relationship Extraction Using LLM, Knowledge Graph Construction. Besides, it has

1. **Community Detection:** Partition graph into hierarchical communities of closely related entities.
2. **Summary Generation:** Generate report-like summaries for each community using an LLM.
3. **Query Processing:** For a given query, use community summaries to generate partial answers in parallel (map), then combine into a final global answer (reduce).

Here's a sample code to use GraphRAG via CLI.

```
# !pip install graphrag

# assume the data files, e.g., txt, are
  stored in data folder
graphrag init --root ./data

# assume that you already configured LLM,
  e.g., openai
# build the index
graphrag index --root ./data

# query using --method global or local.
graphrag query \
--root ./ragtest \
--method global \
--query "What are the top themes in this
  story?"
```

Or you can use a different implementation of the LangChain library via "pip install langchain-graphrag."[5] This code snippet below from ch14\graphrag_langchain.py shows the workflow: it loads a text file, constructs the GraphRAG index, and then queries it to get a summarized answer based on the community summaries.

```
index_creator = GraphIndexCreator(llm=llm)
chain = GraphQAChain.from_llm(llm,
  graph=graph, verbose=True)
chain.run("What is the relationship between
  Harry Potter and Voldemort")
```

A similar implementation from DataStax[6] is available in ch14\graphrag_datastax.py:

```
retriever = GraphRetriever(
    store=vector_store,
    # Define edges based on document metadata
    edges=[("name", "location")],
)
```

14.3.4.2 *LightRAG and MinRAG*

LightRAG integrates graph structures for enhanced indexing and retrieval (a dual-level of retrieval strategy) into conversional RAG.[7] It utilizes graph structures with vector representation for efficient entity and relationship retrieval. Extensive testing shows that LightRAG significantly

---

[5]https://github.com/ksachdeva/langchain-graphrag.

[6]https://github.com/datastax/graph-rag/tree/main/packages/langchain-graph-retriever.

[7]During indexing, text is mapped into a heterogeneous graph of entities, relations, and chunk nodes, which is stored alongside vector embeddings. At query time, LightRAG performs (1) low-level retrieval to fetch precise entity- and relation-centric evidence and (2) high-level retrieval to surface broader, topic/community-level context. The two evidence streams are fused to provide the generator with semantically rich but compact context, enabling multi-hop reasoning while controlling token budget.

outperforms baseline models in speed and accuracy, ensuring its adaptability and effectiveness in dynamic content environments.[8] MinRAG has some further optimizations for the small LLMs. A sample code is available in ch14\graphrag_lightrag.py.

14.4 Concluding Remarks

GraphRAG's effectiveness heavily depends on the quality of the underlying KGs. Constructing comprehensive and accurate KGs requires significant resources, and most of the publicly accessible ones remain incomplete or inconsistent. Without a unified ontology, transferring planning algorithms across domains becomes difficult. Moreover, building KGs from raw text presents a trade-off between detail and efficiency: preserving fine-grained information results in large, unwieldy graphs, while compact graphs risk losing critical insights. Adding to this, computational costs are high when LLMs are used for knowledge summarization, and methods such as Open Information Extraction (OpenIE) often introduce noise, irrelevance, or gaps that limit graph connectivity.

Another major limitation lies in efficiency and scalability. GraphRAG introduces considerable computational overhead because it must handle graph-structured data, including nodes, triples, and paths. The process inflates prompt size, sometimes exceeding tens of thousands of tokens, which can degrade relevance and introduce redundancy. Managing large-scale graphs with millions of nodes and edges is especially challenging, given the input size restrictions of LLMs. Many current GraphRAG implementations are optimized only for smaller graphs, leaving industrial-scale applications difficult to manage. Additionally, some methods under-exploit graph structures, failing to fully leverage the potential benefits of graph-based reasoning.

---

[8]Methodologically, LightRAG emphasizes efficiency and freshness: its graph structures guide retrieval so fewer (but more relevant) chunks are sent to the LLM, and an incremental update algorithm refreshes indices without full rebuilds. In practice, this yields faster responses with higher contextual fidelity versus flat, vector-only pipelines, and avoids the heavyweight orchestration common in prior GraphRAG designs. An open-source server and API streamline document ingestion, KG exploration, and RAG querying for real deployments. In practice, we noticed that the creation of the graph might not complete if some low-performance LLMs are chosen.

Reasoning and integration challenges further complicate adoption. LLMs are not naturally suited to process graph-structured data, and converting graphs into text often strips away critical geometric or relational information. Maintaining coherence across large contexts is also problematic due to fixed LLM context windows, which can cause loss of important details. Some GraphRAG approaches rely too heavily on graph structure, introducing noise at the expense of semantic meaning, while others struggle with measuring similarity between text queries and graph data. Knowledge conflicts between sources and difficulty in retaining precise examples or citations add additional friction, reducing interpretability and user trust.

Dynamically updating graphs in GraphRAG systems also faces several key challenges, primarily stemming from the reliance of most current methods on static databases, which makes incorporating evolving knowledge costly and requires re-initialization or retraining. This process leads to significant computational overhead and memory requirements, particularly with large-scale graphs, increasing response latency and prompt length quadratically. Moreover, ensuring consistency and resolving conflicts when integrating new information from diverse sources is a major hurdle. In addition, the current unimodal (text-based) focus of GraphRAG presents obstacles for integrating multimodal data, which would exponentially increase graph complexity and size for dynamic updates.

Finally, GraphRAG suffers from limited evaluation and generalization. Most current implementations are tested on narrow, task-specific datasets without systematic comparison to conventional RAG methods. Evaluation frameworks are predominantly unimodal, focusing on text-based contexts and overlooking multimodal challenges such as images or audio. Existing benchmarks often lack nuanced task differentiation and ignore reasoning complexity, making them insufficient to fully capture GraphRAG's advantages or weaknesses. As a result, the field lacks robust tools to measure performance consistently, hindering both research progress and practical adoption.

By understanding GraphRAG's main limitations and core value propositions, we can apply this technique wisely, which is summarized in Table 14.6 [3, 8]. For example, we use it for tasks requiring multi-hop reasoning and contextual integration,[9] whereas for straightforward fact

---

[9]Its knowledge graph structure enables chaining logic across multiple supporting facts to yield richer, more context-aware results.

Table 14.6.   GraphRAG's application scenarios.

Scenario Type	Suitable for GraphRAG	Unsuitable for GraphRAG
Knowledge Graph Availability	When high-quality, comprehensive, and domain-specific KGs already exist.	When high-quality KGs are absent or costly to construct, leading to noisy or incomplete graphs.
Complex Reasoning	Multi-hop reasoning, relational inference, and queries requiring logical dependencies across entities.	Simple fact retrieval where direct keyword search or standard RAG suffices.
Domain Characteristics	Domains with structured relationships (e.g., biomedical, supply chain, legal compliance).	Domains like mathematics (symbolic reasoning) or ethics (subjective judgments), where symbolic or value-based reasoning dominates.
Query Type	Exploratory or analytical queries needing contextual and relational depth.	Precision-demanding tasks such as multiple-choice, fill-in-the-blank, or multi-select questions where retrieval noise degrades accuracy.
Latency Tolerance	Offline analysis, research, or use cases where higher processing time is acceptable.	Time-sensitive queries that require rapid response.
System Resources	Enterprise-grade infrastructure with sufficient memory, compute, and storage to handle large graphs.	Resource-constrained environments such as edge devices or lightweight cloud instances.
Modalities	Text-based tasks where graph enrichment improves relational understanding.	Multimodal tasks requiring integration of images, audio, or video into graph structures without proper multimodal LM support.
Information Needs	Tasks needing structured context or global summaries with some tolerance for abstraction.	Detailed Q&A requiring fine-grained facts, where summarization-based approaches risk information loss.

retrieval, traditional RAG approaches often outperform it. Consequently, when faced with tasks that demand the integration of multiple information sources and complex logical inference, GraphRAG offers distinct advantages that conventional RAG cannot match [3, 6].[10] Thus, practitioners should reserve GraphRAG for scenarios demanding deep integration of multiple information sources and logical inference, while relying on streamlined, vector-based RAG when only straightforward fact retrieval is needed.

Future research and development in GraphRAG are focused on addressing its current limitations and expanding its capabilities. A critical area is the development of dynamic and adaptive graphs, enabling Graph-RAG systems to efficiently incorporate new entities and relationships in real-time or near real-time, which is essential as knowledge constantly evolves. This involves creating strategies for constructing, updating, and storing graphs dynamically while maintaining efficiency and effectiveness. Closely related to this is the challenge of efficiency and scalability, particularly for large-scale graphs, by optimizing knowledge retrieval and integration processes to reduce computational overhead, memory requirements, and response latency. This includes exploring lossless compression techniques for long contexts to prevent prompt inflation and enhance inference speed.

Another significant direction involves improving knowledge quality and consistency. This entails developing advanced mechanisms for systematic knowledge organization, automated quality refinement, and intelligent knowledge base expansion, ensuring that constructed graphs are concise, relevant, and consistent, even when integrating information from multiple, potentially conflicting, sources. Furthermore, the current uni-modal (text-based) focus needs to evolve towards multimodal data integration, allowing GraphRAG to process and synthesize information from diverse sources like images, audio, and video into cohesive graph structures. This expansion will address the exponential growth in complexity and size that comes with integrating varied modalities.

Finally, enhancing the trustworthiness of GraphRAG systems is a major priority, especially for high-stakes domains. This includes improving reliability by quantifying and calibrating multi-hop uncertainty, developing robust defenses against safety threats like adversarial attacks on graph structures, addressing privacy concerns related to the relational

---

[10]https://github.com/jeremycp3/GraphRAG-BenchGraphRAG-Bench.

nature of graphs and message passing in GNNs, and boosting explainability by generating transparent reasoning paths. The field also requires standard benchmarks for consistent, comprehensive, and objective evaluation across all components of GraphRAG, from graph construction to generation. Additionally, exploring broader applications in various sectors like healthcare, finance, and legal services, and combining GraphRAG with Graph Foundation Models are promising avenues for future research.

References

[1] H. Han and *et al.*, *Retrieval-Augmented Generation with Graphs (GraphRAG)*, arXiv:2501.00309v2, 2025.

[2] Q. Z. Zhang, S. Chen and *et al.*, *A Survey of Graph Retrieval-Augmented Generation for Customized Large Language Models*, arXiv:2501.13958, 2025.

[3] H. Ha, H. Shomer and *et al.*, *RAG vs. GraphRAG: A Systematic Evaluation and Key Insights*, arXiv:2502.11371, 2025.

[4] D. Edge and *et al.*, *From Local to Global: A GraphRAG Approach to Query-Focused Summarization*, arXiv:2404.16130v2, 2025.

[5] B. Peng, Y. Zhu and *et al.*, *Graph Retrieval-Augmented Generation: A Survey*, arXiv:2408.08921v2, 2024.

[6] Y. Xiao and *et al.*, *GraphRAG-Bench: Challenging Domain-Specific Reasoning for Evaluating Graph Retrieval-Augmented Generation*, arXiv: 2506.02404v3, 2025.

[7] Z. Guo, L. Xia, Y. Yu, T. Ao and C. Huang, *LightRAG: Simple and Fast Retrieval-Augmented Generation*, arXiv:2410.05779v3, 2025.

[8] Z. Xiang, C. Wu and *et al.*, *When to use Graphs in RAG: A Comprehensive Analysis for Graph Retrieval-Augmented Generation*, arXiv:2506.05690, 2025.

Chapter 15

Agentic RAG

15.1 Introduction

AI agents are autonomous or semi-autonomous systems capable of perceiving their environment, reasoning about goals, and executing actions to achieve those goals with minimal human intervention. The recent rise of *agentic AI*, i.e., AI systems that orchestrate multiple capabilities, tools, and decision-making loops, marks a shift from static model inference toward dynamic, goal-driven problem solving. This evolution is not merely a technical novelty; it addresses practical needs in complex domains where single-step responses are insufficient, such as financial analysis, regulatory compliance, and research synthesis. Agentic AI offers tangible benefits: the ability to break down tasks into subtasks, select optimal strategies, adapt to evolving contexts, and leverage multiple models or tools, thereby delivering more accurate, explainable, and actionable outcomes.

At its core, an AI agent consists of several essential components: a *perception layer* to receive input, a *reasoning engine* to interpret context and make decisions, a *planning module* to decompose tasks, an *action interface* to execute operations via tools or APIs, and a *memory subsystem* to store and recall relevant information. When these agents are integrated with retrieval-augmented generation (RAG), they evolve into *agentic RAG* systems, which are capable not only of retrieving knowledge from diverse sources but also of dynamically deciding *what* to retrieve, *how* to process it, and *when* to combine results. This agentic layer transforms traditional RAG pipelines into adaptive, multi-stage workflows where retrieval, reasoning, and synthesis are orchestrated intelligently rather than executed linearly [1–3].

The emergence of agentic AI has been accompanied by the development of standardized protocols and toolkits that enable interoperability and modular design. The Model Context Protocol (MCP) facilitates consistent communication between models and external tools. Agent-to-Agent (A2A) protocols define secure, structured exchanges between autonomous agents in multi-agent ecosystems. Agent-GUI (Ag-UI) standards support human-friendly interaction layers for visualizing and controlling agent workflows. On the tooling side, frameworks such as LangChain, LlamaIndex, AutoGen, and Microsoft's Semantic Kernel provide building blocks for creating, chaining, and deploying agents. Meanwhile specialized connectors allow integration with search engines, databases, APIs, and domain-specific applications. Together, these protocols and tools form the foundation for building scalable and interoperable agentic RAG solutions.

Agentic RAG enables a wide spectrum of advanced scenarios that go beyond the capabilities of static RAG pipelines. In multi-document retrieval, agents can dynamically select sources, rank results, and iteratively refine queries to build comprehensive, context-rich answers. For mixed structured and unstructured data processing, specialized agents, such as a text2SQL agent and a doc-analyzer agent, can collaborate under a coordination framework to merge insights from databases, APIs, and document corpora. In human-in-the-loop (HITL) scenarios, agents can pause execution to seek user validation for high-risk decisions or subjective interpretations, ensuring reliability and compliance. Other advanced use cases include real-time monitoring with adaptive knowledge updates, cross-lingual retrieval with translation agents, and compliance auditing with rule-driven decision agents. Collectively, these scenarios illustrate how agentic RAG can bridge the gap between raw data access and actionable, trustworthy intelligence.

15.2 Agent Main Component and Feature

A typical single agent[1] contains several main parts including memory modules, planning modules, reasoning modules, and tools for executing specific tasks, as well as action (executor) and observation (observer), as shown in Figure 15.1.

---

[1]An AI agent operates as an intelligent loop that repeats until a task is completed. We can roughly divide the workflow into 3–4 steps on how it works:

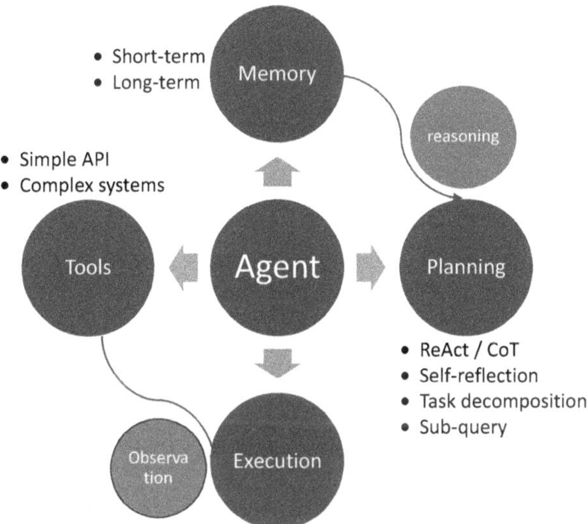

Figure 15.1. A singe agent structure.

Memory

Memory in LLM agents is divided into short-term and long-term categories, enabling the agent to store and recall information essential for user interactions and internal processing. These memory systems help maintain continuity across conversations and allow the agent to refine its responses based on previous interactions.

- A large language model (LLM) decides the next action, outputting a structured directive, typically formatted as JSON, indicating which tool to invoke (a process often called "tool calling").
- Deterministic code executes the tool as instructed.
- The tool's output is added back into the agent's context window.
- The loop repeats until the LLM outputs a special instruction signaling "done," at which point the cycle terminates.

In less formal terms, you can think of this as akin to a while or for loop combined with a switch statement. A functional programming analogy describes it as a foldL: the accumulator represents the evolving context window, while the reducer combines the LLM's decision-making step with control logic to determine which tool to execute next. In essence, an agent is a structured orchestration that alternates between reasoning (via the LLM) and execution (via deterministic code), iteratively enriching the context until the task is resolved.

Planning

The planning module plays a vital role in breaking down complex queries into smaller, manageable tasks, allowing the agent to address each part sequentially. There are mainly two types:

- **Planning without feedback:** Techniques like Chain of Thought and Tree of Thoughts are popular methods for task decomposition, and they can be classified into two types: single-path reasoning and multi-path reasoning. These approaches enable the LLM to generate a detailed plan that addresses the user's question. However, traditional planning without feedback can struggle with long-horizon planning required for solving complex tasks because it lacks mechanisms to refine strategies based on past actions or outcomes.
- **Planning with feedback:** To address this challenge without feedback, planning with feedback introduces mechanisms that allow the LLM to iteratively evaluate and improve its execution plan using historical actions and observations. This iterative process aims to correct and improve upon past mistakes, enhancing the quality of the final results, especially important in complex real-world environments where trial and error are key. Methods like ReAct and Reflexion incorporate reflection or critic mechanisms into the planning process. For example, ReAct combines reasoning and acting by cycling through steps: Thought, Action, and Observation. Observations from the environment provide feedback, helping the model adapt its plan. Human guidance or internally generated feedback can also play a role, improving overall planning flexibility and performance. This capability is further enhanced by tools, ranging from basic APIs like weather services to advanced systems such as code interpreters or specialized data retrieval platforms, enabling the agent to access or compute the necessary information for answering user queries effectively. Below is the simplified pseudo-code:

```
while True:
    # 1. Thought: LLM call
    #     Based on the LLM output, decide
whether a tool call is needed,
    #     which tool to call, and what the
input parameters should be.
    if no_tool_call_needed:
        break
```

```
    # 2. Action: Execute the tool call
    result = call_tool(selected_tool,
input_params)

    # 3. Observation: Get the result of the
tool call,
    #     append it to the prompt/context, and
continue the loop (back to 1)
    prompt.append(result)
```

Those techniques bolster an LLM agent's reasoning abilities by facilitating iterative processing of thoughts, actions, and observations as shown in Figure 15.2. This method enables the agent to continuously refine its strategies, leading to improved responses over time. This iterative reasoning process equips LLM agents to handle complex problem-solving tasks and deliver more accurate, contextually appropriate answers.

15.2.1 *General Workflow of Single Agent*

Combing all the components above, a simplified step-by-step tool-calling process of an LLM-powered agent is illustrated in Figure 15.3.

When a user submits a request, the agent first constructs a comprehensive prompt that includes the user's query, a list of available tools (with names, descriptions, and schemas), the required call format, and any relevant conversation history. This prompt is sent to the LLM, which "thinks" by generating a plan composed of Thought, Action, and Action Input. If the LLM determines that a tool is needed (indicated by an Action specifying a tool), the agent invokes that tool with the supplied parameters and records the result as an Observation. This observation is appended back into the conversation history, and the enriched prompt is sent to the LLM again for further reasoning. The cycle of *think* → *tool call* → *observe* → *think* continues until the LLM outputs a Final Answer, at which point the agent returns this answer to the user. This process ensures that tool usage is deliberate—the LLM always decides *if* and *when* to call external tools—creating a structured, iterative decision-making flow.

15.2.2 *Workflow vs Agent*

When dealing with a task using RAG, it is best to start with the simplest workflow approach and only introduce complexity (i.e., intelligent agents)

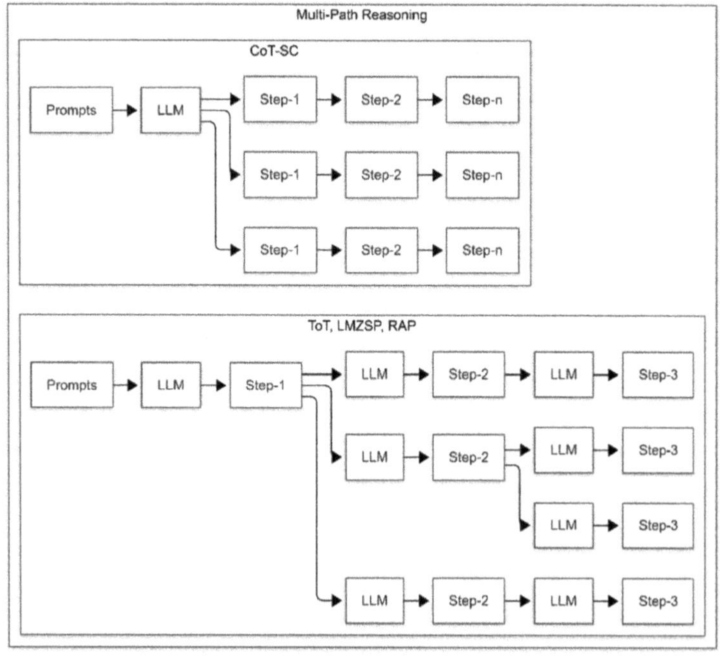

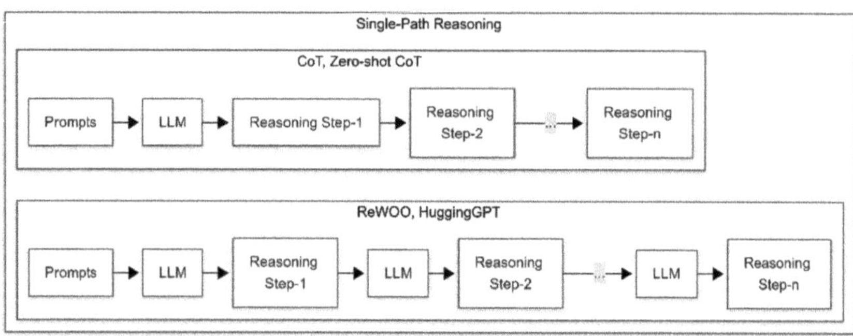

Figure 15.2.　Single-path vs multi-path reasoning [1] [4].

when it truly adds value to it.[2] Chaining together a single call to a large model enhanced with information retrieval or illustrative context samples, is sufficient in many situations. This approach is easier to implement, lower in latency, more cost-effective, and often just as accurate for tasks like summarization, classification, small knowledge lookups, or standard

---

[2]In some literatures, the workflow with LLM and predefined rules+tools is also considered as one types of agents in Level 1 [7].

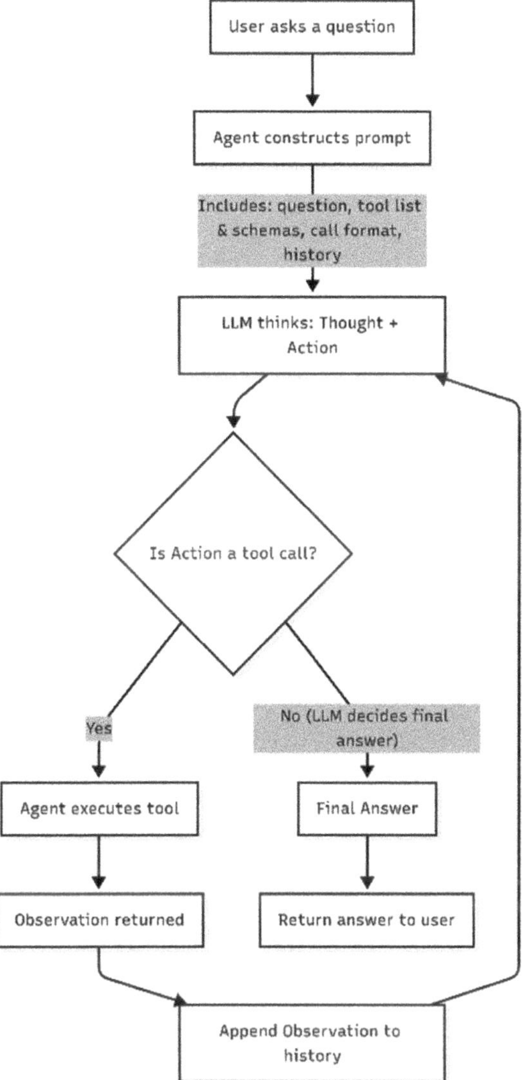

Figure 15.3. Agent workflow.

question-answering. While agent achieves better performance for complex tasks, it usually brings increased latency and compute expense. In general, we should evaluate task complexity and measure performance/cost before jumping toward an agent approach, or even adopt progressive enhancement. Figure 15.4 compares the workflow and agent.

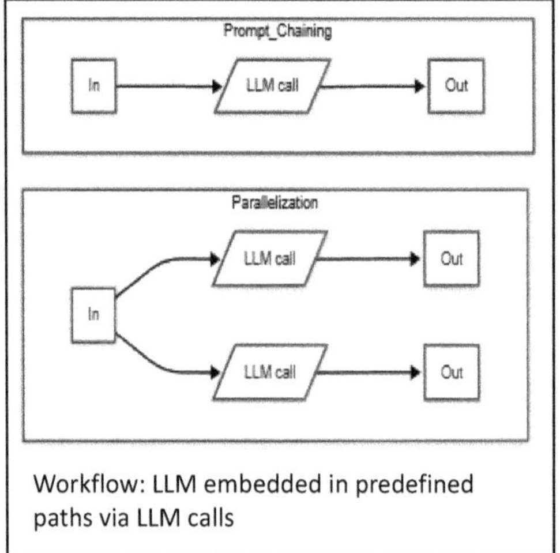

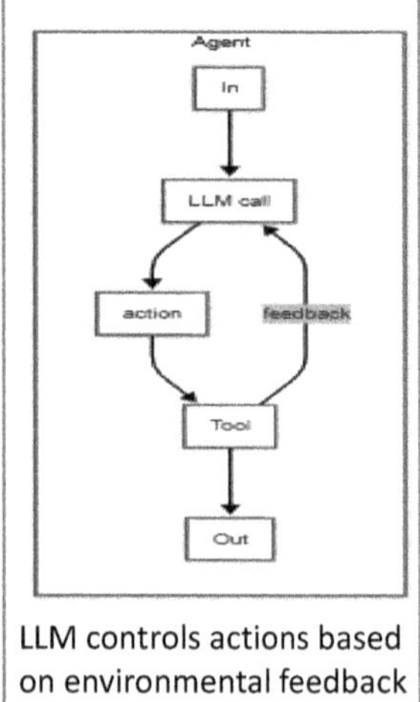

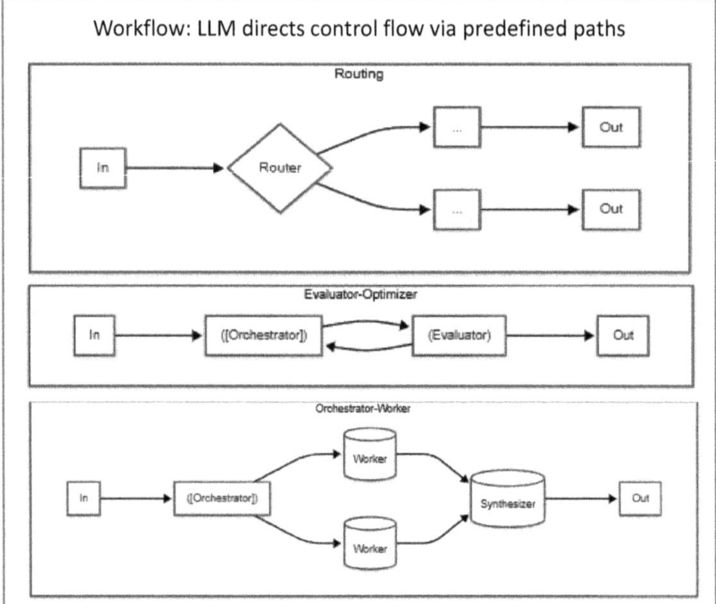

Figure 15.4.   Workflow vs agent.

- **Workflows** shine when your task is something like "scrape a site →
  extract specific data → format and store it." i.e., when tasks require
  clear multi-step processes with little need for flexible intelligence.
- **Intelligent agents**, in contrast, are better suited to tasks like
  "research competitive products, evaluate features, and draft strategic

recommendations," where autonomous planning, open-ended reasoning and tool usage bring real value.

15.2.3 *Single vs Multiple Agents*

Multi-agent systems distribute workload across specialized agents, each focusing on a specific function or domain. This modularity fosters scalability, flexibility, resilience, and robustness—attributes that monolithic single-agent systems often lack. Multi-agent systems enhance efficiency and adaptability, particularly in environments requiring parallel processing, domain partitioning, or fault tolerance. Transitioning from a single-agent to a multi-agent architecture becomes necessary when:

- **Task Complexity Increases:** When responsibilities such as database querying, analysis, visualization, and report generation become too broad for one system, splitting tasks among agents improves clarity, maintainability, and reliability.
- **Diverse Data Sources:** When workflow spans relational databases, document stores, and graph data, specialized agents optimized per source streamline processing and improve accuracy.
- **High Scalability Needs:** For large-scale, mission-critical systems, the modularity of multiple agents supports parallel execution, fault tolerance, and dynamic orchestration—features lacking in single-agent setups.
- **Limitations of Single-Agent Context Management:** Single agents may struggle with context overload or confusion among many tools, leading to degraded performance.

Usually, we use single-agent systems when tasks are limited in scope, simple, and well-defined, with a path of minimal complexity and operational overhead. On the other hand, we shall adopt multi-agent systems when workflows grow complex, require specialized handling across diverse domains, demand scalability, or benefit from parallelism and resilience. The detailed comparison is listed in Table 15.5.

15.3 Agentic RAG Framework

The integration of these components and techniques allows LLM agents to operate across a range of applications from customer service and data analysis to more creative tasks like writing assistance or interactive media.

Table 15.1.   Single vs multiple agent system.

Factor	Single-Agent Systems	Multi-Agent Systems
Structure	One agent handles all tasks.	Multiple agents, each specialized.
Resource Use	Efficient for simple tasks.	Less efficient; higher overhead, but better at handling complex workloads.
Scalability	Limited with adding features risks context overload.	Highly scalable via modular addition of agents.
Reliability	Predictable behavior, easier to debug.	Better resilience, e.g., failure of one agent doesn't collapse the system.
Coordination	No coordination needed.	Coordination critical, i.e., requires protocols and communication frameworks.
Suitability	Best for bounded, sequential, or isolated tasks.	Ideal for parallel, exploratory, domain-diverse, or evolving workflows.

The potential for LLM agents to transform various sectors is vast, given their ability to not only understand and generate human language but also to reason, plan, and learn from their interactions. By combining agent and RAG into an agentic RAG framework, the resulting system offers broader applicability, enhanced flexibility, and superior performance compared to using each component individually. Table 15.2 compares non-agentic RAG and agentic RAG systems in multiple aspects.

15.3.1 *Agentic Frameworks and Tools*

As RAG systems evolve from passive retrieval pipelines to autonomous, task-driven systems, agentic frameworks play an increasingly important role. These frameworks enable reasoning, decision-making, tool-use, memory, and multi-turn orchestration, transforming RAG into a dynamic, self-improving agent capable of interacting with tools, APIs, and users. Many agent frameworks and tools are available in the markets including the general purpose agents, e.g., openai agent, langchain/langgraph and llamaindex, and specific agents, e.g., CrewAI, Phidata and AutoGen. Table 17.5 and Table 17.6 in supplementary material list some popular frameworks and tools. Table 15.3 further compares the three main frameworks with their unique features and applicable scenarios.

Table 15.2.   Agentic vs non-agentic RAG.

Feature/Aspect	Non-agentic RAG	Agentic RAG
Prompt Setup	Dependent on manual crafting of prompts.	Automatically adjusts prompts based on contextual needs.
Contextual Awareness	Static, lacks real-time context integration.	Dynamically integrates ongoing conversation history into decision-making.
Efficiency & Overhead	Prone to inefficiency with potential for high overhead.	Optimizes retrieval to enhance efficiency and reduce unnecessary operations.
Process Complexity	Separate tools and models needed for complex reasoning.	Streamlines complex reasoning and tool integration, removing the need for separate models.
Decision Autonomy	Governed by static rules without real-time adjustments.	High autonomy, making context-aware decisions on information retrieval and processing.
Information Retrieval	Limited to initial queries without follow-up actions.	Proactively seeks additional information as needed during retrieval.
Adaptability & Responsiveness	Restricted to predefined queries and static responses.	Highly adaptable, revising strategies based on immediate feedback and observations.
User Interaction	Frequently requires user input for guidance.	Minimizes user need for interaction by proactively managing tasks.
Error Management	Basic error handling capabilities.	Advanced error correction and robust recovery mechanisms in place.

Table 15.3.   Agent comparison.

Framework	Unique Features	Applicable Scenarios
LangChain/ LangGraph	**Graph-based orchestration** with state management. Excellent for defining complex, cyclical workflows and multi-agent interactions.	**Complex Reasoning Tasks:** Ideal for problems that require iterative refinement, like multi-step question answering, conversational chatbots that need to search multiple times, or self-correcting RAG systems.

(Continued)

Table 15.3.   (*Continued*)

Framework	Unique Features	Applicable Scenarios
LlamaIndex	**Deep integration with data sources and indexing strategies:** Provides a rich ecosystem for ingesting and preparing data specifically for RAG.	**Data-Centric Applications:** Best suited for applications where the primary challenge is dealing with a vast, diverse, and messy corpus of data. Think enterprise knowledge bases, document analysis, and research assistants.
OpenAI Swarm (Conceptual)	**Collaborative, multi-agent approach** using a "swarm" of specialized agents. Focuses on task decomposition and delegation.	**Domain-Specific Expertise:** Great for problems that benefit from a division of labor. For example, a medical diagnosis assistant where one agent specializes in symptoms, another in treatment options, and a third in patient history.

15.3.2 *Taxonomy of Agentic Frameworks for RAG*

Agentic RAG frameworks can be categorized into the following four major layers, based on their design principles and functionalities:

1. Agent-Orchestration Layer

Responsible for managing multiple agents, tools, and workflows. Enables complex reasoning, dynamic routing, and subtask decomposition.

Category	Libraries/Tools	Capabilities
Graph-Based Agents	LangGraph (by LangChain), CrewAI, AutoGen	Node-level orchestration, conditional routing, tool/agent chaining, async loops
Swarm/ Team-Based Agents	OpenAI Swarm, Autogen, CrewAI	Parallel agents with role-specific tasks, peer-to-peer messaging, collaborative decision-making
Workflow DSL	LangChain Expression Language (LCEL), PromptChainer, SupaAgents	Declarative RAG workflows, context-aware flow control, dynamic agent/task injection

2. Memory & Context Layer

Handles memory persistence and context fusion across retrieval rounds or interactions.

Category	Libraries/Tools	Capabilities
Short-Term Memory (STM)	LangChain Memory, LangGraph State, LlamaIndex Context	Manages context window, keeps chat history or retrieved chunks
Long-Term Memory (LTM)	ChromaDB, Qdrant, Weaviate, LlamaIndex Vector Store	Persistent storage of documents and user history
Memory-Augmented Retrieval	RAG-Fusion, Memory-RAG	Recall-enhanced prompting, relevance-aware retrieval conditioned on past tasks

3. Planning & Reasoning Layer

Enables abstract reasoning, task decomposition, and self-reflection within RAG.

Category	Libraries/Tools	Capabilities
ReAct Loop	LangChain AgentExecutor, LangGraph, LlamaIndex Agent	Chain-of-thought reasoning with tool-use
Self-Reflection & Critique	AutoGPT Feedback Nodes, ReWOO, Self-RAG, ToolLLM	Improves output by self-critiquing, retrying with refined reasoning
Task Planning	Plan-and-Solve, GraphRAG, LAMAR	Breaks down queries into retrievable sub-questions or logical flows

4. Tool Integration Layer

Connects agents with external systems like APIs, databases, or file systems.

Category	Libraries/Tools	Capabilities
Tool Libraries	LangChain Tool, FunctionAgent, Autogen Tools, LlamaIndex Tools	Tool definitions, input/output schema enforcement
Text-to-SQL/ API agents	OpenAI Agent SDK, vanna, AgentKit, ReAct + OpenAPI	Enables querying of structured data or triggering complex actions
Knowledge Graph Agents	GraphRAG, KG-RAG, LlamaIndex KG Index	Multi-hop reasoning over structured data (RDF, Neo4j, etc.)

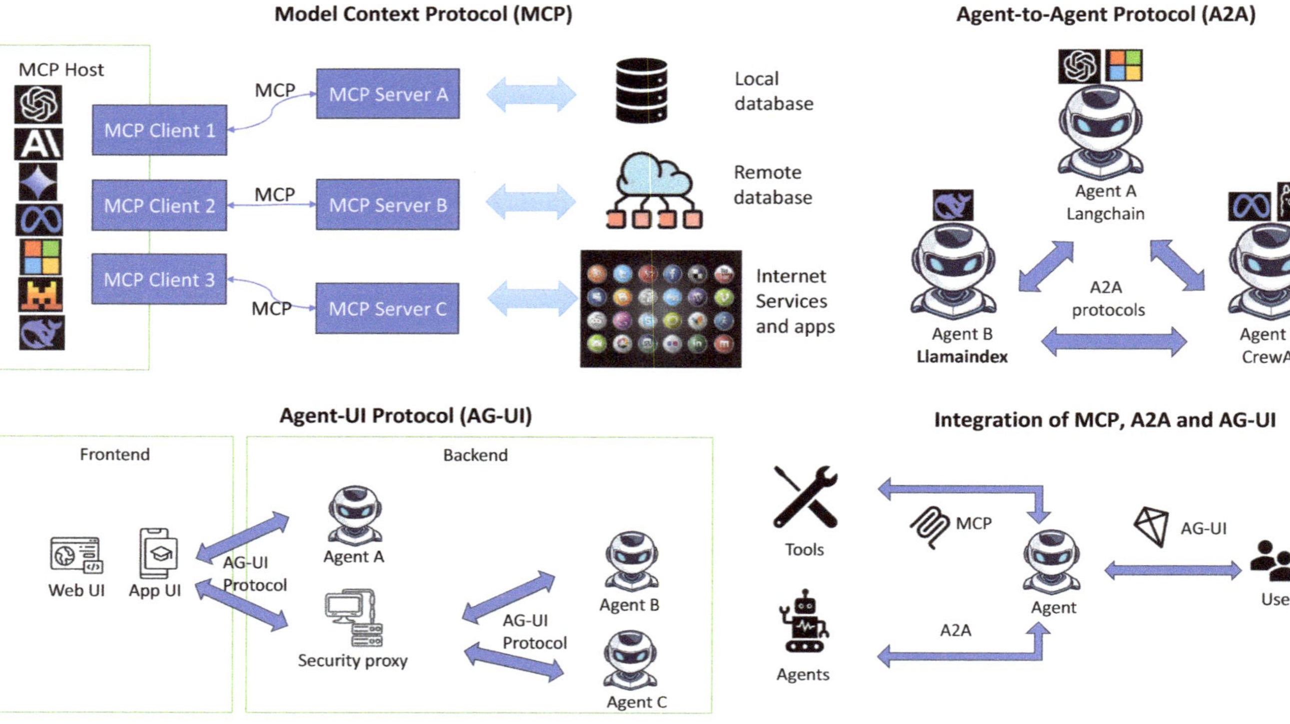

Figure 15.5. Three agent protocols and their integration.

15.4 Agentic Protocols

Although most of the aforementioned agents integrate multiple tools, they cannot interact with each other and cannot build smart bridges among different LLMs. Therefore, some universal protocols dedicated for the agents were developed. There are many types of protocols, but the three main types are: 1) Agent to tools; 2) Agent to agent; 3) Agent to UI.

15.4.1 *Agent to Tools: Model Context Protocol (MCP)*

MCP is an open protocol that standardizes how applications provide context and tools to LLMs, solving the "NxM problem" of custom integrations between models and data sources. It can be viewed as a significant extension to function calling discussed in Chapter 12.[3] Initially developed by Anthropic and released in late 2024, it has since gained broad adoption, with support from organizations like OpenAI, Microsoft (Semantic Kernel), JetBrains, and community contributors.

MCP addresses the challenge of LLM isolation by allowing models to access real-time data. For example, live weather updates or recent GitHub commits: instead of relying solely on static, pre-trained information. This capability not only enhances the models' responsiveness but also dramatically simplifies the integration process for developers, reducing the typical complexity from an N×M to an N+M problem through standardized tool schemas. Furthermore, MCP's interoperability across multiple LLMs, including Claude, GPT, and LLaMA, minimizes vendor lock-in and provides a consistent framework for advanced, context-aware tool usage. For more detailed information, one can consult the official MCP specification or OpenAI's MCP integration guide.

MCP serves as a universal adapter for AI by offering a standardized interface that enables large language models (LLMs) to connect with external tools and data sources in a manner similar to USB-C. This

---

[3] MCP and function calling share many commonalities. Both of them enable LLMs to trigger external functions, APIs, or services beyond pure text generation. Both rely on schema-like definitions (e.g., JSON input/output) so that the LLM knows how to call the function and interpret the results. Both approaches try to reduce hallucinations and make tool use more deterministic by enforcing structured interfaces instead of relying only on free-form text. However, their differences are also obvious, which is compared in Table 15.4.

Table 15.4.   Comparison of MCP and function calling.

Aspect	MCP (Model Context Protocol)	Function Calling (Native LLM Feature)
Scope	A general interoperability protocol for LLM <-> tools communication. Supports multiple tools, agents, and environments.	A model capability to call predefined functions exposed by the developer during inference.
Standardization	Defines a protocol-level contract (like gRPC/HTTP for LLMs), independent of any single vendor.	Tied to specific model vendors (OpenAI, Anthropic, etc.) and implemented differently across them.
Extensibility	Can orchestrate many tools, services, and agents, with metadata exchange (e.g., discovery, permissions, schemas).	Limited to functions provided at inference time; orchestration logic usually must be implemented outside the model.
Interoperability	Designed for cross-platform and multi-agent ecosystems, so any MCP-compatible client/server can talk.	Usually single-vendor and session-scoped, making portability harder.
Complexity	More complex: requires MCP servers, message exchange, and infrastructure. Best suited for enterprise-scale or agentic AI.	Lightweight: define functions inline with the model API call; best for simple tool use.
Governance & Security	Built-in support for policies, access control, auditing, and sandboxing in multi-tool settings.	Basic guardrails only; depends on app logic to enforce governance.
Use Case Fit	Suitable for agent frameworks, marketplaces, multi-LLM orchestration, enterprise RAG/ agent systems.	Suitable for single-app integrations, e.g., "LLM can call a weather API or database query function."

standardization allows models to integrate seamlessly with databases, APIs, GitHub, Slack, and more without requiring custom one-off scripts, thereby minimizing redundant development efforts. MCP is maintained by Anthropic yet remains community-driven with SDKs available in multiple popular languages such as Python, TypeScript, Java, C#, and Kotlin. Moreover, the MCP ecosystem includes over 1,000 community-built servers for platforms like PostgreSQL, Slack, and Docker, as well as official integrations for services such as Stripe and GitHub.

MCP adopts a client-server architecture inspired by the Language Server Protocol (LSP), where user-facing applications, such as desktop clients and IDEs, act as hosts, and lightweight servers expose a variety of tools and data, including weather APIs and code repositories. The system employs JSON-RPC 2.0 to support both local (STDIO) and remote (HTTP combined with Server-Sent Events) communication. OpenAI has integrated MCP into its Agents SDK to enhance tool augmentation capabilities; this allows agents to access external tools, such as filesystem operations or API calls, within multi-step workflows while keeping or Streamable HTTP the process secure. Additionally, MCP embeds user consent mechanisms and privacy guardrails, including read-only queries and sandboxing, to maintain control over sensitive operations. To implement a MCP framework, FastMCP is a recommended python library for building MCP applications, providing the fastest path from idea to production.[4] Another interesting project is MCP-agent,[5] which implements effective agent building patterns in a composable way.

15.4.2 *Agent2Agent (A2A)*

A2A is an open-source framework developed by Google designed to streamline agent-to-agent communication and collaboration in 2025. It provides a standardized infrastructure that enables heterogeneous agents to interact seamlessly by adopting common protocols and interface specifications. This unified approach removes interoperability barriers between different AI systems, fostering a more cohesive ecosystem where distinct agents can exchange information and coordinate their actions effectively.

The architecture of A2A is built around modularity and scalability. By using well-defined communication protocols, such as JSON-RPC 2.0, A2A ensures that agents can easily be integrated into a larger multi-agent system, e.g., whether they are task-specific models, data retrieval modules, or other service components. The framework offers flexibility for developers to extend and customize agent behavior without altering core functionalities, thus promoting rapid prototyping and iterative

---

[4]https://github.com/jlowin/fastmcp.
[5]https://github.com/lastmile-ai/mcp-agent.

development. Detailed documentation, sample code, and integration guides provided in the repository further simplify the process of building and deploying these interconnected systems.

The significance of A2A lies in its potential to revolutionize how AI systems are constructed and deployed. By enabling dynamic interactions among agents, the framework supports complex workflows that require multi-step reasoning and real-time problem-solving capabilities. This capability is particularly valuable in domains such as conversational AI, automated decision-making, and real-time data analysis, where coordinated action among diverse models can lead to more intelligent and adaptive systems. Ultimately, A2A helps reduce development overhead and accelerates innovation by providing a robust foundation for building scalable, interoperable, and intelligent multi-agent architectures.

15.4.3 *Agent to User Interaction (AG-UI)*

Modern AI-agent stacks have no universal "language" for connecting back-end agents to front-end apps: LangGraph, CrewAI, Dify and others each ship their own WebSocket formats, JSON schemas, and status conventions. Any UI migration therefore requires bespoke adapters, leading to brittle code and high maintenance cost. The Agent-User Interaction (AG-UI) protocol was created to eliminate this fragmentation by specifying a single, lightweight event stream that every agent and every client can understand.

Architecturally, AG-UI sits between the UI and one or many agents. A front-end component speaks the protocol directly or through a Secure Proxy that routes requests to multiple agents. Each agent, whether self-hosted or third-party, emits AG-UI-compliant events, while the proxy enforces authentication and multi-tenant isolation. Because the protocol is purely event-based, it can be layered over server-sent events (SSE), WebSockets, or HTTP/2 streams without locking the stack to any vendor or runtime.

The interaction pattern is simple: the client sends a POST to initialize a session, then opens a streaming channel to receive JSON-encoded events. Standard event types include TEXT_MESSAGE_CONTENT for token-level output, TOOL_CALL_START / TOOL_CALL_END for tool execution traces, STATE_DELTA for shared-state updates, and AGENT_ HANDOFF to transfer control among cooperative agents. Crucially, the same channel is bidirectional—the UI can push new events or context back to the agent at any time, enabling real-time, iterative heartbeat planning rather than one-shot calls.

Key features follow naturally: (1) Portability and framework-agnostic: the same UI code works with multiple agent frameworks, e.g., LangGraph today and CrewAI; (2) Real-time bi-directional streaming with back-pressure handled by the SSE/WebSocket/webhook substrate; (3) Extensibility: new event kinds can be added without breaking older clients; (4) Open-source SDKs in TypeScript and Python that expose typed helpers (RunAgentInput, Message, Context, Tool, State) for rapid adoption. Collectively, AG-UI gives developers a REST-like standard for live agent interaction, cutting integration effort while delivering consistent, framework-agnostic user experiences.

The MCP, A2A, and AG-UI protocols are generally used for different aspects and compliment to each other; Table 15.5 compares their main

Table 15.5.   Comparison of agent protocols.

Comparison Dimension	MCP	A2A	AG-UI
Positioning	Standardizes how models/agents connect to external **tools and data** ("tool augmentation").	Standardizes **inter-agent communication and collaboration** (discovery, delegation, task lifecycle).	Standardizes **agent – frontend** interaction via a streaming, event-based UI contract.
Transport Protocol	Primarily uses JSON-RPC 2.0 over stdio or HTTP/S with Server-Sent Events (SSE).	Uses JSON-RPC 2.0 over HTTP/S. Supports SSE for streaming and webhooks for push notifications.	Transport-agnostic, but commonly uses SSE or WebSocket for real-time event streaming.
Data Unit	JSON-RPC messages for Requests, Results, and Notifications.	Task/Message: Structured JSON containing task types, parameters, and artifacts. Message containing one or more "Parts" (e.g., TextPart, FilePart), and Artifacts for final output.	Structured JSON Events. Defines 16 standard event types (e.g., TOOL_CALL_ START, STATE_DELTA).

(*Continued*)

Table 15.5.   (*Continued*)

Comparison Dimension	MCP	A2A	AG-UI
Core Functions	• Tool registration/ discovery. • Schema description. • Tool invocation & result return. • Multi-agent/ multi-tool orchestration.	• Agent discovery & authentication. • Task initiation & assignment. • Artifact transmission. • Multi-turn messaging & state management. • Long-running task support.	• Capturing user events. • Rendering agent responses. • UI-agent state synchronization. • Supports rich interactions (text, buttons, forms). • Enables human-in-the-loop workflows and interruptible execution.
Typical Scenarios	Code assistants calling compilers, search tools, or local/cloud file systems.	Multi-agent collaborative problem-solving (e.g., knowledge reasoning, distributed tasks).	Embedding agents into IDEs, web apps, mobile apps, or live agent UIs for real-time interaction.

features and Figure 15.5 illustrates them. In practice, one agent can simultaneously:

- exchange messages with another agent via A2A,
- serve the user interface through AG-UI, and
- invoke external tools exposed by an MCP server.

Each protocol addresses a distinct need and the three operate in a complementary fashion.

- AG-UI manages human-in-the-loop interaction and streaming UI updates.
- A2A enables inter-agent communication and collaborative task execution.
- MCP standardizes tool calls and context handling across different models.

15.4.4 *Adoption Roadmap*

From the enterprise adoption point of view, we can therefore develop a clean, enterprise-oriented roadmap from basic LLM → RAG → explicit Agent → MCP-enabled LLMs → Agent-to-Agent (A2A) as shown in Table 15.6. Usually, the most reliable step beyond stand-alone LLMs is to add RAG so answers are grounded in current and proprietary sources with provenance, as established by the original RAG work and subsequent practice. From there, explicit agent patterns such as ReAct (reason-and-act) and Toolformer (self-learned tool use) underpin multi-step planning and tool orchestration; these boost capability but also introduce more LLM calls, latency, and new failure modes that demand good observability and guardrails. To simplify integrations at scale, organizations increasingly standardize tool/data access with the MCP, introduced by Anthropic and now supported on Windows, reducing bespoke connectors while still requiring strong security, consent, and rate/latency controls. As agent apps proliferate, interoperability becomes the next bottleneck; the A2A protocol, addresses discovery, messaging, and delegation across vendors, while Google's AP2 extends this with trusted, policy-bound payments between agents. In practice, adopt A2A only when multiple independent agents truly need to collaborate across teams or partners; otherwise, standardizing tool access with MCP and disciplined agent orchestration (e.g., reliability loops as promoted by LangGraph) usually provides the best balance of capability, cost, and governance.[6] Figure 15.6 further provide a decision workflow to choose the right stage.

15.5 Typical Agent-based Scenarios

15.5.1 *Agents in Multi-Document Retrieval*

This section introduces an agent-based multi-document retrieval case using OpenAI Agent, which operates within the RAG system described below.

---

[6]Jun Xu, From LLMs to Agent Meshes: A Pragmatic Enterprise Roadmap (LLM → RAG → Agents → MCP → A2A), https://medium.com/@xujun0628/from-llms-to-agent-meshes-a-pragmatic-enterprise-roadmap-llm-rag-agents-mcp-a2a-36ee8b3cd025, Oct 2025. Accessed on 11/11/2025.

Table 15.6.   The 5 stages of GenAI adoption.

Stage	Core Capability & Strengths	Typical Enterprise Scenarios	Limitations/When to Avoid
1. Stand-alone LLM	Prompt → text from the model's parametric knowledge. Fast to pilot; zero infra	Drafting, translation, summarization, generic FAQs, ideation	Knowledge cutoff, no citations, can hallucinate; no private data/tools
2. External Knowledge base (Web + RAG)	Browse/search + RAG to ground answers in current/private corpora Fresh, citeable answers; controllable knowledge scope	Policy/KB assistants, customer support, product & compliance Q&A with provenance	Retrieval quality/latency bound by indexing, chunking, reranking; still needs guardrails
3. Explicit Agents	Plan → tool/API calls → observe → reflect loops (multi-step. Handles multi-tool, multi-step tasks; transparent trajectories orchestration)	"Analyze docs, build charts, then draft an email"; "Search, compare, and compile a brief"; semi-autonomous ops with human-in-the-loop	More moving parts; higher cost/latency; new failure modes (tool choice, prompt injection)
4. Smarter LLMs + standardized tool access à MCP	Native tool/function calling + MCP servers/ clients for consistent connectors. Simpler integration; auditable registries/ permissions; broad ecosystem support (Windows/Copilot/IDEs)	Unifying many internal tools/ DBs; OS/IDE integrations; reducing glue code	Ecosystem maturity varies; still probabilistic; require security & consent layers
5. Agent-to-Agent (A2A) mesh	Inter-agent discovery, messaging, and delegation across teams/vendors; optional commerce via AP2 for payments Avoids "agent silos"; enables capability composition; standardizes trust/identity	Enterprise agent ecosystems (procurement, finance, travel, ITSM); partner integrations; cross-domain workflows	Governance, identity/ consent, and tracing get harder; added coordination latency

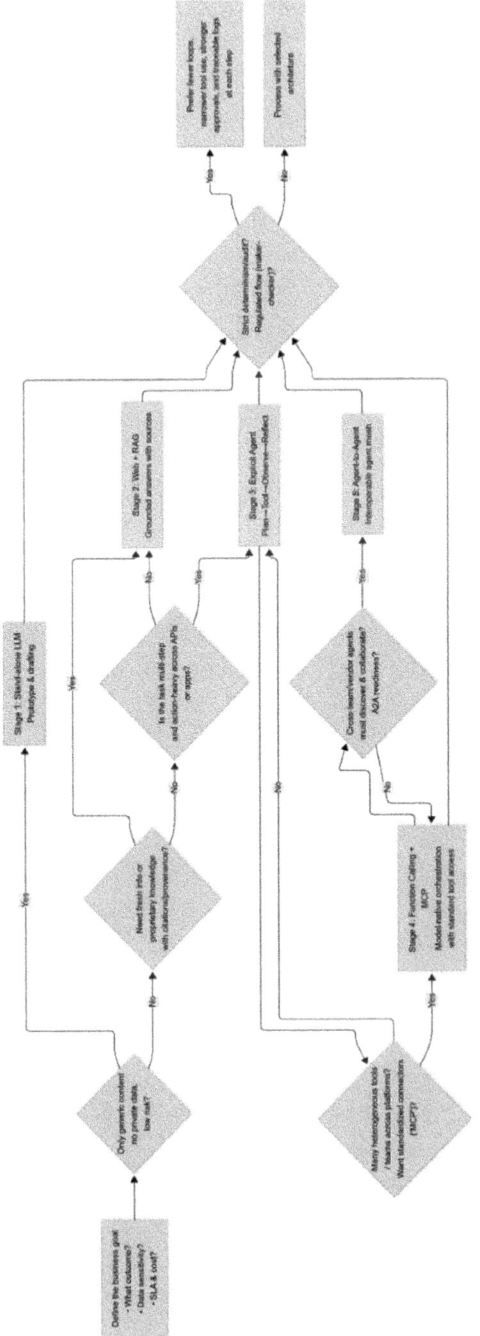

Figure 15.6. Decision workflow for stages.

OpenAI Agent

OpenAI Agents now unify core LLM capabilities and tool use.[7] With Responses, you can invoke built-in tools (notably web search, file search over vector stores, code interpreter, and computer use) and also register custom function tools defined by JSON-schema, avoiding much of the bespoke glue previously needed to integrate external systems. Conversation state can be handled server-side by storing responses and chaining turns via previous_response_id, while the Agents SDK provides higher-level sessions (persistent memory), handoffs (agent-to-agent delegation), and guardrails for validation. These primitives reduce custom infrastructure while still enabling deep connections to external data and services (including via MCP connectors).

Furthermore, the OpenAI Agent class provided by LlamaIndex builds on this foundation by fusing OpenAI's function-calling capabilities with its own ChatEngine (which supports contextual, multi-turn conversations) and QueryEngine (which facilitates knowledge retrieval). This integration enables the agent to perform multi-step reasoning within a conversational flow, retrieving relevant information and executing function calls as needed, thus mimicking advanced human-like problem-solving. The result is a more autonomous agent capable of handling complex workflows, from dynamic data analysis to context-aware decision-making, all while sustaining coherent dialogue with users.

Multi-Document Agents Scheme

In a sophisticated multi-document agent setup, an agent (OpenAIAgent) is initialized for each document, capable of document summarization and classic QA mechanics. A top agent is responsible for routing queries to document agents and synthesizing the final answer. Each document agent uses two tools: a vector store index and a summary index. Based on the routed query, the document agent decides which index to use. For the top agent, all document agents act as tools.

This advanced RAG architecture involves many routing decisions by each agent, allowing for the comparison of different solutions or entities described in various documents and their summaries, along with classic

---

[7]Note that OpenAI announced the Assistants API deprecation with a planned sunset on Aug 26, 2026; new builds should target Responses + Agents SDK and follow OpenAI's migration guidance.

single document summarization and QA mechanics. This approach covers the most common chat-with-collection-of-docs use cases.

The multi-document agents scheme involves both query routing and agentic behavior patterns. The benefit of this approach is the ability to compare different solutions or entities described in different documents and their summaries. However, the drawback of such a complex scheme is its slower performance due to multiple back-and-forth iterations with the LLMs inside the agents. LLM calls are always the longest operations in a RAG pipeline, as search is optimized for speed by design. For large multi-document storage, some simplifications to this scheme may be necessary to make it scalable.

The implementation using llamaIndex[8] is built on a two-level hierarchy:

- **Document Agents:** Each individual document is assigned a dedicated agent. This agent is equipped with two query engines: a vector index for semantic search and a summary index for summarization. These are converted into tools that an OpenAI Function Calling agent can use, allowing it to intelligently choose the best way to retrieve information from its specific document.
- **Top-Level Agent:** This is a retriever-enabled OpenAI agent that orchestrates the entire system. It uses a tool retriever to identify the most relevant document agents (tools) for a given user query. It also incorporates a Cohere reranker to improve the filtering of candidate documents, ensuring the most pertinent information is selected. Additionally, a query planning tool is dynamically created to handle complex queries that require synthesizing information from multiple documents.

15.5.2 *Agents in Mixed Database and Document Processing*

In this subsection, we intend to build an agentic RAG that can help to process both database (tabular RAG) and document (traditional RAG). In particular, we use some HR data to answer HR questions, e.g., questions on HR policies and the inquisitor's own information from the database.

---

[8]https://docs.llamaindex.ai/en/stable/examples/agent/multi_document_agents-v1/. Accessed on 14/09/2025.

It creates two agents, one tool_df is processing the queries on a database, while the other tool_doc is handling the policy documents. Then we apply OpenAIAgent from llama_index.agent.openai to build agent workflow. tool_df is built on pandasai, a Python platform that enables inqueries to your data in natural language. It helps non-technical users to interact with their data in a more natural way, and it helps technical users to save time and effort when working with data. tool_doc uses a basic SimpleDirectoryReader and automerging_query_engine introduced earlier to handle documents. The corresponding code is available in ch15\ agent_hybrid.py. Below are the code snippets for the tooling and the agent setup:

```python
import json
from typing import Sequence, List
from openai.types.chat import
 ChatCompletionMessageToolCall

# for llama_index version >=0.10
from llama_index.llms.openai import OpenAI
from llama_index.core.llms import ChatMessage
from llama_index.core.tools import BaseTool,
 FunctionTool

tool_doc = FunctionTool.
 from_defaults(fn=chat_response_doc)
tool_df = FunctionTool.
 from_defaults(fn=chat_response_df)

query_engine_tools = [
    QueryEngineTool(
        query_engine=tool_df,
        metadata=ToolMetadata(
            name=f"individual_KPI",
            description=(
                "useful for when you want to
answer queries about my own KPI, budget,
target and/or actual performance, Gross
Profit (GP) Margin, Net Income Margin, SSG
of Revenue Contribution."
            ),
```

```
            ),
        ),
      QueryEngineTool(
          query_engine=tool_doc,
          metadata=ToolMetadata(
              name=f"company_report",
              description=(
                  "useful for when you want to
  answer queries about the company Lenovo's
  annual reports and statements."
              ),
          ),
      ),
  ]

from llama_index.agent.openai import
  OpenAIAgent
agent = OpenAIAgent.from_tools(query_engine_
  tools, verbose=True)
```

15.5.3 *Agent with Human-in-The-Loop*

As enterprise-grade agents gain autonomy, a central concern arises: how can we trust them when they perform high-risk actions? For instance, an AI agent might autonomously delete a crucial database record or initiate a large financial transfer: mistakes here could be catastrophic. Human-in-the-Loop (HITL) technology emerges as a vital safety mechanism. HITL enables a human to step in during key decision points, giving final approval before any irreversible step is taken. This architecture helps preserve the efficiency and autonomy of AI while maintaining safety and control. Common high-risk use cases include database management (e.g., confirming deletions), financial systems (e.g., approving large transfers), and medical diagnostics (e.g., final sign-off on treatment recommendations). While HITL enhances trust and safety, it also introduces serious technical challenges: triggering human review without undermining agent autonomy, accurately recognizing when intervention is needed, pausing and resuming agent state cleanly, and meeting stringent response-time requirements to avoid bottlenecks.

An enterprise-grade HITL architecture typically includes the following key modules:

- **User Input Layer** – Captures the human command or request (e.g., "Delete this database record"), which seeds the downstream workflow.
- **AI Agent Core Processing Layer** – Treats the task as a flow graph (as in LangGraph), with nodes like "call_model" (begin execution), "detection node" (risk assessment), "execute_users" (dangerous, irreversible actions), and "END" (completion). The "execute" nodes are flagged as high-risk.
- **Breakpoint Mechanism** – Acts as the system's "brake," supporting both static breakpoints (predefined pause points) and dynamic ones (triggered at runtime by risk detection). If the agent reaches a critical node or flags risky behavior (e.g., large transaction, sensitive update), execution halts before proceeding.
- **Human Approval Layer** – A human reviewer assesses the state presented, then chooses "approve" to resume or "reject" to halt.
- **Checkpointer / State Manager** – Captures a snapshot of the agent's execution context at pause time, preserving all relevant variables and state so that after human input, execution can resume seamlessly.
- **External Tools & Datastores** – The AI agent may interact with APIs, databases, or file systems (e.g., delete records, call payment endpoint). These are the "execution battlefield."
- **Security, Monitoring & Audit** – Supports logging of every decision and action, fine-grained access control, risk scoring, compliance checks, and real-time monitoring to ensure integrity and traceability.

A typical HITL workflow follows two possible routes as shown in Figure 15.7:

- **Normal (Direct) Flow:** The human inputs a command. The AI agent processes it, runs through the detection and execution nodes and directly completes the task—no interruption.
- **HITL-Intervention Flow:** The AI agent detects a sensitive or high-risk operation (e.g., tool call to delete data). A breakpoint is triggered; the system pauses at the checkpoint, saving the full execution state. A prompt is sent to the human reviewer, who inspects the context and then either approves or rejects the operation. If approved, the agent

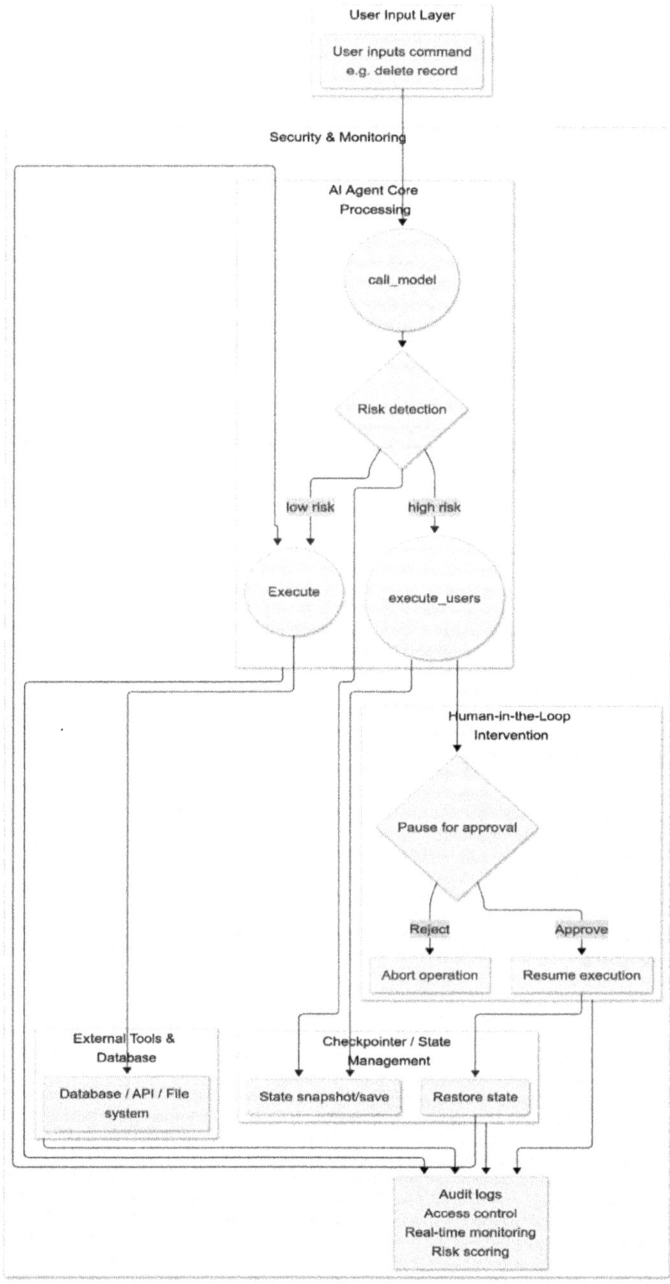

Figure 15.7.   Agent with HITL.

resumes from exactly the saved state and completes execution. If rejected, the workflow halts and returns an appropriate message.

This dual-path design achieves efficiency (autonomous execution when risk is low), safety (human control over risky steps), flexibility (both static and dynamic breakpoints), and state continuity (via checkpoint/ resume). The result is a robust, enterprise-ready AI Agent framework— especially when implemented with tools like LangGraph, whose graph-based flow, dynamic breakpoints, and memory-saver checkpointer deliver seamless human-AI collaboration. An example will be discussed in the next subsection with text2sql.

15.5.4 *Agent with Text2SQL and Human Interrupt*

Text2SQL was discussed in the previous chapter, which may be valuable for non-technical users, particularly those who may not be familiar with database query languages but need access to insights hidden within large datasets. However, it may not always provide the right answer in many cases. Agents can be further designed to bridge the gap between natural language input and structured SQL queries. As powerful as these agents are, they still benefit from human oversight, especially when dealing with critical or sensitive data operations. There are several reasons why incorporating human judgment into the process is essential:

1. **Validation:** Humans can validate whether the generated SQL accurately reflects the intent behind the natural language question.
2. **Error Prevention:** Complex queries might contain logical errors or unintended side effects. A human reviewer can catch these before execution.
3. **Security:** Preventing unauthorized or potentially harmful operations like DELETE or UPDATE without explicit approval.
4. **Interpretation:** Some queries may require domain-specific knowledge or business logic that an AI agent might not fully grasp without context.

In the associated code "ch15\agent_sql.py," we show how to use langgraph to develop such an agent with HITL interruption. The general workflow is illustrated in Figure 15.8.

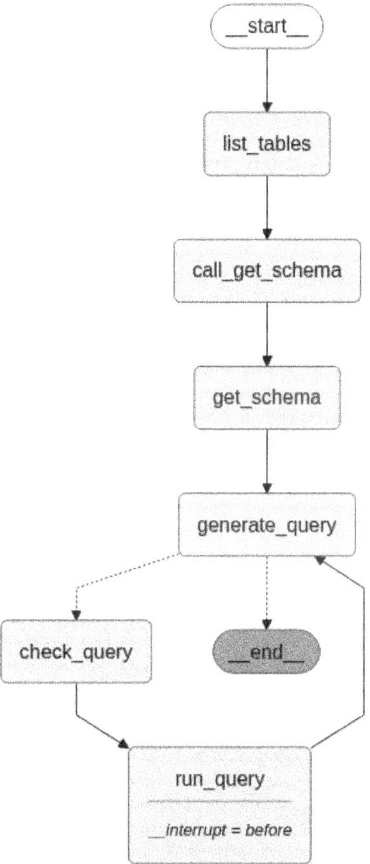

Figure 15.8.   Text2SQL agent with HITL.

Step 1: Initialize Language Model and Tools

We begin by initializing a chat model and connecting it to the SQL database through a toolkit:

```python
from langchain_community.utilities import
  SQLDatabase
db = SQLDatabase.from_uri("sqlite:///./data/
  Chinook.db")

from langchain_community.agent_toolkits import
  SQLDatabaseToolkit
toolkit = SQLDatabaseToolkit(db=db, llm=llm)
tools = toolkit.get_tools()
```

These tools include functions such as sql_db_query, sql_db_schema, and others that help the agent interact with the database.

Step 2: Define Agent Nodes and Graph Structure

We define nodes for handling tasks like listing tables, retrieving schema information, generating queries, and checking them for correctness:

```python
from langgraph.graph import StateGraph, END
builder = StateGraph(MessagesState)

# Add nodes
builder.add_node(list_tables)
builder.add_node(call_get_schema)
builder.add_node(get_schema_node)
builder.add_node(generate_query)
builder.add_node(check_query)
builder.add_node(run_query_node)

# Define edges
builder.add_edge(START, "list_tables")
builder.add_edge("list_tables",
  "call_get_schema")
builder.add_edge("call_get_schema",
  "get_schema")
builder.add_edge("get_schema",
  "generate_query")
builder.add_conditional_edges("generate_
  query", should_continue)
builder.add_edge("check_query", "run_query")
builder.add_edge("run_query", "generate_query")
```

This graph structure ensures that the agent follows a logical sequence when processing user queries.

Step 3: Enable Human Intervention

To incorporate human review before executing any query, we set up memory checkpointing and configure the agent to pause execution at specific points:

```python
if HITL:
    from langgraph.checkpoint.memory import
  MemorySaver
    memory = MemorySaver()
    agent = builder.compile(checkpointer=memory,
  interrupt_before=["run_query"])
else:
    agent = builder.compile()
```

With this configuration, the agent will pause right before executing a query and wait for human approval.

Step 4: Execute and Review Queries
When a user submits a query, the agent processes it through the defined workflow:

1. **List Tables** – Identifies available database tables.
2. **Get Schema** – Retrieves schema details for relevant tables.
3. **Generate Query** – Constructs an initial SQL query based on the user's question.
4. **Check Query** – Validates the query for potential issues (e.g., syntax, performance).
5. **Human Approval** – Waits for confirmation before proceeding.
6. **Run Query** – Executes the approved query and returns results.

```python
If the human approves the query, the agent
  continues execution:
python
for step in agent.stream(
    {"messages": [{"role":"user","content":
  question}]},
    config,
    stream_mode="values",
):
    step["messages"][-1].pretty_print()

if HITL:
    try:
```

```
            user_approval = input("Do you want to
go to execute query? (yes/no): ")
    except Exception:
        user_approval = "no"

    if user_approval.lower() == "yes":
        for step in agent.stream(None,
config, stream_mode="updates"):
            print(step)
    else:
        print("Operation cancelled by user.")
```

Building aText2SQL agent with human intervention provides a robust solution for translating natural language questions into actionable SQL queries while ensuring safety and accuracy. By leveraging LangGraph's stateful execution model and integrating a simple yet effective human validation step, developers can create intelligent database assistants that combine the power of AI with the precision of human oversight.

15.5.5 *Agents with Large Number of Tools or MCP servers*

Modern language-agent systems rely on agentic tooling or MCP to expose external tools (APIs, functions, databases) to LLMs. As the number of (MCP-defined) tools grows into the hundreds or thousands, naïvely embedding every tool description in the prompt exceeds context limits and wastes tokens that should be reserved for reasoning.

"Prompt bloat" inflates both the size and cognitive complexity of the prompt. Faced with a long, partially redundant tool list, the LLM must first understand each option and then decide which to call. This increases latency, raises the risk of hallucinating non-existent APIs, and degrades accuracy, especially when tools have overlapping capabilities.

RAG-MCP combines RAG with MCP tooling. Instead of pre-loading every tool, all tool schemas, examples, and metadata are stored in an external semantic index. For each user query, a lightweight retriever ranks the tools by semantic similarity and returns only the top-k candidates. These few descriptions are injected into the prompt (or passed via function-call metadata), shrinking context length and focusing the model's attention. Below is a sample procedure.

1. **Encode & Retrieve:** The user's task description is embedded; a vector search over the MCP index selects the k most relevant tools.
2. **Validate (optional):** Candidate tools are sanity-checked by issuing synthesized test calls to confirm compatibility.
3. **Inject & Plan:** The best tool (or small set) is supplied to the LLM, which now plans and executes without distraction.
4. **Iterate:** In multi-turn dialogues, the retriever is re-run only when needed, keeping prompts lean while allowing the tool registry to scale indefinitely.

Agent with Cross-domain Experience

The traditional agents struggle to generalize across tasks and domains. A Reason-Retrieve-Refine framework was proposed to address its inefficiency and error-proneness. At its core, it introduces a shared agent knowledge base (AGENT-KB) [5] that captures both high-level problem-solving workflows and fine-grained step-level execution experiences. This experiential memory enables agents to learn from past successes and transfer strategies to new challenges. The framework employs a two-stage Teacher-Student reasoning model, where the student improves planning using workflow-level knowledge, while the teacher iteratively refines those plans using step-level experience to address execution failures. This architecture significantly enhances agent performance on complex tasks, achieving up to a 16–19% improvement on benchmarks like GAIA [6] and SWE-bench [7].

15.5.6 *Deep Research*

Deep Research is usually an agent-driven research assistant designed to autonomously carry out multi-step investigations. Unlike simple retrieval tools, it can break down complex queries into sub-questions, conduct iterative searches across hundreds of sources, analyze the findings, and synthesize them into a structured and comprehensive report. OpenAI and Perplexity describe it as an AI agent capable of working at the level of a human research analyst.

Its workflow involves planning, autonomous information retrieval, reasoning, and synthesis. The LLM acts as an agent: first decomposing a query into manageable tasks, then retrieving information via search engines or plugins, analyzing and reasoning over results, and iterating until sufficient evidence is collected. Reinforcement learning (RLHF) is often applied to improve multi-step reasoning. The system finally consolidates the findings into a report, often with citations and a transparent "chain of thought" interface.

Deep Research is well-suited for scenarios requiring comprehensive reports, e.g.,

- **Market research and competitive analysis:** Compiling industry trends and rival strategies.
- **Academic research:** Assisting with literature reviews.
- **Professional analysis:** Legal, policy, and financial reports with data-driven conclusions.
- **Consumer or personal decisions:** Supporting major purchases or in-depth personal queries.

It excels at automating the tedious process of gathering and synthesizing vast information, saving significant time. However, results may still contain factual errors or reasoning flaws, requiring human verification. It is also resource-intensive compared to simpler tools. Table 15.7 compares the differences between deep search (introduced in Chapter 12), deep research and the traditional RAG techniques.

Table 15.7.   Comparison of deep research, deep search and RAG.

Dimension	Deep Research	Deep Search	Traditional RAG
Core Goal	Autonomously complete multi-step research and generate comprehensive reports.	Provide deeper, more relevant, and comprehensive search results.	Combine retrieval with generation to produce accurate, evidence-grounded answers.
Technical Approach	LLM acts as an autonomous agent: plans tasks, performs iterative retrieval, reasoning, and synthesis.	Semantic query expansion, wide-ranging retrieval, re-ranking, and AI-driven summaries. Usually use advanced RAG techniques.	At least two stage process: retrieval from knowledge base (vector/ keyword search) + generation by LLM.

Table 15.7.   (*Continued*)

Dimension	Deep Research	Deep Search	Traditional RAG
Outputs	Structured reports with conclusions, data, and citations.	Curated and ranked search results, sometimes with AI summaries.	Natural-language answers or generated content with references to retrieved documents.
Typical Applications	Market research, academic literature reviews, policy/ legal/financial reports, consumer decision support.	Complex or ambiguous queries, professional research (patents, papers), long-tail or personalized searches.	Enterprise Q&A, customer service bots, professional assistants (legal, medical, finance), content creation, personalization.
Strengths	Automates comprehensive research; saves time; outputs detailed, structured reports; transparent reasoning process.	Improves relevance and completeness of results; handles complex/ambiguous queries; reduces user effort in filtering information.	Produces more accurate and trustworthy answers; reduces hallucinations; dynamically updates knowledge without retraining the model.
Limitations	High computational cost; slower execution; possible factual errors or reasoning bias; still requires human verification.	Slower than traditional search; resource-intensive; may misinterpret ambiguous queries; higher implementation cost.	Quality depends on retrieval accuracy; requires maintaining external knowledge bases; integration complexity in large-scale systems.

There are many implementations are available in the markets, from the commercial OpenAI, Claude, Gemini, Kimi, Doubao, Perplexity, Manus and Grok to the open source langchain and Nvidia QIA research assistant.[9] A simple, configurable, fully open source langchain implementation works across many model providers, search tools, and MCP

---

[9]A benchmark leaderboard is available at https://github.com/NVIDIA-AI-Blueprints/aiq-research-assistant. More implementations can be found in https://github.com/DavidZWZ/Awesome-Deep-Research.

servers,[10] and another similar one using langgraph and gemini is also available at github.[11] Some other comprehensive open solutions include GPT Researcher,[12] OpenManus,[13] OpenClaw[14] and deer-flow.[15]

15.6 Concluding Remarks

In this chapter, we have explored the foundations and evolution of Agentic RAG, framing it as the natural progression from traditional RAG systems toward autonomous, goal-driven AI workflows. We delved into its core components, from perception and reasoning modules to memory and tool interfaces, and outlined how these coalesce into an agentic pipeline. We discussed prominent protocols (such as MCP, A2A, Ag-UI) and tool ecosystems that enable flexible, interoperable agentic systems. Finally, we illustrated diverse application scenarios from multi-document retrieval and hybrid data processing to human-in-the-loop operations, which demonstrate how agentic RAG empowers dynamic, context-aware solutions.

Several advanced methodologies are emerging that promise to augment Agentic RAG further. For example, Graph-R1 [8], an agentic GraphRAG framework trained via end-to-end reinforcement learning, harnesses lightweight hypergraph structures and turns retrieval into an agent-environment interaction optimized through rewards. Another innovation, MAO-ARAG (Multi-Agent Orchestration for Adaptive RAG), empowers a planner agent to dynamically sequence RAG executors, such as query reformulators and generators, balancing quality with computational cost using reinforcement learning [9]. Moreover, ReasonRAG introduces process-level supervision reinforced by a structured dataset (RAG-ProGuide), enhancing stability and sample efficiency in agentic training by rewarding fine-grained steps like query generation and evidence extraction [10]. These techniques offer promising pathways to make Agentic RAG more adaptive, efficient, and scalable.

Looking ahead, Agentic RAG is poised for transformative growth.

---

[10]https://github.com/langchain-ai/open_deep_research.

[11]https://github.com/google-gemini/gemini-fullstack-langgraph-quickstart.

[12]https://github.com/assafelovic/gpt-researcher.

[13]https://github.com/FoundationAgents/OpenManus.

[14]https://github.com/openclaw/openclaw.

[15]https://github.com/bytedance/deer-flow.

- **Enterprise Platforms, Modular Infrastructure & No-Code Accessibility:** The path to production deployment increasingly lies in scalable, modular platforms. Solutions such as AWS AgentCore, n8n and Dify offer enterprise-grade infrastructure with features like long-running workflows, memory, observability, and secure tool integration. Complementing these are no-code and low-code AI agent builders designed for business users and domain experts, not only developers. Examples include:
  - Microsoft Copilot Studio, enabling users to build agents for tasks like handling onboarding or sending internal emails via a low-code interface.
  - Salesforce Agentforce, offering pre-built agents for service, marketing, and sales workflows.
  - Google's Vertex AI Agent Builder (within Agentspace/Agent2Agent ecosystem), supporting no-code agent creation with interoperability.
  - Workflow automation tools like Appian, Dify, and n8n, offering drag-and-drop or template-based agent design. For instance, n8n lets users build context-aware agents that scrape web content, summarize it, and interact with a knowledge base, all without coding.

  These platforms lower the barrier to deployment, enabling rapid prototyping, user-driven customization, and accelerated adoption across organizations.

- **Efficient Memory and Semantic Caching Layers:** Production systems benefit from optimized memory architectures to reduce latency and cost. Techniques such as semantic caching (storing query-response contexts for reuse) have proven effective in reducing redundant LLM calls and improving responsiveness. Advanced memory systems like Zep [11] (a temporal knowledge-graph engine) and SHIMI [12] (a hierarchical memory index for decentralized agents) further enhance retrieval accuracy, latency, and multi-agent coordination in real-world applications.

- **Modular Multi-Agent and Specialized Architectures:** Agentic RAG systems are increasingly adopting modular architectures composed of specialized agents, each trained for domains such as legal, financial, or support workflows. These agent "teams" can perform parallel subtasks and collaboratively compose the final output, mirroring organizational coordination. Frameworks like LangChain, LlamaIndex, and AutoGen already enable such designs in production workflows. Incorporating personalization within Agentic RAG is increasingly viable. PersonaAI [4], for example, combines RAG with secure user modeling to craft personalized digital avatars while maintaining privacy constraints. On-device personalization is emerging through frameworks that enable fine-tuning LLMs locally with sparse annotations and small memory footprints, balancing user-specific strengths with privacy preservation. These privacy-conscious approaches can be embedded into multi-agent systems, enabling agents to utilize personal knowledge, encrypted vector stores, and local context, particularly important for sensitive domains and edge deployments. AgentFlow [13][16] is a trainable, modular agent framework that solves complex, tool-using tasks by coordinating four specialized modules: Planner, Executor, Verifier, Generator, over a shared memory and toolset. Unlike one-shot or offline approaches, it optimizes "in the flow" of multi-turn interaction to improve planning and tool use.

- **Enterprise Data Foundations & Governance Mechanisms:** The strength of agentic RAG hinges on high-quality, well-managed data, spanning CRM, e-commerce, and support systems. Modern enterprise document management systems (DM), integrated with metadata and knowledge-graph representations, empower agents to ground outputs in accurate, contextual intelligence. Without this foundation, RAG agents may underperform regardless of their sophistication. At the same time, trustworthy deployment requires robust governance via monitoring agent behavior, enforcing access controls, and capturing human feedback. These governance measures are indispensable in regulated domains like healthcare or finance.

Looking forward, Mixture-of-Agents (MoA) methodology can leverage the collective strengths of multiple LLMs for immediate efficiency

---

[16]AgentFlow formalizes problem solving as a multi-turn Markov Decision Process (MDP) and trains the Planner online with Flow-GRPO: an on-policy RL method that broadcasts a single trajectory-level outcome reward to every turn, applies token-level PPO-style updates with KL regularization, and uses group-normalized advantages for stability, turning long-horizon optimization into tractable single-turn updates. By learning how to plan and call tools during live interaction rather than baking all capability into a monolithic model, AgentFlow shows a path to smaller, cheaper systems that generalize better across diverse tools and tasks. Open resources (paper, code, demo, models) further its practical adoption and reproducibility.

and reasoning capability improvement [14] and self-evolving agents may serve as the essential bridge to Artificial Super Intelligence (ASI) [15]. One can conceptualize the evolution as a continuum, i.e., from static language models to foundational task-capable agents, onward to self-evolving agents, and ultimately to autonomous ASI. Central to this progression is the RAG-enhanced agent: by seamlessly retrieving and integrating fresh, external information into their reasoning processes, RAG agents enable continuous knowledge acquisition and expansion over time. This dynamic, self-feeding repository of knowledge empowers agents to adjust their perception, reasoning, and actions based on up-to-date, real-world information, thereby laying the groundwork for escalated autonomy and adaptation.

Self-evolving agents distinguish themselves from earlier AI paradigms through their capacity for evolutionary adaptation at multiple levels. They dynamically refine their internal models, memories, toolsets, and model structural architecture (not just between tasks, but even mid-task) using mechanisms such as reward-based reinforcement, imitation learning, and population-based optimization. Unlike static models, these agents also possess capabilities for active exploration, self-reflection, and self-evaluation. The inclusion of RAG mechanisms further enables them to continuously absorb new knowledge, mitigating hallucinations and ensuring their decisions are grounded in verified, up-to-date information.

This shift from the traditional scaling of fixed models to developing inherently self-evolving agents grounded in real-time knowledge, represents a transformative trajectory toward ASI. By enabling agents to autonomously evolve through active learning, structural refinement, and responsive knowledge integration, they may eventually attain or surpass human-level intelligence across a diversity of tasks. In such a future, ASI emerges not just from model size, but from perpetual adaptation, introspection, and seamless knowledge assimilation.

References

[1] Y. Huang, *Levels of AI Agents: From Rules to Large Language Models,* arxiv:2405.06643v2, 2024.

[2] F. Xing, *Designing Heterogeneous LLM Agents for Financial Sentiment Analysis,* arXiv:2401.05799, 2024.

[3] Y. Yu, H. Li, Z. Chen, Y. Jiang, Y. Li, D. Zhang, R. Liu, J. W. Suchow and K. Khashanah, *FinMe: A Performance-Enhanced Large Language Model*

Trading Agent with Layered Memory and Character Design, arXiv:2311. 13743, 2023.

[4] X. LI, P. JIA and *et al., A Survey of Personalization: From RAG to Agent,* arXiv:2504.10147v1, 2025.

[5] X. Tang and *et al., AGENT KB: Leveraging Cross-Domain Experience for Agentic Problem Solving,* arxiv 2507.06229v2, 2025.

[6] G. Mialon, C. Fourrier, C. Swift, T. Wolf, Y. LeCun and T. Scialom, *GAIA: A Benchmark for General AI Assistants,* arXiv:2311.12983, 2023.

[7] C. E. Jimenez, J. Yang, A. Wettig, S. Yao, K. Pei, O. Press and K. Narasimhan, *SWE-bench: Can Language Models Resolve Real-World GitHub Issues?,* arXiv:2310.06770v3, 2024.

[8] H. Luo and *et al., Graph-R1: Towards Agentic GraphRAG Framework via End-to-end Reinforcement Learning,* arXiv:2507.21892, 2025.

[9] Y. Chen and *et al., MAO-ARAG: Multi-Agent Orchestration for Adaptive Retrieval-Augmented Generation,* arXiv:2508.01005, 2025.

[10] W. Zhang and *et al., Process vs. Outcome Reward: Which is Better for Agentic RAG Reinforcement Learning,* arXiv:2505.14069, 2025.

[11] P. Rasmussen, P. Paliychuk, T. Beauvais, J. Ryan and D. Chalef, *Zep: A Temporal Knowledge Graph Architecture for Agent Memory,* arXiv: 2501.13956, 2025.

[12] T. Helmi, *Decentralizing AI Memory: SHIMI, a Semantic Hierarchical Memory Index for Scalable Agent Reasoning,* arXiv:2504.06135, 2025.

[13] Z. Li and *et al., In-the-Flow Agentic System Optimization for Effective Planning and Tool Use,* arXiv:2510.05592, 2025.

[14] J. Wang, J. Wang, B. Athiwaratkun, C. Zhang and J. Zou, *Mixture-of-Agents Enhances Large Language Model Capabilities,* arxiv:2406.04692, 2024.

[15] H.-a. Gao, J. Geng and *et al., A Survey of Self-Evolving Agents: On Path to Artificial Super Intelligence,* arXiv:2507.21046v2, 2025.

Chapter 16

Conclusion

16.1 Introduction

Since GPT-4's debut in 2023, the LLM field has shifted from the "bigger is better" scaling-laws mindset to a multidimensional strategy centered on **efficiency**, **reasoning**, and **agency**. GPT-4 extended context windows and improved reliability, but its closed-source nature and the absence of a GPT-5 in 2024 triggered reflection on scaling's limits. Although GPT-5 was finally launched in August 2025, hailed for its "PhD-level" intelligence and marked improvements in coding, writing, and hallucination reduction, it nonetheless fell short of AGI benchmarks and sparked debate over realistic expectations for "general intelligence."

Three pressures therefore have emerged: the need for architectural efficiency to handle longer contexts cost-effectively; the need for true multi-step reasoning beyond sheer size; and the push toward agentic AI, where models take action based on a combination of reasoning and environment interaction. Enhancement in computational efficiency will help with improvements in reasoning, although there are other factors that lead to the limitations of the reasoning and planning abilities of LLM's and related large reasoning models (LRMs). Improvements in reasoning will continue to underpin the practical expansion of agent capabilities.[1]

---

[1] There are many debates on the existing of the true reasoning capability of LLMs. In my view, LLMs and humans share a broad commonality: both can be framed as predictive systems that generate hypotheses and update them with evidence, but they differ profoundly in substrate, grounding, and control. In humans, leading accounts such as

Efficiency innovations have largely come from sparse architectures and attention mechanism advances. Mixture-of-Experts (MoE) designs which have been used by the major providers of LLM models for several years already, have more recently been championed by DeepSeek, Qwen, and Minimax. These designs achieve massive total parameter counts but activate only a fraction per token, cutting costs while retaining capacity. In parallel, attention bottlenecks are being addressed with methods like DeepSeek's Multi-Head Latent Attention (compressing KV caches), Minimax's Lightning Attention (linear complexity with periodic full attention), and Qwen's Grouped Query Attention. Open-source labs disclose these techniques to compete on cost-performance, while closed-source leaders use efficiency gains to power proprietary high-value features, creating divergent strategic positions. A recent DeepSeek-OCR

---

dual-process theories and predictive-processing posit interacting fast/slow systems and pervasive prediction, yet these frameworks remain debated and incomplete; recent brain-wide recordings further show that decision signals are distributed across much of the brain rather than localized, underscoring that human reasoning is not yet mechanistically unified [8]. In contrast, LLMs are trained for next-token prediction over text, and their "reasoning" often manifests as probabilistic pattern completion that can nonetheless display human-like content effects such as belief bias.

In practice, LLM reasoning is elicited and strengthened by scaffolds and search: chain-of-thought prompting encourages intermediate steps, self-consistency samples multiple paths and aggregates them, tree-of-thoughts explores branches explicitly, and instruction-tuning/RLHF shapes preferences for stepwise, helpful outputs; newer "reasoning" models further allocate compute to deliberation before answering [13]. Tool-augmented approaches such as ReAct and PAL offload sub-problems [12] to retrieval or code execution, improving reliability on math, logic, and knowledge tasks [10]. A recent test-time method, Deep Think with Confidence (DeepConf), adds confidence-aware filtering, and yields majority-vote systems that are both more accurate and cheaper to run [11]. Despite these advances, LLMs still face limitations in compositional generalization, causal modeling, and factuality, with persistent "hallucinations" documented across surveys and benchmarks [12].

A pragmatic outlook recognizes two points: first, there is no definitive mechanistic account of human reasoning today; second, prediction is plausibly one pathway of human reasoning, making machine-native probabilistic methods legitimate rather than derivative. Accordingly, build LLM systems that optimize for calibrated, verifiable reasoning: combine generator–verifier loops, branch-and-bound search (e.g., ToT), and tool use (ReAct/ PAL); log uncertainties and prefer abstention over brittle guesses; and evaluate with outcome-based metrics alongside plausibility. Given the distinct mechanisms of brains and models, convergence should be functional, not structural: engineer dependable pipelines that fuse prediction, external checks, and domain grounding, rather than forcing LLMs to mimic human cognition.

proposed a novel method for compressing long contexts by converting text into high-resolution images ("optical 2D mapping") and decoding them with a vision-language model. Early results report roughly 10× token reduction and about 97% decoding accuracy at lower compression in the ICDAR 2023 dataset, directly targeting tokenization inefficiency in long-prompt workloads and latency costs [1].

Reasoning advances, even with their well-documented limitations, have moved compute into the inference stage, with models like OpenAI's o-series and v5, Anthropic's Claude 4, Google's Gemini 2.5, and Qwen3. These models offer so-called "thinking" modes, but that term can be misleading. These models are not "thinking" and reasoning in ways we might think they are, based on our knowledge of human thinking processes. Also, while the model might show intermediate steps or "chain-of-thought" to the human user, this may not necessarily be how the model is internally processing the information and making its inferences. Even with these limitations, it is still the case that Chain-of-Thought (CoT) reasoning and "thinking budgets" deliver large performance jumps in complex tasks like AIME math problems. Reinforcement learning (RL) has evolved from alignment tuning to teaching reasoning strategies, exemplified by DeepSeek-R1's RL-first pipeline and Minimax's CISPO algorithm. This shift redefines scaling laws to include "test-time compute" and creates new hardware, algorithmic, and business model considerations, as reasoning becomes a differentiable product feature.

Agency capabilities apply reasoning to real-world actions. OpenAI's o3/o4-mini and Anthropic's Claude 4 integrate tool use, code execution, and even GUI control, enabling autonomous multi-step problem solving. These agentic functions mark the transition from static problem-solvers to dynamic actors in both digital and, potentially, physical environments. Benchmarks are adapting accordingly: traditional NLP tests like MMLU are saturated, and reasoning/agent benchmarks (AIME, GPQA, SWE-bench) now define state-of-the-art, with specialization emerging by task domain.

Looking ahead, trends point toward embodied intelligence and world models (which are closing the loop between perception, reasoning, planning, and action) and exploration of post-Transformer architectures. While no architecture has supplanted Transformer yet, research in state space models, normalization tweaks, and long-context efficiency is active. The field's new foundation rests on the triad of efficiency (MoE, novel attention), reasoning (inference-time computation, RL-driven logic), and agency (tool and environment interaction), signaling a mature phase where the most competitive systems will not simply be the largest, but the most efficient, thoughtful, and capable of acting.

Owning to those advances, the past two years have seen a remarkable shift in how enterprises talk about AI "productionization." On one hand, consulting firms and large investment bank analysts (with their own self-interest in getting business to spend money on these efforts) estimate the potential productivity upside of corporate AI at roughly US $4.4 trillion a year,[2] while Stanford's 2025 AI Index shows inference costs for GPT-3.5-level performance falling more than 280-fold in just twenty-three months.[3] Yet Gartner still finds that barely half of prototypes ever reach production and that almost a third of generative-AI projects are abandoned for lack of clear value or spiraling cost.[4] In short, building a model is easy; turning it into business impact is hard.[5]

A first and very human pain point is **"novelty without utility."** Several organizations bought Microsoft Copilot licenses for thousands of staff, only to watch usage crater once the early buzz wore off; Australia's Treasury, for example, ran a 14-week trial after which most volunteers said the tool felt "less useful than expected."[6] Analysts note a similar

---

[2]McKinsey & Company, The economic potential of generative AI: The next productivity frontier, https://www.mckinsey.com/capabilities/mckinsey-digital/our-insights/the-economic-potential-of-generative-ai-the-next-productivity-frontier, Jun 2023, Accessed on 09/08/2025.

[3]Stanford HAI, The 2025 AI Index Report, https://hai.stanford.edu/ai-index/2025-ai-index-report, Apr 2025, Accessed on 09/08/2025.

[4]Gartner, Gartner Predicts 30% of Generative AI Projects Will Be Abandoned After Proof of Concept By End of 2025, https://www.gartner.com/en/newsroom/press-releases/2024-07-29-gartner-predicts-30-percent-of-generative-ai-projects-will-be-abandoned-after-proof-of-concept-by-end-of-2025, Jul 2024, Accessed on 09/08/2025.

[5]From a widely accepted strategic viewpoint in enterprise technology adoption, we shall not consider AI success in project level. It should be understood as a core enterprise infrastructure and a catalyst for organizational transformation. Sustainable AI value emerges when it is embedded into data platforms, compute environments, governance mechanisms, and operating processes, rather than deployed through isolated use cases. Treated as infrastructure, AI becomes reliable, scalable, and reusable across the organization. More importantly, it reshapes how work is designed and executed, driving workflow re-engineering, decision augmentation, and new operating models. This perspective positions AI as a long-term strategic capability that underpins continuous improvement and enterprise-wide innovation, rather than a short-lived technology initiative.

[6]The Commonwealth of Australia, Evaluation of a trial of generative artificial intelligence (Copilot) in The Department of the Treasury, https://evaluation.treasury.gov.au/sites/

skepticism elsewhere: large buyers still need to be "convinced" that Copilot meaningfully boosts daily work. The cure is to anchor roll-outs in a single pain-killing workflow (say, drafting client emails directly from CRM data) and publish hard time-saved metrics so that colleagues pull the tool into their routine instead of having it pushed upon them.

The second pain point is the **"accuracy paradox."** In regulated sectors, AI outputs cannot be trusted blind; many professionals in law, risk and tax worry about the need for stricter oversight, and U.S. securities regulators list model-risk management and supervisory controls as critical adoption hurdles. Consequently, every LLM answer often triggers a mandatory human check, so labor hours are not saved, and in some cases, this checking is more time intensive than having the human generate the result. A more resilient pattern is to let the model decide *what* should happen (intent classification) and hand the *how* off to deterministic bots: classic RPA can execute rule-bound steps with far greater auditability.

Third comes **"compute first, use-case later."** Many firms have rushed to build on-prem GPU clusters for "future" AI agents, only to see the racks idling once pilot enthusiasm fades. Training or inference in the public cloud is financially safer until a workload's ROI is proven, argues Pluralsight's 2025 framework for cloud-versus-on-prem AI, and a hybrid strategy can cut overall costs for bursty demand while keeping sensitive data local.[7] Falling hardware prices make later repatriation attractive, but only after the business case is nailed. Meanwhile, as reasoning (deep research) model is generally superior over non-reasoning model, which leads to more token and more computational power.

Real-world stories underline both the pitfalls and prizes. UK law firm Shoosmiths put a £1 million bonus pool on the table if employees log one million Copilot interactions, turning usage into a collective KPI and surfacing "shadow AI" work that staff previously hid.[8] Elsewhere, market

---

evaluation.treasury.gov.au/files/2025-02/evaluation-generative-artificial-intelligence.pdf, Feb 2025, Accessed on 09/08/2025.

[7]Axel Sirota, Cloud AI vs. on-premises AI: Where should my organization run workloads? https://www.pluralsight.com/resources/blog/ai-and-data/ai-on-premises-vs-in-cloud, Mar 2025, Accessed at 27/07/2025.

[8]Shoosmiths, Shoosmiths offers £1m bonus pot for 1 million AI prompts, https://www. shoosmiths.com/insights/news/shoosmiths-offers-1m-bonus-pot-for-1-million-ai-prompts, Apr 2025, Accessed at 27/07/2025.

watchers note that the most successful deployments embed generative models *inside* existing applications rather than shipping yet another stand-alone chat window.

Finally, the frontier is shifting from LLM "co-pilots" to fully fledged **agentic workflows**. Tools such as Google full agentic platforms and Anthropic Claude 4 family let LLM agents drive a virtual Windows desktop to automate legacy software without human clicks, and vendors like UiPath are repositioning RPA suites as orchestration backbones for those agents. Gartner analysts describe this convergence as the route from "robotic process automation to agentic AI," echoing research on intelligent-automation patterns that marry AI decision-making with rule-based execution. Together, these cases suggest that sustainable AI productionization is less about the next flashy model and more about stitching models into accountable, value-visible business processes. Those observations and principles are also applicable to RAG.

16.2 The Rise of RAG: From 1.0 to 2.0

2024 marked the "Year of RAG," where RAG exploded into the mainstream, reshaping how LLMs interact with real-world knowledge. In 2025, RAG evolved beyond its early limitations, transitioning from a niche tool for text augmentation to a versatile framework powering cross-industry application. While some initially believed that long-context windows (a hotly debated feature in early 2024) might replace traditional RAG, reality proved otherwise. Even the largest context windows can't load *all* enterprise data, and LLMs remain prone to hallucinations without external grounding.

The maturation of MLOps/LLMOps architectures, which are coupled with advances in vector databases, embedding models, and multimodal capabilities, has fueled RAG's rapid evolution. Weekly research papers now flood arXiv and major conferences, pushing RAG from its experimental "1.0" phase into a sophisticated "2.0" era. Today's RAG isn't just about patching LLM weaknesses. It's a full-stack engineering paradigm for dynamic knowledge integration.

16.2.1 *Key Challenges: Why RAG 2.0 Isn't Perfect*

Even with smarter retrievers, longer contexts, and agentic loops, RAG still carries some hard limits. These issues can be *mitigated* but not fully eliminated, and they're why disciplined engineering matters more than chasing "RAG 2.0" or "RAG 3.0" labels.

- **Multimodal ingestion fidelity is brittle:** Most enterprise knowledge lives in PDFs, slide decks, and reports with dense tables and charts. Extracting faithful structure (tables, multi-column layouts, figures) remains a major pain point; flattening or OCR errors quietly degrade retrieval and grounding.
- **Retrieval–intent mismatch ("semantic gap") persists:** Abstract or multi-hop questions still trip up pure vector search; classic sparse methods miss paraphrases. Hybrid retrieval with re-ranking helps, but it's not a silver bullet and adds complexity.
- **Longer context windows aren't a cure-all:** Models use long contexts unevenly: evidence in the *middle* of a long prompt is often ignored, so "just stuff more text in" doesn't guarantee grounding.
- **Hallucinations are reduced, not removed:** RAG cuts confabulations, but legal/medical evaluations and surveys still find ungrounded or mis-attributed answers in production tools. Expect improvement, not perfection.
- **Latency vs. accuracy is a real trade-off:** Cross-encoder or LLM-based re-rankers boost relevance but cost milliseconds and compute; production guidance often favors shallower or cascaded rankers under tight SLAs.

  Also, search backends are tuned for millisecond queries over indexed content not whole PDFs so orchestration adds overhead you must budget for.
- **Security risks are structural, not cosmetic:** Indirect prompt injection is an inherent risk when models read external content; no single filter "fixes" it. On top of that, knowledge-base poisoning (even a single document) can taint answers or multi-hop reasoning. Defense must be layered.
- **Freshness has an operational cost:** Keeping indices current is non-trivial: changes require re-chunking, re-embedding, and re-indexing. Stale vectors quietly rot answer quality, especially in real-time domains.
- **"Citations" aren't always faithful:** A model can cite a correct source yet still have formed its answer elsewhere ("post-rationalization"). Evaluating faithfulness not just correctness remains an open problem with emerging but incomplete benchmarks.

16.2.2 *Pragmatic Principles: Avoiding the Hype-cycle*

To avoid drowning in complexity, keep these rules below in mind as the practical principles in for production:

- **Don't Fix What Isn't Broken:** If your TEXT2SQL pipeline works, don't replace it with RAG just to chase trends. Focus on problems that *need* retrieval augmentation.
- **Resist "Shiny Object Syndrome": Start Simple and Iterate:** GraphRAG? Multimodal RAG? Cool names, but start simple. A retail company should validate basic RAG on a single product catalog before overhauling everything. For example BM25+semantic retrieval with clean, well-structured data. Proven workflows show that even basic systems provide quick validation and a foundation for further enhancement. Then start to adopt robust chunking strategies and two-stage ranking.
- **RAG Is a Smart Search Engine, Not a Panacea:** Treat RAG as a *business-driven* toolkit, not a magic solution. For instance, a travel agency's RAG needs clear intent routing: *Is this a visa query, a flight search, or a hotel review?* Nail the logic, then build.
- **Open Source ≠ Production-Ready:** Frameworks like LangChain are great for prototyping, but production demands customization. One fintech team learned this the hard way when their RAG bot leaked data through unvetted connectors.
- **Optimize the 80% First:** Don't obsess over edge cases. Spending months fine-tuning rerankers for rare scenarios? Save that effort for life-or-death use cases.
- **Document Parsing: Good Enough Beats Perfect:** Striving for 100% document fidelity is a fool's errand. Focus on extracting *actionable* insights, not pixel-perfect reconstructions.
- **Balance Latency, Cost, and Accuracy:** Invest in evaluation & monitoring to understand inference latency, using GPU acceleration where possible. Design APIs with streaming support and resilience to prioritize user experience.
- **Apply Security & Risk Protocols:** Protect your pipeline: consider privacy, compliance, and potential attack vectors. Stay aligned with enterprise security standards throughout your RAG stack.

16.3 Future Directions: Where to Place Your Bets

Many promising techniques on RAG have been provided in the previous chapters. Below, some key directions are further highlighted.

Agentic RAG: For certain types of business applications, especially those involving multi-step business process execution, agentic RAG systems would be good choices. The systems like Self-RAG and Adaptive RAG, close the loop between retrieval and action. And it will become a core component inside a genetic agentic system. Picture a financial analyst's AI agent that autonomously retrieves market data, flags inconsistencies, and drafts reports, i.e., all while learning from its mistakes [2–3].[9]

Unified Multimodal Understanding: GPT-4o/5 and Gemini 2.5 hint at a future where RAG handles text, images, and audio seamlessly. But challenges remain: How do you align a chart's visual data with its textual analysis? Early adopters are experimenting with cross-modal rerankers, but true unification is still years away [4].

Unified-Foundation Model vs Specific Model: Enterprises may start to build some industry-specific models, and later might converge on a *single, general-purpose foundation model* that can be lightly personalized for any task, rather than maintaining dozens of models. Analysts see this shift as the natural outcome of scale economics, i.e., larger models learn more transferable abstractions, while new techniques such as mixture-of-experts routing, parameter-efficient fine-tuning and adaptive inference allow the same backbone to run lean when workloads are small and to scale up automatically for harder problems. At the same time, we may expect domain-specific models targeted at the edge devices and mobile devices with limited infrastructure resources. With proper knowledge distillation, compression and network trucking techniques, the models enable local efficient deployment.

---

[9]Don't overgeneralize Andrew Ng's claim that "LLMs will be agents, or they will be nothing," which applies when you need true end-to-end automation, but many business tasks don't demand agentic capabilities.

Personalized RAG: Imagine a streaming platform tailoring recommendations by blending your watch history with critics' reviews—*without* compromising privacy. Recent work like *"A Survey of Personalization: From RAG to Agent"* explores injecting user preferences at every stage, from adaptive prompts to reinforcement learning [5–6]. The challenge lies balancing customization with data scarcity and privacy walls.

Continuous-learning RAG: This advanced RAG framework capable of incrementally updating its external knowledge and retrieval mechanisms without requiring full model retraining. It maintains a living, evolving knowledge repository: as documents and user feedback arrive, the retriever's index and reranking logic are updated, often using knowledge distillation or curriculum-based fine-tuning, enabling the system to retain past competency, adapt to new domains, and remain responsive to real-world changes. Meanwhile, it may combine RAG with dynamic memory banks, knowledge-graphs and multi-agent incremental learning to reduce hallucinations, improve reasoning, and dynamically incorporate new structured knowledge. This trend may evolute into self-learning RAG via self-evolving agents [7].

TrustRAG: Safety by Design: Tools like **TrustRAG** use clustering and self-assessment to filter malicious content with a certain level of explainability. For instance, a journalism RAG might flag conflicting sources before generating a summary. Red teaming, as practiced by DeepMind, will become standard because securing RAG isn't optional [8].

Note that RAG 2.0 isn't just an upgrade. It's a whole paradigm shift in how machines interact with knowledge. We're moving:

- From *static text dump* to *dynamic, multimodal reasoning*, when the knowledge is no longer one-dimensional, and neither is retrieval.
- From *one-shot queries* to *self-correcting agentic loops*, which means better answers with fewer hallucinations—because the agent can catch itself before it drifts.
- From static RAG pipeline to self-improving automation, e.g., models will auto-annotate poorly retrieved content to finetune retrievers.
- From cloud to edge RAG, e.g., on-device retrieval (e.g., smartphones) will bypass cloud latency.

The road ahead will see labels like "3.0" or "4.0," but the core truth remains: RAG succeeds when it solves real problems with simplicity and rigor. The winners won't be those chasing the latest framework, but those mastering the art of balance between innovation and pragmatism, ambition and caution.

As we navigate this frontier, remember: Sometimes, the smartest systems are the ones that know when to slow down and more carefully and deliberately consider what must be done and how to do it, daresay, even "think." The human cognitive architecture has demonstrated the power of simultaneously having both a very fast and efficient perceptual and pattern matching approach (Type 1 processing) as well as a much slower and more effort and reasoning intensive approach (Type 2 processing) [9]. It looks as if LLM's and RAG supplementation are moving in this same direction of realizing this need for these dual processes, and the resulting ability to think fast and think slow.

References

[1] H. Wei, Y. Sun and Y. Li, *DeepSeek-OCR: Contexts Optical Compression,* arXiv:2510.18234, 2025.

[2] Y. Gao and *et al.*, *Retrieval-Augmented Generation for Large Language Models: A Survey*, arxiv:2312.10997, 2023.

[3] F. Xing, *Designing Heterogeneous LLM Agents for Financial Sentiment Analysis*, arXiv:2401.05799, 2024.

[4] Y. Ding and *et al.*, *A Survey on RAG Meets LLMs: Towards Retrieval-Augmented Large Language Models*, arxiv:2405.06211v1, 2024.

[5] X. LI, P. JIA and *et al.*, *A Survey of Personalization: From RAG to Agent,* arXiv:2504.10147v1, 2025.

[6] Z. Zhang and *et al.*, *Personalization of Large Language Models: A Survey,* arXiv:2411.00027v2, 2025.

[7] H.-a. Gao, J. Geng and *et al.*, *A Survey of Self-Evolving Agents: On Path to Artificial Super Intelligence,* arXiv:2507.21046v2, 2025.

[8] H. Luo and L. Specia, *From Understanding to Utilization: A Survey on Explainability for Large Language Models*, arXiv:2401.12874v1, 2024.

[9] K. E. Stanovich and R. F. West, Individual difference in reasoning: Implications for the rationality debate?, *Behavioral and Brain Sciences*, vol. 23, no. 5, pp. 645–65, 2000.

[10] International Brain Laboratory, D. Angelaki and *et al.*, A brain-wide map of neural activity during complex behaviour, *Nature,* vol. 645, pp. 177–191, 2025.

[11] S. Yao, J. Zhao, D. Yu, N. Du, I. Shafran, K. Narasimhan and Y. Cao, *ReAct: Synergizing Reasoning and Acting in Language Models,* arXiv: 2210.03629v3, 2023.

[12] Y. Fu, X. Wang, Y. Tian and J. Zhao, *Deep Think with Confidence,* arXiv:2508.15260, 2025.

[13] H. Liu, Y. Ding, Z. Fu, C. Zhang, X. L. Liu and Y. Zhang, *Evaluating the Logical Reasoning Abilities of Large Reasoning Models,* arXiv:2505.11854, 2025.

[14] L. Gao and *et al.*, *PAL: Program-aided Language Models,* arXiv:2211.10435v2, 2023.

[15] J. Wei and *et al.*, *Chain-of-Thought Prompting Elicits Reasoning in Large Language Models,* arXiv:2201.11903v2, 2023.

Index